About Island Press

Since 1984, the nonprofit Island Press has been stimulating, shaping, and communicating the ideas that are essential for solving environmental problems worldwide. With more than 800 titles in print and some 40 new releases each year, we are the nation's leading publisher on environmental issues. We identify innovative thinkers and emerging trends in the environmental field. We work with world-renowned experts and authors to develop cross-disciplinary solutions to environmental challenges.

Island Press designs and implements coordinated book publication campaigns in order to communicate our critical messages in print, in person, and online using the latest technologies, programs, and the media. Our goal: to reach targeted audiences—scientists, policymakers, environmental advocates, the media, and concerned citizens—who can and will take action to protect the plants and animals that enrich our world, the ecosystems we need to survive, the water we drink, and the air we breathe.

Island Press gratefully acknowledges the support of its work by the Agua Fund, Inc., The Margaret A. Cargill Foundation, Betsy and Jesse Fink Foundation, The William and Flora Hewlett Foundation, The Kresge Foundation, The Forrest and Frances Lattner Foundation, The Andrew W. Mellon Foundation, The Curtis and Edith Munson Foundation, The Overbrook Foundation, The David and Lucile Packard Foundation, The Summit Foundation, Trust for Architectural Easements, The Winslow Foundation, and other generous donors.

The opinions expressed in this book are those of the author(s) and do not necessarily reflect the views of our donors.

TIDAL MARSH RESTORATION

Tidal Marsh Restoration

A Synthesis of Science and Management

Edited by Charles T. Roman and David M. Burdick

 **ISLAND**PRESS

Washington | Covelo | London

Library of Congress Cataloging-in-Publication Data

Tidal marsh restoration : a synthesis of science and management / edited by Charles T. Roman, David M. Burdick.
    p. cm. — (The science and practice of ecological restoration series)
  ISBN 978-1-59726-575-1 (hardback) — ISBN 1-59726-575-6 (cloth) — ISBN 978-1-59726-576-8 (paper)
1. Salt marsh restoration. 2. Salt marsh ecology. I. Roman, Charles T. (Charles True) II. Burdick, David M.
  QH541.5.S24T53 2012
  578.769—dc23

                                                                              2012014219

Printed on recycled, acid-free paper

Manufactured in the United States of America
10 9 8 7 6 5 4 3 2 1

Keywords: Island Press, tidal marsh, tidal restoration, restoration ecology, ecological restoration, salt marsh, tidal wetlands, ecosystem services, coastal wetlands, adaptive management, ecological monitoring

CONTENTS

W. Gregory Hood
Skagit River System Cooperative (La Conner, Washington)

Charles A. Simenstad
School of Aquatic and Fishery Sciences, University of Washington

Ecosystem restoration is a simple concept: return degraded, dysfunctional ecosystems to their former healthy and functional conditions. Yet, just as the process of restoring a sick person to health is complicated and problematic, so is ecosystem restoration. It requires an understanding of how the original healthy ecosystem was formed and maintained, what the causes of degradation were, and what might be the best restoration techniques to employ—directly analogous to the practice of medicine, which requires understanding the anatomy and physiology of the patient, disease diagnosis and etiology, and treatment through medicines, surgery, or behavioral prescriptions. Restoration scientists and managers are ecosystem physicians; our patients are dysfunctional ecosystems and landscapes. Just as physicians must integrate and apply principles of physiology, genetics, biochemistry, microbiology, and parasitology to address practical problems in human biology, those engaged in restoration need to integrate and apply principles of ecology, hydrodynamics, geochemistry, geomorphology, and engineering to solve practical environmental problems. However, while medicine has been practiced for as long as people have been injured or sick, and modern medical practice can be traced back to Hippocrates (ca. 460–ca. 377 BC), Galen (AD ca. 129–ca. 199), Avicenna (980–1037), and Vesalius (1514–1564), restoration science is a comparatively young discipline, and anthropogenic environmental degradation and species extinctions are primarily postindustrial problems. Restoration ecology in the Americas probably originated in the mid-1930s, when Aldo Leopold's family and the US Civilian Conservation Corps replanted tallgrass prairie on degraded Wisconsin farmland, but the governmental response was not significant until after the initial passage of the Clean Water (1972), Clean Air (1970), and Endangered Species (1973) Acts. In addition to introducing the notion that ailing ecosystems could be healed, Leopold contributed two fundamental advancements in restoration ecology—development of an environmental technology and a template for ecological research. He recognized that the process of reassembling, repairing, and adjusting ecosystems can lead to profound insights into their structure and function.

This book summarizes some of our still early attempts to understand responses to environmental degradation and ecological restoration as observed in tidal marshes in New England and Atlantic Canada, while also referencing relevant restoration efforts from other regions. Case studies, supported by chapters on related disciplinary considerations, clearly illustrate the technological practice of restoration in tidal marsh systems. Beyond simple description, these examples demonstrate the utility of restoration in testing the robustness of ecological theory as applied to practical problems in ecosystem management. Investment in restoration science and monitoring is generally a small fraction of restoration project costs, yet the findings described in this volume illustrate how much practical value can be derived from even such marginal investment. The discussion of social, political, and bureaucratic concerns reminds us that scientific theory alone cannot sustain the practice of ecosystem restoration: social and political specialists are necessary members of the restoration team, and restoration scientists must themselves develop social and political skills to supplement their scientific expertise. While these lessons from New England and Atlantic Canada have broad applicability, we hope they will inspire other regional synopses from areas with a diversity of tidal wetland restoration approaches, namely Chesapeake Bay and the Carolinas, the Gulf of Mexico and the Mississippi Delta, coastal California and San Francisco Bay, the Pacific Northwest's Puget Sound and Columbia River estuary, and other regions. Comparisons of such synopses could reveal important regional differences in restoration ecology practices and performance. For example, the New England and Atlantic Canada tidal wetland restoration approaches described in this volume focus on local hydrologic exchange rather than landscape-scale processes and estuarine gradients, which are often at the center of restoration practice in Puget Sound and the Columbia River estuary. This difference may be due to a strong management focus on recovery of threatened anadromous salmon in the Pacific Northwest where historical salmon habitat has been lost throughout an extensive landscape, ranging from headwater streams to estuarine gradients in river deltas, to fringing salt marshes, lagoons, and other coastal landforms providing rearing habitat along juvenile salmon migration routes to the open ocean. Similarly, a Mississippi Delta perspective on tidal wetland restoration would presumably emphasize marsh subsidence to a greater degree than in other regions, while a California perspective might emphasize impacts on tidal wetlands from freshwater diversion and urbanization.

Reading these accounts will give coastal wetland managers and restoration scientists greater confidence in the resilience of emergent coastal wetlands that are under stress. Although there are many legitimate reasons to question the hypothesis that marshes follow restoration trajectories in multistressed landscapes, such as extensively urbanized estuaries, the consistent appearance of progressive and often rapid trajectories toward more natural states increases confidence in the feasibility and even the predictability of these efforts. Expansion of the predominantly structural metrics (vegetation development) to more socially relevant indicators of marsh function (performance of fish, avifauna, or nutrient cycling processes) may more closely represent the ecosystem goods and services that can motivate social support for and investment in restoration. Greater incorporation of long-term reference wetland studies and retrospective historical ecology, as called for throughout this book and as demonstrated in other coastal regions (e.g., the National Estuarine Research Reserves, the Louisiana Coastwide Reference Monitoring System, the Puget Sound River History Project, and the Historical Ecology Program at the San Francisco Estuary Institute), would ultimately anchor these trajectories more firmly into ecological science and theory.

While often undervalued by funding agencies, long-term restoration monitoring and adaptive management are fundamental to improving the performance of a specific site or, more generally, to further develop the practice of tidal marsh restoration. To return to our medical analogy, restoration without monitoring is like surgery without follow-up to see if the intervention was successful or if any complications arose that may need additional treatment. Such shortsighted practice would never be tolerated in medicine. The contributors to this book demonstrate restoration successes and emphasize the value of monitoring and science to tidal wetland restoration; hopefully their work will encourage greater agency commitment to the funding of monitoring. It is our hope that this first regional coastal restoration synopsis will inspire others and will ultimately lead to greater understanding of how we can more effectively revive our ailing coastal ecosystems.

The authors are gratefully acknowledged for their contributions to this book, for sharing their wealth of knowledge on salt marsh restoration, and for their patience throughout the long process of completing this edited volume. We also thank our respective mentors for introducing us to salt marsh ecosystems early in our careers, with a particular emphasis on science-based support for management decision making: Franklin Daiber, William Niering, and Scott Warren; Graham Giese and Irving Mendelssohn. Stimulating and informative discussions with our students and colleagues, while we conducted fieldwork or attended meetings, have provided the foundation for many of the ideas and concepts offered in this book. Thank you.

The chapters in this book have benefited from critical peer reviews, for which we extend thanks to the following colleagues: Britt Argo, Wellesley College; David Bart, University of Wisconsin; Kirk Bosma, Woods Hole Group; Christopher Craft, Indiana University; R. Michael Erwin, US Geological Survey; Joan LeBlanc, Saugus River Watershed Council; Bryan Milstead, US Environmental Protection Agency; Gregg Moore, University of New Hampshire; Pamela Morgan, University of New England; Lawrence Oliver, US Army Corps of Engineers; Lawrence Rozas, National Oceanic and Atmospheric Administration; Stephen Smith, National Park Service; Megan Tyrrell, National Park Service; Kerstin Wasson, Elkhorn Slough National Estuarine Research Reserve; Cathleen Wigand, US Environmental Protection Agency.

Special thanks are extended to Amanda Meisner and Robin Baranowski, students at the University of Rhode Island, for their attention to editorial details as the manuscript was being produced, and to Roland Duhaime for dedicated assistance with the graphics. It was also a pleasure working with Barbara Dean and Erin Johnson, of Island Press, as they expertly guided us through the process.

Finally we offer sincere thanks to our families for their always present support.

PART I
Introduction

Chapter 1

A Synthesis of Research and Practice on Restoring Tides to Salt Marshes

CHARLES T. ROMAN AND DAVID M. BURDICK

The structure and ecological function of salt marshes are defined by many inter-acting factors, including salinity, substrate, nutrient and oxygen availability, sedi-ment supply, and climate, but hydrology (the frequency and duration of tidal flooding) is a dominating factor (e.g., Chapman 1960; Ranwell 1972; Daiber 1986). When tidal flow is restricted there can be dramatic changes to physical and biological processes that affect vegetation patterns, fish and avian communities, and biogeochemical cycling, among others. Throughout the developed coastal zone, roads and railroads that cross salt marshes often have inadequately sized bridges and culverts that restrict tides (fig. 1.1). Tide gates are also a common fea-ture, eliminating or dramatically restricting flood tides from entering salt marshes but allowing for some drainage on the ebb tide. Other tide-restricting practices that have been ongoing for centuries include impoundments for wildlife manage-ment purposes (Montague et al. 1987) and diking and draining to facilitate graz-ing and agriculture (Daiber 1986; Doody 2008). Diking is particularly extensive in Atlantic Canada (Ganong 1903), Europe (Davy et al. 2009), and the United States (e.g., Delaware Bay, Sebold 1992; San Francisco Bay, Nichols et al. 1986).

With tidal restriction there are often dramatic changes in vegetation as salt and flood-tolerant species of the salt marsh are displaced by plants typically found in fresher and drier conditions. Under regimes of tidal restriction, *Spartina*-dominated (cordgrass) marshes in the northeastern United States have been in-vaded by the aggressive *Phragmites australis* (common reed), often in dense monocultures, and other less salt-tolerant herbaceous and woody species (e.g., Ro-man et al. 1984; Burdick et al. 1997; Crain et al. 2009). *Phragmites* marshes, when compared to short-grass *Spartina* meadows, reportedly do not provide suitable habitat for birds, especially those that typically nest in salt marshes (Benoit and

3

FIGURE 1.1. Tide-restricted salt marsh, Herring River, Cape Cod, Massachusetts. (a) Tide-restricting road/dike and culverts at the mouth of the estuary. (b) *Spartina alterniflora* salt marsh dominates downstream of the tide restriction. (c) The invasive *Phragmites australis* occurs immediately upstream of the tide restriction. (Photos courtesy of Charles Roman)

Askins 1999; DiQuinzio et al. 2002). Fish abundance, species composition, and food web support functions are altered by tidal restriction when compared to tide-unrestricted systems (e.g., Dionne et al. 1999; Able et al. 2003; Raposa and Roman 2003; Wozniak et al. 2006). Feeding, reproduction, and nursery function can be much reduced or eliminated based on studies documenting the response of the dominant East Coast marsh fish, *Fundulus heteroclitus* (mummichog), to *Phragmites* invasions (Able and Hagan 2000; Able et al. 2003; Hunter et al. 2006). Tidal restriction can result in significant subsidence of the sediment surface and acidification of salt marsh soils, with subsequent declines in marsh primary production and export (e.g., Anisfeld and Benoit 1997; Portnoy 1999). Water quality concerns, especially low levels of dissolved oxygen in tide-restricted marshes, have been reported with detrimental effects on estuarine fauna (Portnoy 1991).

The practice of restoring tidal flow to degraded tide-restricted salt marshes has been actively pursued for decades. In Delaware Bay (New Jersey) over 1700 hectares of salt marsh that had been diked and cultivated for salt hay are now undergoing tidal restoration (Weinstein et al. 1997; Philipp 2005). Similarly, restoration efforts through the natural or deliberate breaching of dikes are under way in the United Kingdom and other parts of Europe (Pethick 2002; Wolters et al. 2005; Davy et al. 2009), Bay of Fundy (Byers and Chmura 2007), San Francisco Bay (Williams and Faber 2001; Williams and Orr 2002), the Pacific Northwest (Thom et al. 2002), and elsewhere. Along populated coasts, managers are also engaged in programs to restore tidal flow to degraded salt marshes by removing tide gates and enlarging culverts, bridge openings, and other flow restrictions, with numerous examples from the northeastern United States (Warren et al. 2002; Crain et al. 2009), southeastern US (NOAA Restoration Center and NOAA Coastal Service Center 2010), Pacific US (Zedler 2001; Callaway and Zedler 2004), Australia (Williams and Watford 1996; Thomsen et al. 2009), and other regions.

Purpose and Book Organization

To help guide future restoration efforts throughout the coastal zones of the world and to advance restoration science and management, this edited volume compiles, synthesizes, and interprets the current state of knowledge on the science and practice of restoring tidal flow to salt marshes. This book focuses on the New England and Atlantic Canada region, where the practice of restoring tidal flow to salt marshes has been ongoing for decades, accompanied by extensive multidisciplinary research efforts. However, the book is far from limited in regional scope; the contributing authors incorporate relevant literature from other regions to complement and support the information base developed in New England and Atlantic Canada.

This book will serve as a valuable reference to guide managers, planners, regulators, environmental and engineering consultants, and others engaged in planning, designing, and implementing individual projects or programs to restore tidal flow to tide-restricted or diked salt marshes. Those involved in restoration science will find the technical syntheses, presentation of new concepts, and identification of research needs to be especially useful as research and monitoring questions are formulated and as research findings are analyzed, interpreted, and reported. Perhaps this book will inspire undergraduate and graduate students to pursue careers in coastal habitat restoration from the restoration science or resource management perspective.

The book is divided into six major parts—Introduction, Synthesis of Tidal Restoration Science, The Practice of Restoring Tide-Restricted Marshes, Integrating Science and Practice, Communicating Restoration Science, and a Summary. Following this introductory chapter, the second part of the book synthesizes the extensive literature that is available on the hydrologic, biogeochemical, and biological (vegetation, nekton, birds) responses of salt marshes to tidal restoration. The focus is on the New England and Atlantic Canada region, but, as noted, the chapters also provide broader geographic perspectives. There is an emphasis on trajectories of change throughout the restoration process. Each chapter closes with recommended research needs aimed at improving our understanding of marsh responses to tidal restoration.

Coastal managers from local, state, and federal agencies and conservation organizations have an extraordinary knowledge base on the practice of salt marsh tidal restoration. The third part of the book provides a rather unique opportunity for those at the forefront of facilitating tidal restoration projects to offer insight on the challenges of developing and maintaining salt marsh restoration programs, to highlight project achievements, identify monitoring and adaptive management approaches, and discuss the essential role of partnerships. Some agencies in the New England and Atlantic Canada region have been engaged in tidal restoration projects, with dedicated programs, for over two decades. Other programs are newly emerging, and some offer no formal restoration program but present the structure used to implement successful projects. The chapters present a broad range of lessons learned, which are transferable to agencies or organizations that are developing programs (or leading individual projects) aimed at tidal restoration of coastal wetlands.

Part IV of the book integrates science and practice, with chapters on the role of monitoring, adaptive management, and documentation of ecosystem services as multiparameter tools to evaluate trajectories of restoration. Another chapter offers ecosystem-based simulation models—that go beyond predicting hydrologic responses to restoration—as an informative methodology to aid in the regional prioritization of restoration sites, to guide the design of projects, and to facilitate communication of restoration objectives and anticipated outcomes. The final chapter in part IV presents modifications to tide-restricting infrastructure (e.g., modified tide gates) that can be used to achieve a desired hydrologic condition.

Part V contains four case studies focused on successes and challenges associated with advancing tidal restoration projects to the public, regulatory, and stakeholder audiences. Each chapter discusses the role of interdisciplinary science, hydrologic and ecological modeling, and effective communication to address societal concerns (e.g., flood protection, mosquito control, water quality, altered habitat) that are associated with tidal restoration projects.

The book closes with a summary of the state of science with regard to tidal restoration of salt marshes and the application of this knowledge to the implementation or practice of salt marsh restoration. Enhancing our ability to understand, predict, and plan for the response of tide-restored salt marshes to accelerated rates of sea level rise was a recurring theme throughout this edited volume and is appropriately a focus of the final chapter.

A Justification for Tidal Restoration Initiatives

Within coastal zones of the world, salt marshes, mangroves, and other ecosystem types have been destroyed due to filling and dredging operations, sometimes at alarming proportions. In the New England region it is estimated that 37 percent of salt marshes have been lost, while in urban centers, like Boston, salt marsh loss is even greater (81 percent) (Bromberg and Bertness 2005). Within the Canadian Maritimes there has been an estimated 64 percent loss of coastal wetlands, mostly attributed to agricultural reclamation, and along the Pacific US coast there is reportedly a 93 percent loss of coastal marsh, with the urban San Francisco Bay dominating the loss statistic (Gedan and Silliman 2009). In addition to these losses, coastal wetland habitat is degraded by tidal restrictions, impoundments, diking, ditching, invasive species, storm water discharge, nutrient enrichment, and other factors. Combined with losses, habitat degradation has impacted the ability of once vibrant coastal marshlands to support fish and bird populations, provide storm protection, sequester carbon, contribute to water quality maintenance, and provide open space for recreation and aesthetics. Reintroducing tidal flow to tide-restricted salt marshes represents a technique that can be successfully implemented to restore the functions of degraded salt marshes and enhance resilience to climate change effects. It is our hope that this book will provide stewards of the coastal zone with the scientific foundation and practical guidance necessary to implement effective and necessary tidal restoration initiatives.

REFERENCES

Able, K. W., and S. M. Hagan. 2000. "Effects of Common Reed (*Phragmites australis*) Invasion on Marsh Surface Macrofauna: Response of Fishes and Decapod Crustaceans." *Estuaries* 23:633–46.

Able, K. W., S. M. Hagan, and S. A. Brown. 2003. "Mechanisms of Marsh Habitat Alteration due to *Phragmites*: Response of Young-of-the-Year Mummichog (*Fundulus heteroclitus*) to Treatment for *Phragmites* Removal." *Estuaries* 26:484–94.

Anisfeld, S. C., and G. Benoit. 1997. "Impacts of Flow Restrictions on Salt Marshes: An Instance of Acidification." *Environmental Science and Technology* 31:1650–57.

Benoit, L. K., and R. A. Askins. 1999. "Impact of the Spread of *Phragmites* on the Distribution of Birds in Connecticut Tidal Marshes." *Wetlands* 19:194–208.

Bromberg, K. D., and M. D. Bertness. 2005. "Reconstructing New England Salt Marsh Losses Using Historical Maps." *Estuaries* 28:823–32.

Burdick, D. M., M. Dionne, R. M. Boumans, and F. T. Short. 1997. "Ecological Responses to Tidal Restorations of Two Northern New England Salt Marshes." *Wetlands Ecology and Management* 4:129–44.

Byers, S. E., and G. L. Chmura. 2007. "Salt Marsh Vegetation Recovery on the Bay of Fundy." *Estuaries and Coasts* 30:869–77.

Callaway, J. C., and J. B. Zedler. 2004. "Restoration of Urban Salt Marshes: Lessons from Southern California." *Urban Ecosystems* 7:107–24.

Chapman, V. J. 1960. *Salt Marshes and Salt Deserts of the World*. New York: Interscience Publishers.

Crain, C. M., K. B. Gedan, and M. Dionne. 2009. "Tidal Restrictions and Mosquito Ditching in New England Marshes." Pp. 149–69 in *Human Impacts on Salt Marshes: A Global Perspective*, edited by B. R. Silliman, E. D. Grosholz, and M. D. Bertness. Berkeley: University of California Press.

Daiber, F. C. 1986. *Conservation of Tidal Marshes*. New York: Van Nostrand Reinhold.

Davy, A. J., J. P. Bakker, and M. E. Figueroa. 2009. "Human Modification of European Salt Marshes." Pp. 311–35 in *Human Impacts on Salt Marshes: A Global Perspective*, edited by B. R. Silliman, E. D. Grosholz, and M. D. Bertness. Berkeley: University of California Press.

Dionne, M., F. T. Short, and D. M. Burdick. 1999. "Fish Utilization of Restored, Created, and Reference Salt-Marsh Habitat in the Gulf of Maine." Pp. 384–404 in *Fish Habitat: Essential Fish Habitat and Rehabilitation*, edited by L. R. Benaka. American Fisheries Society Symposium 22. Bethesda, MD: American Fisheries Society.

DiQuinzio, D. A., P. W. C. Paton, and W. R. Eddleman. 2002. "Nesting Ecology of Saltmarsh Sharp-Tailed Sparrows in a Tidally Restricted Salt Marsh." *Wetlands* 22:179–85.

Doody, J. P. 2008. *Saltmarsh Conservation, Management and Restoration*. Springer.

Ganong, W. F. 1903. "The Vegetation of the Bay of Fundy Salt and Diked Marshes: An Ecological Study." *Botanical Gazette* 36:161–86.

Gedan, K. B., and B. R. Silliman. 2009. "Patterns of Salt Marsh Loss within Coastal Regions of North America." Pp. 253–65 in *Human Impacts on Salt Marshes: A Global*

Perspective, edited by B. R. Silliman, E. D. Grosholz, and M. D. Bertness. Berkeley: University of California Press.

Hunter, K. L., D. A. Fox, L. M. Brown, and K. W. Able. 2006. "Responses of Resident Marsh Fishes to Stages of *Phragmites australis* Invasion in Three Mid-Atlantic Estuaries." *Estuaries and Coasts* 29:487–98.

Montague, C. L., A. V. Zale, and H. F. Percival. 1987. "Ecological Effects of Coastal Marsh Impoundments: A Review." *Environmental Management* 11:743–56.

Nichols, F. H., J. E. Cloern, S. N. Luoma, and D. H. Peterson. 1986. "The Modification of an Estuary." *Science* 231:567–73.

NOAA Restoration Center and NOAA Coastal Services Center. 2010. *Returning the Tide: A Tidal Hydrology Restoration Guidance Manual for the Southeastern United States.* Silver Spring, MD: National Oceanic and Atmospheric Administration.

Pethick, J. 2002. "Estuarine and Tidal Wetland Restoration in the United Kingdom: Policy versus Practice." *Restoration Ecology* 10:431–37.

Philipp, K. R. 2005. "History of Delaware and New Jersey Salt Marsh Restoration Sites." *Ecological Engineering* 25:214–30.

Portnoy, J. W. 1991. "Summer Oxygen Depletion in a Diked New England Estuary." *Estuaries* 14:122–29.

Portnoy, J. W. 1999. "Salt Marsh Diking and Restoration: Biogeochemical Implications of Altered Wetland Hydrology." *Environmental Management* 24:111–20.

Ranwell, D. S. 1972. *Ecology of Salt Marshes and Sand Dunes.* London: Chapman and Hall.

Raposa, K. B., and C. T. Roman. 2003. "Using Gradients in Tidal Restriction to Evaluate Nekton Community Responses to Salt Marsh Restoration." *Estuaries* 26:98–105.

Roman, C. T., W. A. Nicring, and R. S. Warren. 1984. "Salt Marsh Vegetation Response to Tidal Restriction." *Environmental Management* 8:141–50.

Sebold, K. R. 1992. *From Marsh to Farm: The Landscape Transformation of Coastal New Jersey.* Washington, DC: US Department of the Interior, National Park Service, Cultural Resources.

Thom, R. M., R. Zeigler, and A. B. Borde. 2002. "Floristic Development Patterns in a Restored Elk River Estuarine Marsh, Grays Harbor, Washington." *Restoration Ecology* 10:487–96.

Thomsen, M. S., P. Adam, and B. R. Silliman. 2009. "Anthropogenic Threats to Australasian Coastal Salt Marshes." Pp. 361–90 in *Human Impacts on Salt Marshes: A Global Perspective*, edited by B. R. Silliman, E. D. Grosholz, and M. D. Bertness. Berkeley: University of California Press.

Warren, R. S., P. E. Fell, R. Rozsa, A. H. Brawley, A. C. Orsted, E. T. Olson, V. Swamy, and W. A. Niering. 2002. "Salt Marsh Restoration in Connecticut: 20 Years of Science and Management." *Restoration Ecology* 10:497–513.

Weinstein, M. P., J. H. Balletto, J. M. Teal, and D. F. Ludwig. 1997. "Success Criteria and Adaptive Management for a Large-Scale Wetland Restoration Project." *Wetlands Ecology and Management* 4:111–27.

Williams, P., and P. Faber. 2001. "Salt Marsh Restoration Experience in San Francisco Bay." Special issue, *Journal of Coastal Research* 27:203–11.

Williams, P. B., and M. K. Orr. 2002. "Physical Evolution of Restored Breached Levee Salt Marshes in the San Francisco Bay Estuary." *Restoration Ecology* 10:527–42.

Williams, R. J., and F. A. Watford. 1996. "An Inventory of Impediments to Tidal Flow in NSW Estuarine Fisheries Habitats." *Wetlands (Australia)* 15:44–54.

Wolters, M., A. Garbutt, and J. P. Bakker. 2005. "Salt-Marsh Restoration: Evaluating the Success of De-embankments in North-west Europe." *Biological Conservation* 123:249–68.

Wozniak, A. S., C. T. Roman, S. C. Wainright, R. A. McKinney, and M. J. James-Pirri. 2006. "Monitoring Food Web Changes in Tide-Restored Salt Marshes: A Carbon Stable Isotope Approach." *Estuaries and Coasts* 29:568–78.

Zedler, J. B., ed. 2001. *Handbook for Restoring Tidal Wetlands*. Boca Raton, FL: CRC Press.

PART II

Synthesis of Tidal Restoration Science

The preceding introductory chapter provided a brief overview of the environmental consequences of salt marshes subject to restricted tidal exchange caused by roads, railroads, dikes, and other infrastructure, followed by recognition that the practice of restoring tidal flow to these degraded marshes has been successfully pursued worldwide. The chapters in this part of the book synthesize the extensive scientific literature that is available on the impacts of tide restriction on salt marshes, and moreover, on the responses to tide restoration. The geographic focus of the science synthesis chapters is New England and Atlantic Canada—a region where tidal flow restoration has been ongoing for decades, documented through an abundant multidisciplinary literature. But it is important to note that the chapters strive to incorporate relevant literature from other regions, as well as providing discussion on the applicability of the findings beyond the region. The science synthesis chapters emphasize trajectories of change throughout the tidal restoration process and close with recommended research needs to further our understanding of marsh responses to tidal restoration.

MacBroom and Schiff (chap. 2) offer a review of tidal marsh hydrologic concepts followed by a synthesis of hydraulic modeling (ranging from simple to complex three-dimensional models) used to predict hydrologic responses to various scenarios of tidal flow restoration. Anisfeld (chap. 3) discusses biogeochemical aspects of salt marshes under regimes of tide restriction and subsequent tide restoration, including pore water/sediment salinity, redox and sulfide, nutrients, metals, and others. Given a foundation on the physical and biogeochemical factors related to tidal restoration responses, the remaining science synthesis chapters focus on biological responses. Smith and Warren (chap. 4) explore the factors that influence plant communities during the restoration process, while Chambers and coauthors (chap. 5) focus on the ecology of *Phragmites australis* (common reed), often a dominant invader of tide-restricted salt marshes and a target of restoration efforts. Raposa and Talley (chap. 6) conduct a meta-analysis of tidal restriction impacts on nekton communities (free-swimming fish and decapod crustaceans) and responses to tidal restoration. This part of the book closes at a higher trophic level, with a synthesis of avian community responses to tidal restoration (Shriver and Greenberg, chap. 7).

Chapter 2

Predicting the Hydrologic Response of Salt Marshes to Tidal Restoration

The Science and Practice of Hydraulic Modeling

JAMES G. MACBROOM AND ROY SCHIFF

The hydraulic gradients caused by tides are the primary source of physical energy in coastal salt marshes. The salt marsh ecosystem is driven by the interaction of tidal and freshwater hydrology, hydraulics, and sediment processes that determine water depth, duration of inundation, and amount of sediment erosion and deposition. The movement of water through tidal creeks and over marshes also establishes local water quality such as salinity, temperature, and dissolved oxygen as freshwater and saltwater mix.

Many tidal marshes have modified hydrologic processes that alter habitat and ecological interactions due to changes in tide levels, tidal prism, and salinity levels (e.g., Roman et al. 1984; Environmental Agency 2008). The origin of degraded salt marshes is often the constriction or blockage of channels that restrict tidal flow and alter tide levels. Tidal barriers modify flow, water surface elevation, flood volume, salinity, sediment transport rates, and the movement of aquatic organisms. The vast storage and conveyance typical of a natural marsh are reduced with increasing frequency and severity of tidal barriers, a common condition in salt marshes, especially within developed watersheds. Tidal barriers can include undersized culverts, tide gates, sluiceways, bridges, and other types of structures.

A key facet of most marsh restoration projects is the return toward natural hydrologic processes; thus hydraulic modeling is an analysis and design element essential to restoring a salt marsh. Modeling of the marsh and structures, in conjunction with investigating marsh channel morphology and equilibrium conditions, enables reduction or elimination of flow restrictions to return the appropriate tide ranges and storm surges, which in turn allow natural (passive) restoration.

The analysis and prediction of hydraulics within a tidal marsh, with its network of channels and complex flow patterns, is one of the most complicated challenges faced by hydraulic engineers. This chapter discusses analysis of tidal marsh hydraulics using analog, empirical, mathematical, and physical models. Important objectives of hydraulic modeling include accurately representing the combination of tidal exchange and storm surge to predict flow depth and velocity over a range of flow magnitudes. Model results may be directly used to identify changes in upland flooding, sediment transport, aquatic habitat, marsh vegetation, salinity levels, and fish passage under a range of restoration alternatives. Many of these challenging tasks require multiple models and interdisciplinary data collection to establish relationships to marsh hydraulics.

Hydrologic and Hydraulic Concepts Relevant to Modeling Salt Marshes

Hydraulic modeling of salt marshes includes the characterization of the tidal prism, tidal action, the marsh water budget, and flow types. The tidal prism and runoff are typical inputs to the model, while the model output includes flow types, hydraulics, and the resulting water budget.

Tidal Prisms

Various definitions of *tidal prism* exist that will guide modeling of the marsh over a range of conditions (PWA 1995). Models should consider the range of tidal prisms as well as storm conditions so that proposed tide restoration alternatives for the marsh can be investigated over a range of conditions.

- *Mean tidal prism* is the volume of water in the estuary between the elevations of mean high water and mean low water. Mean high water is approximately the bankfull channel stage that often includes inundation of low marsh plains.
- *Spring tidal prism* is the volume of water between the annual mean spring high and spring low tides.

Tides in Marshes and Rivers

High tide levels in marshes are usually lower than in open coastal waters because the hydraulic roughness associated with tidal creeks and the marsh surface limits the tidal surge (e.g., Aubrey and Speer 1985). Friction delays tidal exchange, and the marsh generally does not have time to fill before tides begin to fall in the open coastal waters. The hydraulic roughness leads to a marsh high tide that lags behind open-water high tide. High tide levels may vary throughout the marsh, with the lowest high tide levels generally occurring in the most hydraulically remote areas. The interior tide elevations in marshes and channels are influenced by channel bed and bank friction, channel conveyance, vegetation, sediment bars, freshwater runoff, and artificial restrictions.

The elevation of the marsh surface is driven by tidal variations and freshwater inflow and the associated patterns of sediment erosion and deposition. In older marshes, the marsh plain reaches an equilibrium surface elevation between mean high water and mean spring high water, and salinity will influence vegetation up to the elevation of maximum astronomical plus meteorological (storm) tides.

Rising tides push saltwater into marshes, tidal channels, and freshwater rivers. The dense, cold saltwater that flows beneath warmer and less saline water often creates stratified conditions. The flood tide blocks marsh and river discharges by creating an underlying wedge of saltwater causing water levels to rise and local river flow to reverse and head inland. Tidal influence in rivers may extend far inland beyond the limit of saltwater intrusion due to the "backwater" effect.

Tidal Marsh Water Budget

Water is a conservative substance and, within a marsh system, the conservation of mass states that the summation of water inflow minus the summation of water outflow equals the change in water held in storage. As the volume of water stored in a marsh increases or decreases, the elevation of water also increases or decreases due to perennially saturated soils. The ability to account for inflow, outflow, and changes in water storage with corresponding water elevations forms the basis of most hydrology and hydraulic models for marshes.

Sources of freshwater to marshes include precipitation, groundwater, and overland runoff, while some tidal marshes also receive substantial quantities of freshwater from streams and rivers. Freshwater inflow rates may be estimated from US Geological Survey (USGS) gauging stations, local river gauging, regional regression equations such as the USGS National Flood Frequency (NFF) model, and hydrologic computer models such as the US Army Corps of Engineers Hydrologic Engineering Center–Hydrologic Modeling System (HEC-HMS) or the Natural Resources Conservation Service (NRCS) Technical Release 20 (TR-20).

The primary source of water in salt marshes is the ocean, and water surface levels are driven by the rise and fall of coastal tides. Saltwater floods into channels and low-lying marsh surfaces during rising tides. It fills tidal channels as the tide rises, and it occasionally spreads over the marsh surface at the peak of high tide. The saltwater, plus available excess water from freshwater sources, drains out of the marsh during the ebb tide.

The average flow through a tidal system is the tidal prism plus freshwater runoff volume divided by the time duration between mean high water and mean low water (modified from the US Department of Transportation's [USDOT]

Hydraulic Engineering Circular No. 25 [HEC-25], after Neill 1973). The maximum tidal flow can be approximated by the following:

$Qmax = \pi\ (P)\ T^{-1}$

Qmax = Maximum discharge, cfs

$\pi = 3.14$

P = Tidal prism volume, cubic feet

T = Time duration, sec

Peak flow rate is assumed to occur midway between high and low tide. As a rule of thumb, the maximum discharge is about three times the average tidal flow.

Flow Types

Many types of water flow can occur in open channels such as tidal creeks and ditches. The appropriate classification of flow types helps one to select the appropriate type of hydraulic analysis or model used to predict water velocity, elevation, and direction. The primary flow classifications involve spatial characteristics, temporal characteristics, stratification or density, and energy.

Uniform flow occurs when the water profile is parallel to the bed and the depth is approximately constant along the longitudinal center line. Tidal channels typically contain nonuniform flow where the water depth varies with distance along the channel length, such as where cross-sectional area or longitudinal slope varies. Nonuniform flow is called gradually varied flow when the flow depth changes slowly with distance along the channel, and nonuniform flow is called rapidly varied flow when the flow depth changes over a short distance of channel.

Steady flow occurs when the water flow rate is constant over a specified time period that enables the water depths at a cross section to be constant and in equilibrium. In unsteady flow the water depth at a cross section varies with time as the discharge rate changes, such as during a storm event or tidal cycle. Unsteady flow modeling is often required to accurately represent the dynamic flow environment in salt marshes.

Homogeneous, or well-mixed, flow occurs when the water density is constant, a common condition for shallow fresh or marine waters. Stratified flow occurs when water density varies in horizontal or vertical directions. Horizontal stratification often occurs in estuaries as less saline water draining from the coast flows next to more salty water, creating longitudinal boundaries. These boundaries are often visible on the water surface as areas of shear and eddying. Vertical stratification occurs as density varies with depth due to salinity and temperature gradients.

For example, freshwater flowing into a marsh flows over the underlying tidal prism containing more salty and cold water. Most analyses of shallow tidal marshes and channels assume homogeneous flow for simplicity. This level of detail is typically suitable for exploring restoration alternatives.

The Simmons ratio is the volume of the river flow per tidal cycle divided by the volume of tidal inflow (i.e., the tidal prism). The ratio can be used as an indicator of potential stratification and dilution. When the ratio is 1.0 or greater, a highly stratified system can be expected. Ratios less than 0.1 correspond to a well-mixed, unstratified system with limited freshwater (USACOE 1991). Tidal systems with constrictions and high velocities also tend to be well mixed.

The ratio between inertia (velocity) forces and gravity (water depth) is defined by the Froude number (F). Values less than 1 indicate that gravity forces dominate and flow is deep and smooth (i.e., subcritical flow). Water elevations in subcritical flow are influenced by downstream water depths and energy levels. Froude numbers greater than 1 indicate inertial forces dominate with higher velocities, more turbulence, and shallower water depth (i.e., supercritical flow). Only select models can analyze supercritical flow. Upstream water depths are independent of downstream conditions during supercritical flow. Some models are capable of representing subcritical and supercritical flow at the same time (i.e., mixed flow).

Data Collection

Some level of field inspection, survey, geomorphic assessment, and gauging is required to gather information to run hydraulic models of salt marshes. The level of data collection influences the level of detail possible in the modeling and ultimately the accuracy of prediction of the marsh hydraulics.

Field Inspection and Surveys

It is essential for hydraulic modelers to inspect and become familiar with the details of tidal marshes and creeks prior to beginning model setup. Hydrographic inspections usually include both observations and direct measurement of tide elevations, high water marks, flow velocity, discharge rates, freshwater runoff sources and magnitude, salinity levels, and water column stratification by salinity or temperature. The modeler must also estimate friction coefficients for the channels and marsh surfaces, which requires an investigation of substrate type, marsh vegetation, and accumulated organic material. Structures and tidal restrictions must be inspected to determine their size and function. Boundary conditions need to be identified, and cross-section survey points of channel reaches and junctions should be initially located.

Tidal marshes often have pools or intermittently flooded pannes that become isolated from tidal channels at low tide, creating instability in dynamic models. The location, elevation, depth of water, and likely inundation period of the isolated areas that transition between wet and dry during the model run need to be measured to inform a decision to include them in the model or not. Isolated tidal pools that make up a small fraction of the total storage in the marsh during low flow are often eliminated from the model. Similarly, areas of high ground surrounded by low-lying marsh should be field identified as they may become islands during periods of high water, thus reducing flood storage.

Conventional elevation surveys with levels, theodolites, lasers, or Global Positioning System (GPS) are often required to measure channel cross sections, longitudinal profiles, structure size and elevations (e.g., levees, dikes, etc.), and marsh plain topography. To accommodate public safety, ground-based surveys should also locate surrounding infrastructure such as nearby buildings, roads, and utilities that may be prone to flooding and limit flood storage. Due to the low relief of marshes, even very small (i.e., 3 centimeters) changes in elevation can have a substantial impact on flood storage, flow patterns, and resulting ecological communities.

Remotely collected data (e.g., Light Detection and Ranging [LIDAR] data, aerial photographs, topographic maps, vegetation maps) accompanying the field surveys can allow for rapid survey of elevations and location of features. Ground-truthing and expanding remotely sensed data with on-the-ground field survey are essential data collection steps to create an accurate model geometry and project design.

Geomorphic Assessment

Geomorphic field studies are performed to identify channel patterns, mosquito ditch networks, channel substrate types and erosion thresholds, channel bank heights and stability, the extent of active floodplain, ebb or flood tide sediment bars, deltas, shoals, mud flats, and evidence of erosion or deposition. The geomorphic assessment should comment on past marsh form and processes using historic aerial photographs and existing knowledge of the site, current marsh and channel dynamics as observed in the field, and anticipated changes in the future. Probing the marsh surface and channel bed with soil augers or steel rod soil samplers provides information on substratum (i.e., peat, cohesive fine grain deposits, and gravel or till premarsh materials) around the marsh to gain an understanding of erosion potential during a range of flows. Evidence of marsh recession or growth should be recorded.

Flow rates in tidal channels are dependent on the tide range and prism. The geomorphic assessment should include a survey of fluvial geomorphic indicators (e.g., bankfull width, depth, sinuosity, stream order, substrate type, cross-sectional area, and others) at channel cross sections.

Gauging Tides and Floods

One or more gauges are installed to continuously record water surface elevations based on survey to a known benchmark. Water surface elevation data provide key information on the mixing of tidal cycling and freshwater inflow that the modeler is attempting to accurately represent. Short-term gauging (e.g., one week to several months) characterizes the nature of flood conveyance through the marsh under different lunar phases in the tidal cycle and illustrates the impact of hydraulic restrictions, especially when multiple gauges are deployed upstream and downstream of restricting structures and near the model boundary conditions. Comparing short-term gauge data to published data from an existing long-term gauge and the marsh surface elevation can provide a useful picture of marsh flood dynamics.

The National Oceanic and Atmospheric Administration and other agencies operate a network of tide gauges along coastal areas of the United States (http://co-ops.nos.noaa.gov/). Long-term gauge data are useful for verifying the tidal datum (e.g., North American Vertical Datum of 1988, Mean Sea Level, and Mean High Water) and key hydrologic levels. Records from long-term tide gauges contain information that can be used for reference to short-term gauge data and during modeling.

Additional Data Collection

Marsh studies may include more advanced data collection beyond standard stream gauging. Velocity profiles can be recorded with acoustic Doppler current profilers to directly calibrate and validate model results. Acoustic Doppler meters can also measure surface waves to document coastal storm surge.

Temperature, conductivity, and salinity are regularly used to study water quality at various locations in the marsh. These data are important to evaluate changes to water quality following implementation of a restoration project. Conductivity-temperature-depth sensors can characterize stratification and mixing in the water column. Turbidity and total suspended solids are sometimes measured to estimate sediment storage and transport.

Tidal Hydraulic Modeling Approaches

There are several types of models and procedures that can be used to assess salt marsh hydrology, flow restrictions, and restoration. They can be organized in terms of four design techniques—analog, empirical, mathematical, and physical. Analog and empirical models can be used for small projects where site constraints and environmental risks are limited. More commonly, these model types are used to initiate design for further analysis in one or more mathematical models. Mathematical models are the most common for evaluating marsh hydraulics. Due to the high cost and time for executing, physical models are typically reserved for large studies or applied projects.

Analog Techniques

Analog models are based upon modifying disturbed marshes to replicate reference sites in the same region. This approach is an effective field technique to attempt to restore the tidal regime and its relationship to marsh surface elevations, soil types, salinity, and water quality. If reference physical and chemical conditions can be matched, then it is assumed that ecosystem function can be restored.

Required field observations and measurements of the reference site include the use of tide gauges and topographic surveys and measuring the tidal prism and salinity. The depth, frequency, and duration of marsh inundation are needed to determine the boundary between the high marsh and low marsh and associated habitat and natural communities. Channel flow velocities and substrate types help identify sediment transport characteristics.

Restoration efforts may include removing tidal restrictions to modify flow to achieve desired water surface elevations on the marsh. This also helps adjust salinity, water quality, and sediment loads. Without a detailed geomorphic or mathematical model to predict high tide and flood levels, these variables must be observed and adjusted to set desired marsh submergence while minimizing the risk of flooding to surrounding permanent infrastructure. The use of analog techniques is enhanced by careful monitoring before, during, and after project implementation so that project evaluation and adaptive management can be performed.

One approach to collecting data and using reference sites is based upon the Hydrogeomorphic Approach to assess tidal wetland functions and values (Shafer and Yozzo 1998). This technique was developed for the US Army Corps of Engineers, in consultation with the US Fish and Wildlife Service, Natural Resources Conservation Service, US Environmental Protection Agency, and National Oceanic and Atmospheric Administration, to be used as part of the Clean Water Act Section 404 regulatory program. Another technique for evaluating tidal marshes and potential restoration efforts is based upon the current *Highway Methodology Workbook* used by the New England Division of the US Army Corps of Engineers (NED 1993). Coupled with marsh plain and tide elevations, the foregoing wetland evaluation systems provide the foundations for an analog wetland restoration model.

Empirical Techniques

Empirical techniques, primarily the suite of hydraulic geometry relations, can be used to estimate channel sizes for proposed tidal restoration conditions following naturalization of the tidal prism. This is useful during the initial design phases and to compare results to field measurements during analog methods and during more complex modeling. The concept that tidal channels evolve through sediment erosion and deposition toward an equilibrium size means that exact channel sizing may not be necessary for restored marshes. If channels have ample space, dimensions will adjust based on the combination of tidal cycling and the range of incident freshwater flows (Teal and Weishar 2005). In cases where space for the channel to evolve is limited due to infrastructure, property boundaries, or competing land uses, more detailed channel sizing with mathematical techniques is often required.

Common geomorphic features of tidal channels include variations in plan form (straight, meandering, irregular), natural levees of sediment along their banks, low vertical cohesive banks or angled granular banks, bank seepage (only at low tide), pool and riffle along the profile, bend scour, contraction scour, and mass bank failures. The tidal creek bankfull width and depth, typically located at the marsh surface, increases in the downstream direction, and the bed sediment size varies and is sorted by the distribution of channel velocities. The key geomorphic channel metrics in tidal channels are formed by the relationships between physical and biotic variables such as tide range, sediment size, vegetation, and geomorphic history (PWA 1995).

Langbein (in Myrick and Leopold 1963) showed that the tidal channel width, depth, and cross-sectional area are functions of the tidal prism. Given this relationship between channel dimensions and tidal prism, an early example of empirical geomorphic techniques for analyzing and designing tidal channels was established (PWA 1995). Hydraulic geometry data were collected at thirty-seven cross sections in seven marshes from San Diego to San Francisco Bay over ten years and compared to tidal prism. Channel width, depth, and cross-sectional area were measured at the elevation of mean higher high water, and the tidal prism upstream of each cross section was measured based on channel volume and marsh inundation volume. The data showed a positive correlation (log-log) between channel measurements and tidal prism. The aforementioned relationships should be applicable to both the East Coast and the Gulf Coast since they are based on geomorphic processes, yet local reference data should be used for verification (PWA 1995). In addition, Zeff (1999) studied the morphology of six tidal channels in New Jersey and found that the "at a station" hydraulic geometry was similar to tidal channels in Virginia, Delaware, and California, implying that the geomorphic data are transferable.

An expanded hydraulic geometry data set specifically for design of tidal marsh channels based on data from coastal San Francisco, California, was published by Williams et al. (2002). This updated hydraulic geometry database with large marshes (area up to 5700 hectares) was stratified by the age of the marsh because lag time increases with the age of the marsh. From an expanded data set, hydraulic geometry relationships provide an empirical tool to predict tidal channel dimensions in mature tidal marshes (Williams et al. 2002; Williams and Faber 2004). For example, maximum depth (D) is equal to $0.388\ P^{0.176}$, where P is the tidal prism in cubic meters. Regional hydraulic geometry relations can be prepared as a function of tidal marsh size because the tidal prism is related to marsh size (Williams et al. 2002).

The application of tidal marsh geomorphic empirical models was used in New England on the Oyster River and Black River in Old Saybrook, Connecticut. A series of channel cross sections and upstream marsh areas were measured and related to tidal prism. Tidal prism was estimated by multiplying the measured cross-sectional areas times the channel reach lengths for each reach (Carey et al. 2006). The results indicated that Oyster River and Black River have bankfull channel widths similar to the California data (PWA 1995). An empirical relationship for channel width (and depth) now exists along the Connecticut coast with which to design the dimensions of tidal channels.

The British Environmental Agency has been conducting extensive research on tidal estuaries including salt marshes (http://www.estuary-guide.net/). They recognize that empirical geomorphic modeling including the use of hydraulic geometry relationships based upon discharge or tidal prism is an important design tool. Their geomorphically based approach includes historical analysis of changes, analysis of tide range and species, time series analysis of cycles, sediment budget analysis, and geomorphic analysis of landforms.

Mathematical Modeling

A wide range of mathematical models are available to predict the hydraulics of salt marshes, ranging from simple steady-state mass balance approaches to time-varying two-dimensional algorithms. A review of the theory behind the most common approaches, along with several examples, follows.

LEVELS OF MATHEMATICAL HYDRAULIC MODELS

Mathematical models are based upon application of fundamental theories of fluid dynamics including conservation of mass, momentum, and energy. They range from simple "box" models that assume uniform conditions to complex analytical and numerical models in one (i.e., longitudinal), two (i.e., longitudinal and lateral), or three dimensions (i.e., longitudinal, lateral, and vertical) (table 2.1). The selection of appropriate models requires a clear understanding of project goals and objectives, the level of detail for adequate design, available data, project budgets, and project schedule—all a function of the project's level of complexity. For example, restoration projects with no surrounding infrastructure are simple in

TABLE 2.1

Tidal marsh mathematical hydraulic model levels[1]

Model level	Concept	Data needs	Comments
1 (most simple)	Mass balance Box model Steady state	Marsh area Tide range	Preliminary for planning purposes
2	Mass balance Box model Semisteady flow Uniform water elevation Reservoir routing	Marsh area Stage-storage Tide range Stage discharge	Manual routing ,TR-20, HEC1, HEC-RAS
3	1-D dynamic Routing, unsteady Nonuniform	Marsh cross sections Roughness factors Tide data Sediment transport	Linear systems, branching systems: USGS branch, HEC-RAS, HEC-2, WSPRO, UNET, DYNLET
4	2-D dynamic Routing, unsteady Nonuniform	Marsh topography Element layout Roughness Tide data Sediment transport	Irregular shapes (wet, dry, vectors), RMA2, FES-WMS, River 2-D, FLO-2D, ADCIRC, EFDC, MIKE 21, M2D, HIVEL2D, AdH
5 (most complex)	3-D dynamic	Stratified flow, deep water bodies	RMA-10, UNTRIM, EFDC

[1]Many different computer models are available for evaluating tidal marshes. The programs listed here are mainly public domain software that are readily available and in common use.
TR-20, Technical Release 20, Natural Resources Conservation Service; HEC-1, Hydrologic Engineering Center Flood Hydrograph Package; HEC-RAS, Hydrologic Engineering Center River Anaysis System; HEC-2, Hydrologic Engineering Center Water Surface Profiles; WSPRO, USGS Water Surface Profile Model; UNET, USACE unsteady state hydraulic model; DYNLET, USACE DYNamic Behavior of Tidal Flow at InLETs; RMA2, USACE Resource Management Associates 2; FES-WMS, USGS Finite-element surface-water modeling system; River 2-D, Two-dimensional hydrodynamic model by University of Alberta; FLO-2D, River and floodplain model; ADCIRC, Coastal Circulation and Storm Surge Model; EFDC, USEPA Environmental Fluid Dynamics Code; MIKE 21; flows, waves, sediments and ecology in rivers, lakes, estuaries, bays, coastal areas and seas in two dimensions; M2D, USACE Two-Dimensional Depth-Averaged Circulation; HIVEL2D, flow analysis and hydraulic jump in high-velocity open channels mode/modeling/modelling software—free surface; AdH, Adaptive Hydraulics, USACE Coastal and Hydraulics Laboratory; RMA-10, USAC multidimensional (combining 1-D, 2-D either depth or laterally averaged, and 3-D elements) finite element numerical model written in FORTRAN-77; UNTRIM, unstructured grid version of TRIM.

terms of maintaining public safety as compared to marshes with multiple constrictions and surrounding infrastructure that is subject to flooding.

Mathematical models are classified as being analytical or numeric depending upon the form of calculations being used. Analytical solutions are those in which answers are obtained by direct application of mathematical expressions or equations that represent physical phenomena (USACOE 1991). Manning's equation describing uniform open channel flow is an example of a direct analytical model. Analytical equations exist to define weirs, orifices, culverts, and tide gates. Analytical models can often be solved at several locations manually or with a computer program over a network of channels with associated floodplains. Complex processes are often simplified as empirical coefficients, creating a hybrid model between an empirical and an analytical model. Analytical solutions are relatively simple and rapid compared to numerical solutions.

Numerical solutions used for hydrodynamic analysis have iterative computational procedures to solve mathematical expressions that typically do not have unique solutions (USACOE 1991). Numerical analysis often solves mathematical expressions that are a function of time, often by iteration or approximation. Dynamic models that represent tidal flow conditions must arrive at solutions for short time durations during the flood and ebb tides and thus are numerical intratidal models.

Three principal types of numerical models are finite differences, finite elements, and finite volumes. Finite difference models require finite intervals of time and space with regular grids, while the finite element models proceed step by step through a series of simultaneous equations for discrete elements such as a row of elements in a two-dimensional grid. Finite volumes combine the ability of finite elements to represent complex bathymetries with conservation of volume or mass by finite differences. Numerical modeling is capable of more spatially detailed analysis and is typically more accurate than analytical models, yet numerical models require a large coverage of high-resolution survey.

Mathematical models may represent the project sites in one, two, or three dimensions. One-dimensional models contain cross-sectional data along a single linear scale that extends along the length of the marsh, usually following the main channel. One-dimensional models calculate hydraulic conditions only at measured cross-sectional locations. Two-dimensional models represent the length and width of the marsh and assume that waters are well mixed and velocities are vertically averaged. Two-dimensional models describe hydraulic conditions over a spatial grid of the marsh surface. Three-dimensional models represent length, width, and depth, and are used to characterize complex hydraulics such as stratified flow and vertical turbulence.

One-dimensional hydrodynamic models are powerful tools for marsh assessments because they are useful for both assessment and design of tidal channels

TABLE 2.2

Hydraulic structures and equations

Structure	Equation
Bridge (flowing part full)	US Geological Survey and Federal Highway Administration Bridge equations
Bridge (flowing full)	Orifice equation
Spillway	Weir equation
Pipe and box culverts	Inlet and outlet control equations
Tide gates	Weir and orifice (one-way) equations
Pipes	Manning's equation

and structures. In contrast, two-dimensional models provide superior data on marsh water elevations and circulation but tend to be more complex than necessary for basic designs. An important advantage of two-dimensional models is the ability to represent ponding with nonuniform surface elevations.

Hydraulic Structures

Many tidal marshes and creeks have one or more hydraulic structures at road or railroad crossings, contractions, or water level controls that regulate flow rates (table 2.2). Most hydraulic structures such as bridges, culverts, weirs, and tide gates can be represented by standard one-dimensional engineering equations based upon decades of research. These equations, and others, are incorporated into some of the available models that generally require geometric data, water elevations, and use of empirical tables to estimate various coefficients for wingwall, shape, pier, contraction, and expansion conditions.

Level One: Uniform Mass Balance Approximations

The first and most basic level of hydraulic mathematical analysis represents the marsh as a black box, without longitudinal or lateral flow, with some storage capacity. In box models, the marsh is a simple structure that assumes a steady, uniform horizontal planar water elevation while neglecting internal velocities and hydraulic processes. Box models are most valid in marshes with little to no lag time that rapidly transmit water to and from all areas (e.g., an open water tidal pond or nonvegetated tidal flats with little friction). Results would most closely represent average tidal conditions over long periods of time since the rate of inflow to the marsh remains constant over a sustained period of time.

Box models have been used for many decades and are suitable for preliminary planning studies and for simple calculations in small marshes. A primary limitation of these models is the assumption of homogeneous marsh conditions, so they are not able to assess changes in interior marsh conditions such as different creek configurations, creation of marsh pools, or other marsh restoration parameters.

Box model analysis requires only the volume of the potential tidal prism—the balance of water that would exist in the marsh between mean low water and mean high water. The marsh inlet (channel, culvert, bridge, etc.) must be able to supply this volume during the rising tide cycle, which is about 6.5 hours in the northeastern United States.

For simple and low-risk restoration projects, the required size of culverts or tide gate openings that regulate flow into a marsh can be approximated using steady-state flow assumptions and box models. Culvert capacities along freshwater streams are commonly evaluated using Federal Highway Administration Hydraulic Engineering Circulars no. 5 and no. 10 (FHWA 1985), which provide nomographs for sizing calculation, or the steady-state computer program HY-8 that is available in the public domain from the US Department of Transportation (http://www.fhwa.dot.gov/engineering/hydraulics/software/hy8/). It is necessary to know how much of the tidal prism and discharge has to pass through the culvert over a range of flows and tidal levels to size the structure properly. This is obviously limited to small marshes where one can assume uniform water levels.

The required size of open tide gates for the flood tide can be computed using the orifice equation $Q = CA$ (8) $H^{0.5}$ where Q is the discharge, C is the orifice coefficient, A is the gate cross section, and H is the head differential (SCS 1963).

LEVEL TWO: MASS BALANCE AND LEVEL ROUTING HYDRAULIC ANALYSIS

The second level of mathematical models that can be performed to represent tidal marshes also assumes that the marsh has a uniform-level water surface but that the rate of inflow and outflow varies with time. This simplified semi-steady (i.e., quasi-steady) flow method is a "reservoir" or "level" routing technique. The number of time steps for each iterative computation may vary, and the accuracy increases as smaller time intervals are used. Flow is assumed to be steady during each time step, so multiple time steps are necessary.

Stage-storage-discharge data are necessary to relate flow and marsh storage volume to water surface elevation. Discharge into and out of the marsh at hydraulic control points (e.g., river inflow channel, bridge, culvert, tide gate, etc.) must be known. Calculations must extend over the full range of possible values to accurately approximate hydrologic and hydraulic conditions.

Level routing calculations can be done manually, with a spreadsheet, or with a simple computer program. The computations for each step begin with the corresponding tide elevation and an initial assumed marsh water surface elevation and volume. Since the tidal cycle is about 12 hours long for East Coast marshes, the time steps should be less than or equal to an hour to maintain adequate resolution of both the rising flood tide and the falling ebb tide. The difference between the tide and assumed water surface elevation is the hydraulic energy head used in the flow rate computation for each time interval. The volume of inflow or outflow is computed, and then the resulting incremental volume is added to the initial marsh volume. The new marsh water surface elevation corresponding to the incremental volume is next determined and compared with the trial water elevation. If the trial and computed water elevation match, then the time step is solved and the next time step calculations are initiated. Iterations continue until satisfactory trial water elevations are reached for each time step, including time steps for the ebb and flood tide. If the maximum computed marsh water surface elevation is much lower than the maximum tide elevation, then the marsh may be restricted on the downstream end.

A case example was performed in 1974 for the Morris Creek tidal marsh in East Haven and New Haven, Connecticut. This marsh had a total area of over 81 hectares, with tidal flow restricted by large tide gates. Stage-storage data were compiled from aerial topography maps with 2-foot contour intervals. Actual tide data were obtained from a nearby National Oceanic and Atmospheric Administration tide gauge and synthesized into one-hour time steps. The subsequent mass balance, level routing computations accurately predicted peak interior tide levels as measured at gauges.

LEVEL THREE: ONE-DIMENSIONAL, NONUNIFORM FLOW MODELS

One-dimensional hydraulic models may be operated in steady state representing a single point in time such as a peak flood flow, or in unsteady mode representing changing flows over time. Unsteady flow modeling is common for analyzing marshes so that the superposition of tide and freshwater inflow can be investigated as a function of time. Unsteady flow models often use a numeric computational process.

One-dimensional hydrodynamic (i.e., unsteady) models may be used to economically assess tidal marshes and channels where the flow is primarily linear (i.e., along channels) with limited lateral flow such as high marshes. One-dimensional models are based upon one or more channel reaches, each of which is represented by multiple cross sections. Cross sections define the tidal channel and marsh surface width and depth.

The flow pattern in a one-dimensional model is assumed to be perpendicular to the cross section, with a uniform velocity that is vertically averaged. Roughness coefficients must be assigned to each cross section to calculate energy losses due to friction. All points along a cross section have the same hydraulic energy levels.

One-dimensional models cannot represent lateral forces such as wind, the Coriolis effect, or internal turbulence. One-dimensional models are less expensive to set up because there are fewer data requirements than with more detailed models (Zevenbergen et al. 2004).

The US Geological Survey Branch Network Dynamic Model (BRANCH) program is an early example of a one-dimensional hydrodynamic model that has been used extensively for analysis of tidal channels. It is being replaced in professional practice by the Hydrologic Engineering Center River Analysis System (HEC-RAS), but many past marsh restoration projects that depend on BRANCH and its results are still found. BRANCH was used to model the Connecticut River Estuary (Schaffranek et al. 1981) and Pine Creek in Fairfield, Connecticut (MacBroom 1992).

BRANCH is broadly applicable to a wide range of hydrologic situations with complex boundary and junction conditions (Schaffranek 1987; Schaffranek et al. 1981). This numerical model uses the finite difference approximation of the unsteady flow equations in one dimension and can assess branching channels and tributaries. It does not evaluate hydraulic structures, and thus is of limited value for tide barrier restoration projects.

The US Army Corps of Engineers model, HEC-RAS, is a primary model used today for open-channel analysis. It is a one-dimensional hydraulic program that can be used for both nonuniform steady or nonsteady flow. The unsteady flow algorithm in RAS follows that in a previous model called UNET (Barkau 2001). It also contains subroutines for sediment transport analysis, stable channel design, scour analysis, and water quality modeling.

HEC-RAS is able to analyze water profiles and velocities in both channels and floodplains, plus through bridges, culverts, spillways, gates, and levees. In addition to geometry and flow data, the model requires upstream and downstream boundary conditions. For unsteady flow, boundary conditions can include downstream tide data and upstream stage or flow hydrographs. HEC-RAS can be used for branching flow networks, including islands and split flow.

ONE-DIMENSIONAL NONUNIFORM FLOW CASE STUDY

A one-dimensional unsteady model was assembled to investigate flow restoration, fish passage, and flood mitigation in the Old Orchard Beach and Scarborough salt marsh in southwestern Maine. The technical evaluation of undersized culverts, a railroad embankment running through the marsh, a downstream tide gate, and a small dam was performed with the hydrodynamic HEC-RAS computer model. The model was used to evaluate the water-level alterations due to separate road and railroad culverts and then to test the impact of alternative culvert types and sizes.

Calibration included comparing modeled and known water surface elevations recorded from six short-term gauges and adjusting the model with minor refinements to Manning's N and geometry around structures to smooth flow transitions. An April 2007 nor'easter (fifty-year storm event) during a spring tide was recorded on the gauges.

Water surface elevations determined from the initial hydraulics model runs of the fifty-year flood with a duration of six hours were in good agreement with the observed gauge data. Flooding characteristics at key structures were reproduced in the existing conditions model to within 1 foot of the gauge data, which is an acceptable level of accuracy, considering the inflow hydrographs for the model were synthesized from regression equations and lag time estimates.

Backwatering and flooding upstream of undersized structures were replicated in the model and lasted for approximately 1.5 days during the fifty-year, six-hour storm. The existing conditions model (figs. 2.1a and 2.1b) illustrates a flat longitudinal profile (i.e., water surface elevation versus river station in the direction of flow) in the marsh during high flows illustrating the large amount of local storage capacity. At high flows, the marsh water surface elevation is largely controlled by a series of downstream undersized structures, roadway and railroad embankments, and tide levels that fill storage and slow downstream movement of the freshwater flood peak.

The alternatives analysis exploring culvert enlargements, replacing culverts with bridges, dam modification, and embankment removal indicated that increasing the capacity of undersized structures reduced flooding across the marsh (fig. 2.1c) by allowing improved downstream passage of floodwaters.

LEVEL FOUR: TWO-DIMENSIONAL MODELS

Two-dimensional models are able to predict flow velocities and depths at any point along the length and width of a marsh, accommodating both longitudinal and lateral flow. They assume flow velocities are averaged in the vertical direction and do not account for stratification. Two-dimensional models are a level of mathematical modeling normally used to analyze tidal marshes.

Two-dimensional models represent the tidal marsh and channels in a geometric mesh in the horizontal plane. Each node (element corner) is defined by coordinates allowing its area to be determined, and the volume of water in each element is found by iteration as the water depth varies. The geometric mesh allows analysis of complex marsh shapes and sizes. Having a proper mesh is a critical facet in enabling the model to converge on a successful solution.

The dynamic version of two-dimensional models performs flow computations in a series of time steps. Time intervals need to be very short in order to capture

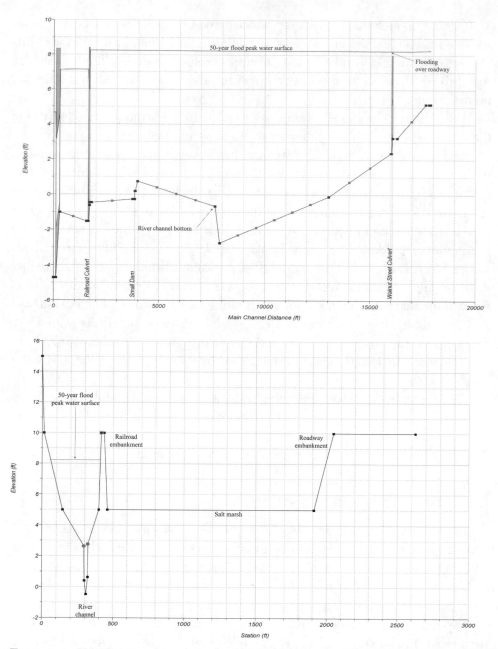

FIGURE 2.1. HEC-RAS model runs for the Old Orchard Beach and Scarborough salt marsh, Maine. (a) Existing conditions flood profile (elevation versus longitudinal river station in the direction of flow with zero at the watershed outlet) of the fifty-year storm event showing flooding in the marsh and overtopping of the upstream roadway/bridge. Note that the water surface during the flood is controlled by undersized culverts at the downstream railroad embankment. (b) Existing conditions cross section (elevation versus lateral river station across flow direction) showing the reduction in floodplain width due to the presence of a railroad embankment.

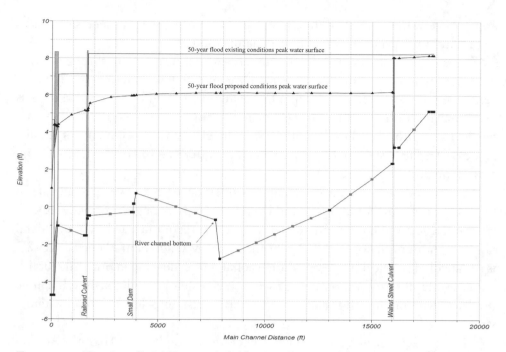

FIGURE 2.1 (Continued). (c) Existing (solid line) and proposed (line with triangles) HEC-RAS flood profiles of the fifty-year event showing reduced flooding in the marsh due to increasing the size of an upstream culvert under a roadway and a downstream culvert under a railroad embankment.

tidal conditions, but small time steps increase computer time, while large time steps cause instability in the trial solutions.

Some of the better two-dimensional models have the capacity to include hydraulic structures as recognizable one-dimensional elements. This facilitates modeling of flow through culverts, pipes, bridges, tide gates, and weirs using conventional one-dimensional coefficients and equations. Some two-dimensional models are limited to subcritical flow only; others have the capacity to accommodate subcritical or supercritical flow. All two-dimensional models can operate with steady or unsteady flow. In many cases, models are first set up and tested in steady flow and then converted to more complex unsteady flow.

Two-dimensional models are a powerful tool for hydrologic analysis of tidal marsh systems and tide restoration modeling. However, their vast capability means that they are complex, and it is not uncommon for a hydrodynamic model to be unstable and unable to reach a solution. Numeric instability occurs when the hydrodynamic model has solution trials that fail to converge within a number of iterations. In order to increase convergence, common techniques include mesh modifications, geometric refinement, variations of eddy viscosity leading up to appropriate values, modified time intervals, and starting with a successful steady-state solution to prime, or stabilize, a dynamic model.

Some hydrodynamic models have difficulty reaching solutions during low tide time steps when many elements are dry. Disconnected elements can lead to convergence failures, such as those due to isolated pannes and pools. Some models use a marsh porosity feature during "wet and drying" cycles. In other models, one has to use "virtual" vertical channels to connect low water elements or modify grades. The latter steps help convergence but introduce minor model inaccuracies in the process.

Many two-dimensional hydrodynamic models have been developed and are in the public domain. For example, the US Army Corps of Engineers model RMA2 is among the most common codes in use and is recommended by the Federal Highway Administration and Federal Emergency Management Agency. River 2D is a Canadian public domain two-dimensional model developed primarily for fish habitat suitability and is now being used in tidal estuaries and marshes. It can be used for dynamic flow but lacks the capacity to use one-dimensional hydraulic structures.

The US Geological Survey, in conjunction with the Federal Highway Administration, developed a public domain two-dimensional hydrodynamic model—the Finite Element Surface Water Modeling System, which has also been used to assess tidal marshes and channels. A nice feature of this model is its incorporation of conventional one-dimensional routines for culverts, tide gates, bridges, and weirs. The author's (J. G. MacBroom) use of this model for salt marshes found that input and output file data are best handled by using commercial software that interfaces with the main program.

RMA2, entitled the Two-Dimensional Model of Open Channel Flows, is a modeling system used for studying two-dimensional hydrodynamics in rivers, reservoirs, bays, and estuaries, with associated secondary models used to predict physical (e.g., sediment transport), chemical (e.g., dissolved oxygen, nutrients), or biological (e.g., bacterial transport) processes using equations for advection, dispersion, diffusion, or decay. Existing and proposed geometry can be analyzed to determine the impact of various project designs on velocity and flow patterns in a water body. The model is also capable of simulating structures such as tide gates and culverts.

RMA2 is a finite element solution of the Reynolds form of the Navier-Stokes equations for turbulent flows. Friction is calculated with Manning's equation, and eddy viscosity coefficients are used to define turbulence characteristics. A velocity form of the basic equation is used, with side boundaries treated as parallel or static (zero flow). The model automatically recognizes dry elements and corrects the mesh accordingly. Boundary conditions may be water surface elevations, velocities, or discharges, and may occur inside the mesh as well as along the edges.

RMA2 computes water surface elevations and horizontal velocity components for subcritical, free surface flow in two-dimensional fields. The model is capable of analyzing both steady-state and hydrodynamic problems. The program has been applied to calculate flow distribution around islands and through and around structures (bridges, culverts, etc.).

The model operates using separate geometry and boundary condition files. The geometry input file defines a system of elements and nodes that are defined with respect to their elevation and spatial location in the water body (x-, y-, and z-coordinates). Material types, such as water, wetland, jetties, outfall structures, and the like, are also defined using the system of elements and nodes. The boundary condition file is used to set hydraulic head and flow conditions at inlet and outlet structures. It is also used to assign Manning's roughness coefficients and viscosity coefficients by material type. Model convergence parameters are specified in the boundary condition file as well.

RMA2 was recently applied to analyze the West River marsh in New Haven, Connecticut, to evaluate water levels, velocities, and flushing under existing and proposed tide restoration conditions. The foundation of the model is a network of grid cells, or mesh, developed to reflect the topographic features of the study area (fig. 2.2). Grid cells are defined to represent changes in bathymetry/topography, slope, and form of the marsh area and principal channels. Flood boundaries are calculated to evaluate site hydraulics (fig. 2.3).

A two-dimensional model was selected because the marsh included separate channels and freshwater inflow, plus a tide gate. The model was used to evaluate (1) a series of potential marsh modifications, including use of a self-regulating tide gate that would increase the tidal prism; (2) the impact of alternative sediment removal options to increase internal flow circulation; and (3) to confirm potential tide elevations at adjacent recreational facilities (Milone & MacBroom, Inc. 2002).

The results indicated that three of the twelve tide gate bays should be retrofitted with the self-regulating tide gates and that this would raise mean high water levels to target elevations.

Hydrodynamic models can also be linked to secondary software that uses the hydraulic data to evaluate sediment transport (SED 2D) and water quality (RMA 10). Coupled models are currently being developed by researchers to evaluate the influence of sea-level rise.

LEVEL FIVE: THREE-DIMENSIONAL MODELS

Several hydrodynamic models exist that compute flow in three dimensions—length, width, and depth. They are used when both width and depth properties of

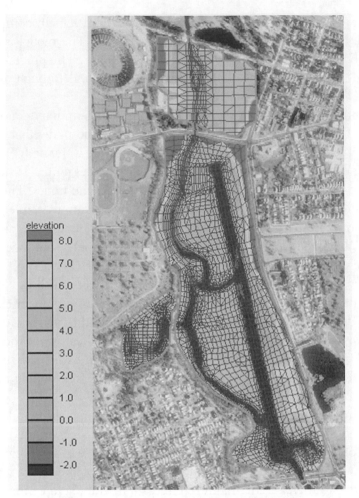

FIGURE 2.2. Two-dimensional model mesh for the West River, New Haven, Connecticut.

water are variables, particularly for highly stratified waters due to density variations caused by salinity, temperature, or sediment load. The principal application of three-dimensional models has been for water quality circulation in lake and deep water harbors or estuaries. They are not normally used for analysis of shallow tidal marshes.

The RMA10 three-dimensional software supported by the US Army Corps of Engineers is one of the better-known models; the popular RMA2 two-dimensional software provides an introduction to its use. It computes water levels and currents, plus temperatures and salinity transport. Some three-dimensional models use a combination of both advection and dispersion equations to distribute water quality parameters.

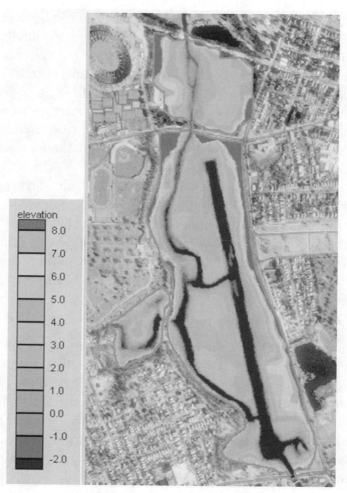

FIGURE 2.3. Two-dimensional model solution with flood limits for the West River, New Haven, Connecticut.

Physical Models

Physical scaled models have been used for many years to represent and analyze tidal hydraulic problems, primarily in enclosed waters such as harbors, bays, and estuaries. Models of coastal processes were most common from 1930 to 1970 before powerful numerical models were available and possible with advances in desktop computing. Physical models of marshes with complex vegetated surfaces, uneven roughness, and shallow flow are difficult due to the large expense of building the model and running simulations with ample monitoring and data recording equipment.

Physical models involve precise topographic data that can be reproduced at a smaller scale in a hydraulic flume. Careful selection of horizontal and vertical model dimensions is performed using similarity equations to scale from field conditions. Scales may be magnified based on the specific project needs. The model's surface roughness and texture must be proportional to fluid properties and field conditions. Models have been used to represent tides, currents, long waves, wind waves, and some sediment processes. The small size of models makes it easier to collect data, and resulting visual observations are informative for observing existing and proposed conditions.

Physical modeling of local restrictions, barriers, and hydraulic structures such as tide gates is more common than full marsh models. Complex topographic and three-dimensional effects are sometimes best represented by physical rather than mathematical models, but they are time consuming, expensive, and often not available for subsequent alternatives analysis. Temporary physical models are often not practical for a specific project given typical project budgets and schedules.

Research Opportunities

Tidal marshes and creeks are complicated hydraulic systems with unsteady flow and varied topography, and often with multiple structures that regulate hydrology and hydraulics. Historical modifications that alter the relationships between tide levels, salinity, and the marsh plain surface can lead to alternation of marsh ecology and even marsh collapse. The hydrologic and geomorphic characteristics of tidal systems can be assessed using one or more analog, empirical, mathematical, and physical modeling techniques to perform redundant assessment and restoration design. Optimum modeling predictions require extensive data collection, including first-hand information of the marsh tidal cycle and freshwater hydrograph.

Continued research and application are needed to establish modeling and design protocols that effectively balance level of detail and cost. There is a need to improve "coupled" models that link tidal marsh hydraulics with salinity, chemical water quality, sediment, and ecological processes.

REFERENCES

Aubrey, D., and P. E. Speer. 1985. "A Study of Non-linear Tidal Propagation in Shallow Inlet/Estuarine Systems: Part I." *Estuarine and Coastal Shelf Science* 21:185–205.

Barkau, R. L. 2001. *UNET, One-Dimensional Unsteady Flow Model, Users Manual.* Davis, CA: US Army Corps of Engineers Hydrologic Engineering Center.

Carey, J., R. Enion, C. Morgan, and J. Westrum. 2006. "Oyster River Restoration, Old Saybrook, CT." Unpublished student paper, Yale University School of Forestry and Environmental Studies, New Haven, CT.

Environmental Agency. 2008. "Understanding and Managing Morphological Changes in Estuaries, Great Britain." The Estuary Guide. http://www.estuary-guide.net/pdfs /morphological_change_guide.pdf.

Federal Highway Administration (FHWA). 1985. *Hydraulic Design of Highway Culverts*. Washington, DC: Federal Highway Administration.

MacBroom, J. G. 1992. "Pine Creek Tidal Hydraulic Study." Pp. 1154–58 in *Hydraulics Engineering: Saving a Threatened Resource—In Search of Solutions: Proceedings of the Hydraulic Engineering Sessions at Water Forum '92*, edited by M. Jennings and N. G. Bhowmik. New York, NY: American Society of Civil Engineers.

Milone and MacBroom Inc. 2002. *West River Memorial Park Tidal Marsh and River Restoration, New Haven, Connecticut*. Cheshire, CT: Milone and MacBroom Inc.

Myrick, R., and L. Leopold. 1963. "Hydraulic Geometry of a Small Tidal Estuary." Professional Paper 422-B. Washington, DC: US Geological Survey.

Neill, C. 1973. *Guide to Bridge Hydraulics, Roads and Transportation Association of Canada*. Toronto, ON: University of Toronto.

Phillip Williams and Associates, Ltd. (PWA). 1995. *Design Guidelines for Tidal Channels in Coastal Wetlands*. Prepared for US Army Corps of Engineers, Waterways Experiment Station, San Francisco, CA.

Roman, C. T., W. A. Niering, and R. S. Warren. 1984. "Salt Marsh Vegetation Changes in Response to Tidal Restriction." *Environmental Management* 8:141–50.

Schaffranek, R. W. 1987. "Flow Model for Open Channel Reach or Network." US Geological Survey Professional Paper 1384.

Schaffranek, R. W., R. A. Baltzer, and D. E. Goldberg. 1981. "A Model for Simulation of Flow in Singular and Interconnected Channels." US Geological Survey Techniques of Water Resources Investigations, Book 7, Chapter C3.

Shafer, D., and D. J. Yozzo. 1998. *National Guidebook for Application of Hydrogeomorphic Assessment to Tidal Fringe Wetlands*. Technical Report WRP-DE-16. Washington, DC: US Army Corps of Engineers.

Soil Conservation Service (SCS). 1963. *Tides and Tidal Drainage in the Northeastern States*. Technical Release EWP-NO6. Upper Darby, PA: US Department of Agriculture Soil Conservation Service.

Teal, J. M., and L. Weishar. 2005. "Ecological Engineering, Adaptive Management, and Restoration Management in Delaware Bay Salt Marsh Restoration. *Ecological Engineering* 25:304–14.

US Army Corps of Engineers (USACOE). 1991. *Tidal Hydraulics*. Engineering Manual EM 1110-2-1607. Washington, DC: US Army Corps of Engineers.

US Army Corps of Engineers, New England Division (NED). 1999. *The Highway Methodology Workbook Supplement: Wetland Functions and Values (NAEEP-360-1-30a)*. Concord, MA: US Army Corps of Engineers.

Williams, P., and P. Faber. 2004. *Design Guidelines for Tidal Wetland Restoration in San Francisco Bay*. Prepared for the Bay Institute, Funded by California State Coastal Conservancy.

Williams, P. B., M. Orr, and N. Garrity. 2002. "Hydraulic Geometry: A Geomorphic Design Tool for Tidal Marsh Channel Evolution in Wetland Restoration Projects." *Restoration Ecology* 10:577–90.

Zeff, M. L. 1999. "Salt Marsh Channel Morphometry: Application for Wetland Creation and Restoration." *Restoration Ecology* 7:205–11.

Zevenbergen, L. W., P. F. Lagasse, and B. L. Edge. 2004. *Tidal Hydrology, Hydraulics, and Scour at Bridges*. Publication no. FHWA-NHI-05-077. Hydraulic Engineering Circular No. 25. Washington, DC: US Department of Transportation, Federal Highway Administration.

Chapter 3

Biogeochemical Responses to Tidal Restoration

Shimon C. Anisfeld

Restoration of tides to a tide-restricted marsh sets into motion significant changes in the biogeochemistry of the marsh. These biogeochemical changes affect the suitability of the marsh for different species of vegetation, nekton, and birds, and set the context for the long-term development of the marsh. In turn, changes in vegetation and nekton can substantially modify the biogeochemistry of the marsh.

This chapter reviews the literature on biogeochemical responses to restoration of tidal flushing. It draws primarily on cases from the northeastern United States but also includes several important studies that were carried out in the Southeast, on the West Coast, and in Europe.

Defining the System

One of the difficulties of studying marsh restoration is the uniqueness of each marsh site. Differences in type of restriction, time since restriction, severity of restriction, freshwater inputs, land use, fill/sediment deposition, and other factors all lead to significant biogeochemical differences among restricted marshes. Responses to restoration would be expected to differ as well. The details of marsh restriction history are often unknown or not reported in the literature; even when they are, it can be hard to know which of the details are important.

To address this variability, the following is a rough classification of restored marsh types:

- *Restored drained sites* This is perhaps the most common scenario for tidal flow restoration, in which tidal flow is being restored to sites where tidal restriction has led to dramatically reduced water levels and salinities.

- *Restored waterlogged sites* These are sites where tidal restriction has created a marsh that is fresher and less frequently flushed but not drier. These sites are waterlogged much of the year with fresh or brackish water (Portnoy and Giblin 1997a) and are now being restored to regular tidal flushing with higher-salinity water.

- *Constructed marshes* These are often included together with true restoration sites and are included in this chapter since the dominant feature—increased tidal inundation—is broadly similar to tidal restoration. The most important difference between constructed and restored sites is probably the substrate (Warren et al. 2002). Most constructed marshes are developing on sandy, low-organic-matter sediment, while restricted marsh soils, despite significant changes (to be discussed), often still qualify as peats. An exception is restoration sites where artificial fill is being removed; these sites would be expected to behave similarly to constructed marshes because the preexisting marsh peat at these sites is typically too low in elevation due to compaction, and the restored marsh is forced to develop on the sandy fill material.

- *"Managed realignment" sites* Managed realignment, also referred to as managed retreat or depoldering, is a technique being widely used in the United Kingdom and the Netherlands for dealing with sea level rise (French 2006). Historically, sea walls were built to protect low-lying coastal land (often salt marsh) from inundation and make it available for agriculture. Now, as sea level rise threatens to overtop those sea walls and flood large areas, coastal managers are breaching the sea walls and often also creating new sea walls in a more landward position. Thus managed realignment restores tidal flow into the lands between the old and new sea walls, with the aim of converting them (back) to salt marsh and intertidal mudflat (though it should be noted that the primary goal of these activities is not habitat creation, but flood protection [Pethick 2002]). This should be functionally similar to removal of dikes or tide gates in "restored drained sites" in the Northeast, with the main difference being the impacts of agricultural land practices (e.g., plowing, which leads to more rapid oxidation of soil organic matter; Dent et al. 1976).

An additional definitional issue has to do with the use of reference sites. Many restoration studies utilize nearby healthy marshes as reference sites to represent the target condition for the restoration. In some cases, restoration is considered successful if and only if the relevant parameters at the restored marsh reach the point where they are statistically indistinguishable from those at the reference marsh. While this approach has some merit, it puts a heavy burden on the choice of an appropriate reference site. Given the high degree of variability among healthy marshes in many important biogeochemical parameters, it may be more appropriate to target a broader suite of reference conditions (Short et al. 2000), rather than trying to match the restored site to one particular reference site.

Each of the following biogeochemical aspects of marsh restoration will be discussed in turn: salinity; redox and sulfide; pH; sediment deposition and accretion; carbon, nitrogen, and phosphorus; bulk density; metals; and other pollutants. Figure 3.1 presents a visual summary of the expected changes in many of these parameters. It is important to note that, while this figure is based on the literature review presented in this chapter, it is a vastly simplified version of the diversity of changes that can happen with restriction and subsequent restoration. It is best viewed as a hypothetical time course of *possible* changes, rather than a definitive statement of how likely this time course is relative to other possibilities. In addition, the figure ignores the spatial variability in these parameters within a given marsh (e.g., with depth in the sediment). The following discussion presents a more nuanced description of different possible responses to restoration, along with an exploration of the supporting evidence.

Salinity

Restricted marshes generally have lowered salinity due to reduced inputs of saltwater, although restriction and reduced tidal flushing can sometimes lead to elevated salinity in situations where freshwater inputs are low and evapotranspiration rates are high. In any case, tidal restoration generally results in relatively rapid return of saltwater into the marsh, as recorded in several different locations:

- Pore water salinity increased dramatically after restoration of two New Hampshire marshes (Burdick et al. 1997).
- Pore water salinity was monitored in 78 percent of Gulf of Maine restoration projects and was found to be indistinguishable from reference marshes after two years (Konisky et al. 2006). However, when tested one year after restoration, salinity was slightly lower than reference (Konisky et al. 2006), indicating that it may take more than a year for pore water throughout the marsh to equilibrate with the flooding tidal waters.
- At the Orplands Farm managed retreat site in the United Kingdom, sediment sodium and magnesium concentrations increased dramatically within a year after flooding with seawater (MacLeod et al. 1999).
- At the Pillmouth managed retreat site (UK), flooding led to an increase in conductivity in both shallow (10 centimeters) and deep (40 centimeters) pore water (Blackwell et al. 2004). However, the increase in the deeper layer was relatively small and gradual, indicating that penetration of

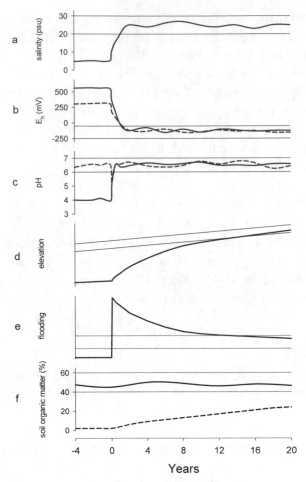

FIGURE 3.1. Possible changes in biogeochemical parameters upon restoration. Each figure shows a hypothetical time course of change from four years before restoration of tidal flow until twenty years after restoration. The range of values in hypothetical reference marshes is shown as lighter solid lines. No attempt is made to illustrate variability along this restoration trajectory.

Note: (a) Salinity. (b) Redox potential in restricted-drained marshes (solid line) and restricted-waterlogged marshes (dashed line). (c) pH in acidified marshes (solid line, more common for restricted-drained marshes) and nonacidified marshes (dashed line, more common for restricted-waterlogged marshes). The dip in pH upon restoration of the nonacidified marshes is a possible feature at some marshes based on Portnoy and Giblin (1997b) and Blackwell et al. (2004). (d) Elevation. Reference levels rise over time as sea level rises. (e) Height of flooding above the marsh surface at mean high water. After the initial increase, the water level decreases as the marsh surface elevation increases more rapidly than relative sea level rise. (f) Soil organic matter content for restricted marshes with high carbon content (solid line) and restricted marshes with low carbon content (dashed line). The former is typical of marshes that have reached an equilibrium with their restricted water level (sensu Anisfeld et al. 1999), with the latter typical of constructed marshes. Many drained marshes may lie somewhere between the two lines.

saltwater was incomplete within the four-month time period of postbreach monitoring. This slow penetration was attributed in part to physical changes caused by saltwater flooding, namely, soil swelling and elimination of cracks.

Redox and Sulfide

Healthy tidal marshes are characterized by a redox system dominated by the sulfate–sulfide pair. Because of abundant inputs of organic matter and poor exchange of oxygen into waterlogged sediments, electron acceptors such as oxygen, nitrate, iron (III), and manganese (IV) are quickly depleted. This results in reducing conditions below the first centimeter or so and leads to the production of phytotoxic sulfide through the reduction of the sulfate that is abundant in saltwater.

For restricted-drained marshes, one of the most significant effects of tidal restriction is the absence of these reducing, sulfidic conditions. Instead, these marshes generally have oxygen penetration and high redox potential down to a depth of 10 centimeters [4 inches] or more. One would expect that reflooding these systems would lead to a decrease in redox potential and renewed production of sulfide.

Restricted-waterlogged marshes, in contrast, can have moderately low redox potential (due to waterlogging), although they tend to have little sulfide production because of the absence of sulfate inputs. Reflooding these systems might be expected to lead to relatively little change in redox potential but higher levels of sulfide.

Unfortunately, redox and sulfide are not routinely reported in restoration monitoring programs in the Northeast. Thus the primary published source of data on these parameters is a greenhouse microcosm study by Portnoy and Giblin (1997b), in which they simulated restoration by flooding sediment cores (from both restricted-drained sites and restricted-waterlogged sites) with seawater. They found that flooding with saltwater led to decreases in redox potential (E_h) for both the restricted-drained and restricted-waterlogged sediments, though the decreases were most dramatic for the restricted-drained peat and for the surface layers, where higher amounts of organic matter led to more rapid depletion of electron acceptors.

Patterns of sulfide production in these greenhouse experiments were complex. Sulfate reduction was probably taking place postflooding in both restricted-drained and restricted-waterlogged peats, but measured pore water sulfide concentrations were quite low in the restricted-drained peats, apparently because an abundance of iron (II) was available to precipitate the sulfide as iron sulfides. In contrast, dissolved sulfide levels in the restricted-waterlogged peats rose to levels comparable to a natural marsh (approximately 5 millimolar).

Measurements of redox potential at managed realignment sites generally corroborate the greenhouse experiments. At both Orplands Farm (MacLeod et al. 1999) and Pillmouth (Blackwell et al. 2004), E_h dropped significantly within a year of flooding, though it continued to drop further in the second year of post-flood monitoring at Orplands Farm.

There is some concern in the managed realignment literature (French 2006) over the potential development of a highly reduced anoxic mud layer at the interface of the old and new sediment. That is, as the old freshwater vegetation is killed by salt and buried by rapid sediment deposition (to be discussed), there is the potential for forming a highly organic layer of decomposing plant material, which may be reducing enough (and at a shallow enough depth) to prevent even marsh vegetation from taking hold. Field evidence documenting this scenario has not yet been obtained.

pH

Restricted marshes, particularly drained sites, can experience acidic conditions due to the oxidation of reduced sulfur that occurs when air enters the previously anoxic sediments (Soukup and Portnoy 1986; Anisfeld and Benoit 1997; Portnoy and Giblin 1997a; Portnoy and Valiela 1997). One would expect that restoration of tidal flow would lead to reflooding of the peat, the reestablishment of anoxic soil conditions, the cessation of sulfide oxidation, and a return to the circumneutral pH typical of healthy marshes.

As was the case for redox and sulfide, there are few data available from the Northeast on pH changes following restoration. In their greenhouse study, Portnoy and Giblin (1997b) found that pore water pH and alkalinity in the restricted-drained peat rose dramatically within three months of flooding (from approximately pH 4 to pH 6.5), while the restricted-waterlogged peat stayed near its initial pH of approximately 6.7, though with a short-term dip to below 6.

Changes in pH have been reported following managed realignment in the United Kingdom. At a site with a preflooding pH of 5.9, MacLeod et al. (1999) observed an increase to 6.5 to 6.9 after flooding. More surprisingly, Blackwell et al. (2004) observed a sharp, but temporary, *decrease* in pore water pH (10 centimeter depth) following flooding, with pH values below 5 for up to nine weeks. This pattern was not found in deeper pore water (40 centimeters) or in shallower sediment pH (2 centimeters). The authors speculate that this acidification may result from patchy sulfide oxidation or from organic acids (or carbonic acid) produced by a pulse of decomposition following salinity-induced plant mortality.

Sediment Deposition and Vertical Accretion

Healthy marshes accrete vertically at roughly the rate of relative sea level rise (RSLR) by accumulating organic matter (primarily from in situ plant production) and inorganic sediment (from tidal deposition). In describing the rate of this process, three different types of measurements should be distinguished: "accumulation" refers to the rate of mass deposition at or near the surface (measured as the *mass* of sediment accumulating above a near-surface marker horizon, grams per square centimeter per year); "accretion" refers to the rate of vertical growth at or near the surface (measured as the *height* of sediment accumulating above a marker horizon, millimeters per year); and "elevation change" also includes processes below the marker horizon such as subsidence or swelling (measured by repeated surveying or with a sediment elevation table [SET; Cahoon et al. 2002], millimeters per year).

Tidal restriction severely disrupts natural elevational processes. Marsh drainage can lead to rapid oxidation, loss of organic matter, and resulting subsidence (Portnoy and Giblin 1997a; Turner 2004). After this initial loss of elevation, some restricted-drained marshes may reach a new equilibrium in which relative water levels are somewhat higher, and organic and inorganic material begins to accumulate again (Anisfeld et al. 1999). In restricted-waterlogged marshes, organic matter oxidation and subsidence may be less than in restricted-drained marshes, but inputs of inorganic material may still be severely reduced (Portnoy and Giblin 1997a).

There is ample evidence from a variety of locations that tidal restoration can lead to rapid accumulation, accretion, and elevation change, especially in the short term, as the marsh "catches up" to the higher water level:

- Morgan and Short (2002) found that accumulation rates at constructed marsh sites in the Great Bay Estuary in New Hampshire (measured as deposition on Mylar disks over three weeks) were very high for newly constructed sites and declined gradually with time (using a space-for-time substitution) to levels comparable to reference marshes.
- Craft et al. (2003) showed that accumulation rates at constructed marshes in North Carolina (measured as deposition on feldspar marker horizons over five months) were higher than reference for more recently constructed sites (one and eleven years old), but were comparable to reference for older sites (twenty-four, twenty-six, and twenty-eight years old).
- At the Orplands Farm managed realignment site, accretion rates were reportedly as high as 41 millimeters per year over the first few years (French 2006), but decreased to an average rate over the first seven years of up to 7.5

millimeters per year (Spencer et al. 2008), which was still more than double the rate of RSLR.

- SET data from the Freiston Shore managed realignment site (UK) showed accretion rates ranging from 1.4 to 50 millimeters per year over the first four years after flooding, although much of this sediment was derived from erosion of seaward salt marshes (Rotman et al. 2008).

- Anisfeld and Hill (unpublished data) measured accretion and net elevation change using artificial marker horizons and a SET at a restored Connecticut marsh (Jarvis Creek, Branford). Twenty-five years after restoration, accretion and elevation change were both approximately 10 millimeters per year, or two to four times the rate at a nearby reference site.

- Boumans et al. (2002) found that the elevation change at two restored marshes in Maine and New Hampshire (measured over a one- to two-year period) was more positive by 20 to 23 millimeters per year than the elevation change at reference sites within the same marshes.

- At the Peazemerlannen site in the Netherlands (Bakker et al. 2002), where an unintentional dike breach in 1973 led to flooding of a former polder, sedimentation led to positive elevation change of approximately 16 millimeters per year over the period 1973 to1996 (based on surveying) and approximately 8 millimeters per year over the period 1995 to 1997 (based on SET measurements).

- After levee breaching, several sites in San Francisco Bay experienced high rates of sediment deposition and major increases in sediment surface elevation, with the most dramatic case being the Warm Springs site, with an elevation change of 5.8 meters over fifteen years (Williams and Orr 2002). Following that increase in elevation, most of Warm Springs had an elevation that was suitable for marsh vegetation, although colonization was proceeding slowly. Other levee breaching sites had much slower sedimentation rates and were too low for plant survival.

- In the ten years following dike removal in the Salmon River marshes, Oregon, elevations in the restored marsh increased by approximately 90 millimeters, compared to approximately 50 millimeters in a reference marsh (Frenkel and Morlan 1991). Over the same time period, accretion (above artificial sand horizons) was approximately 50 to 70 millimeters in the restricted marsh and approximately 30 to 40 millimeters in the reference marsh. These results imply that restoration caused both greater surface sedimentation and an increase in the subsurface volume, perhaps due to soil swelling as a result of higher water levels.

- Anisfeld et al. (1999) compared several restored, restricted, and reference marshes based on radiometric dating of sediment cores. They found that accumulation rates were similar in all three types of marshes, but accretion rates were significantly higher in the restored marsh (approximately 7 millimeters per year compared to approximately 4 millimeters per year in the reference marshes), due to higher porosity. Note that, in this case, the high accretion rates seem to be due primarily to swelling of the soil rather than increased sedimentation.

Does the weight of this evidence mean that restoration sites can always adjust to the higher water level? Of course not. Depending on the degree of subsidence and the amount of RSLR that occurred during the period of restriction, tidal restoration may result in at least temporary *drowning* of the marsh, in which the system is converted to unvegetated intertidal mudflat or subtidal open water. This happened, for example, at Great Harbor Marsh/Lost Lake, Connecticut, where an unintentional tidal restoration resulted in conversion of restricted marsh to open water (Rozsa 1995); one part of this system slowly regained elevation and became marsh, but another portion remains open water. An additional risk of full restoration is the flooding of nearby structures that were built under the assumption that the lower water levels provided by the restriction would endure forever (Roman et al. 1995). If one wants to avoid these risks, it is important to carefully design restoration projects to produce flooding levels that are high enough to lead to desired changes (e.g., restored soil functions, control of invasive plants), but not so high that nearby houses are flooded or the tolerance of low marsh vegetation such as *Spartina alterniflora* (smooth cordgrass) is exceeded.

However, under certain circumstances, it can work well to "aim low"; that is, design a relatively wet system that may be unvegetated for several years but will eventually undergo renewed marsh development: sediment accretion, leading to a reduced hydroperiod, leading to initial colonization by vegetation, leading to further sediment trapping and organic matter accumulation, and so forth. This has been the strategy in San Francisco Bay (Williams and Orr 2002) and, to a lesser extent, at several sites in Connecticut (P. Capotosto, pers. comm.). This approach should be taken only where three conditions apply: there are no nearby structures that would be flooded (or any such structures can be bought out); there is a willingness to accept several years of unvegetated conditions; and hydrologic and sedimentological circumstances are favorable for sediment trapping.

Regarding the last of these conditions, Williams and Orr (2002) identify three factors that they consider to be the most important in determining whether sedimentation and marsh development are likely at breached-levee sites: the availability of suspended sediment; the presence or absence of wind waves that can lead to sediment erosion; and the extent to which tidal exchange is still restricted in a way that can limit sediment delivery. It is likely that at most formerly restricted sites, the hydrology and sedimentology do tend to favor sedimentation, simply because these are generally sheltered sites that at some point did experience sufficient sedimentation for marsh development.

However, current conditions at some sites may be significantly different from the conditions under which the marsh initially developed: RSLR may have increased, sediment availability may have decreased, and other factors may be interfering with the ability of marshes to keep up with RSLR. In the face of predicted accelerations in RSLR (IPCC 2007) and documented—but still not fully understood—cases of marsh drowning (e.g., Hartig et al. 2002), it seems prudent to require careful planning and monitoring in the creation of wetter marshes in the twenty-first century. A tidally restricted *Phragmites*-dominated marsh may be a wiser choice than a "restored" drowned system that is too wet to support marsh vegetation, especially as more and more marshes are lost to drowning. The restoration imperative and the preservation imperative are coming into conflict, and determining how to maximize the quantity and quality of tidal marsh will require creative thinking.

There is also some concern over potential subsidence associated with restoration at waterlogged sites. In their greenhouse experiment, Portnoy and Giblin (1997b) found that flooding of the restricted-waterlogged soils with seawater led to significant subsidence of 60 to 80 millimeters over a short time period of five months. This was apparently due to some combination of salinity-induced plant mortality and sulfate-induced accelerated decomposition. This is potentially of concern for the viability of tidal restoration of restricted-waterlogged sites, since this amount of subsidence could lead to conditions too wet for halophytic vegetation to establish itself. However, it is important to note that the greenhouse experiment did not include tidal delivery of sediment. As shown earlier, this is a huge factor in the elevation dynamics of restored marshes, especially in the short term, and may well mitigate any subsidence. There have been no field reports of subsidence in the period following tidal restoration.

One approach for dealing with excessive flooding in tidal restoration projects is to artificially increase the marsh elevation through the placement of fill material. Cornu and Sadro (2002) evaluated the success of this approach in four marshes in Coos Bay, Oregon, which had received different amounts of fill material as part of a tidal restoration. They found that all sites experienced significant shallow subsidence and elevation loss, apparently due to compaction of the fill material. Subsidence was greatest at the site with the greatest amount of fill. Marsh restoration projects need to take into account the possibility for compaction when designing marsh elevations using fill material.

Carbon, Nitrogen, and Phosphorus

Sequestration of carbon (C), nitrogen (N), and phosphorus (P) in peat is an important function provided by healthy marshes, and restoring that function can be one of the goals of marsh restoration (Andrews et al. 2006, 2008). As already discussed, drained marshes experience oxidation, which leads to loss of organic matter and reduced sediment C concentrations (Portnoy and Giblin 1997a; Portnoy 1999). Anisfeld et al. (1999) found that marshes that had been restricted for more than ninety years showed evidence of having experienced an oxidation phase, resulting in low C levels at depth, but had reached a new equilibrium in which surface C concentrations were not significantly lower than those of reference sites. Even during the oxidation phase, much of the sediment N and P appears to be retained in the sediment (Portnoy and Giblin 1997a).

Restoration might be expected to lead to increased C accumulation, as higher water levels reduce decomposition rates. Evidence from the accretion studies cited earlier tends to suggest, however, that the primary source of new material is inorganic sedimentation and that increases in C accumulation are more minor. Likewise, Anisfeld et al. (1999) found that restored marshes did not have significantly higher sediment organic matter concentrations (percentage) or rates of C accumulation (grams per square centimeter per year) than restricted sites. However, these restricted marshes were the ones already discussed that had apparently reached a new equilibrium and were similar in C content to reference sites. N accumulation rates were slightly higher in these restored marshes than in the restricted and reference sites.

In contrast, studies in *constructed* marshes—with their low initial C concentrations—have generally found that sediment C content increased gradually over time as organic matter accumulated due to high productivity and low decomposition rates. In constructed sites in Great Bay Estuary (Morgan and Short 2002), soil organic matter took fifteen to twenty years to reach reference levels, while in California (Zedler and Callaway 1999), soil organic matter appeared to level off after six years, but at a concentration that was still approximately 25 percent lower than that of reference sites. The California study also measured sediment N, which appeared to increase steadily over time but reached only approximately 60 percent of reference at the end of the study period (eleven years).

Craft et al. (1999, 2003) have carried out extensive studies of C, N, and P dynamics in constructed marshes of different ages, each paired with a nearby reference marsh that is similar in many of the controlling variables. They found that concentrations of C and N tended to increase with time since marsh construction, but that even twenty-eight-year-old marshes were significantly lower in C and N than their reference sites. P behaved differently: levels did not increase over time and constructed marshes were often higher than reference sites, probably due to increased P sorption capacity in the more inorganic soils.

Craft et al. (2003) also found that the organic matter in constructed marsh sediments was generally more labile than in reference marshes (presumably because it is younger), and that C mineralization rates per unit of sediment organic matter were higher in constructed marshes. Despite this, overall C accumulation rates were similar in the constructed and reference marshes, even in the youngest systems. P accumulation rates were also similar across systems, while N accumulation rates were slightly higher in the constructed marshes.

In the short term (weeks to months) after restoration, there is the possibility of loss of stored nutrients from the sediment due to ion exchange with seawater. Portnoy and Giblin (1997b) found that flooding led to sharp increases in pore water ammonium (NH_4^+) and, to a lesser extent, phosphate (PO_4^{-3}) in their greenhouse experiments. Blackwell et al. (2004) found a similarly dramatic increase in pore water NH_4^+ at the Pillmouth managed realignment project; levels remained elevated (up to 15 mg NH_4^+–N/L) for several months.

Bulk Density

Healthy marshes generally exhibit low bulk densities. Drainage tends to cause collapse of pore space, loss of organic matter, and an increase in density (Portnoy and Giblin 1997a), although some restricted marshes may recover and reach a new equilibrium characterized by relatively low bulk densities (Anisfeld et al. 1999). Agricultural use of the drained marsh (common in Europe) can lead to significant compaction (French 2006).

For restricted marshes that do have high bulk densities, restoration of tidal flow should lead to a decrease in bulk density through two processes: swelling of the marsh as it fills with water (Paquette et al. 2004); and an increase in organic matter and associated porosity. The former should be quite fast, while the latter may take decades, as already discussed.

Spencer et al. (2008) have shown that, at managed realignment sites, the relict agricultural surface may persist as an identifiable layer in the soil for a long time. Below that level, density is high and moisture content is low. This may significantly impede water and nutrient movement within the soil, and could potentially affect plant health and marsh development (Spencer et al. 2008; French 2006).

Metals

Healthy salt marshes are generally considered sinks for trace metals, which are bound tightly as organic complexes or as sulfides. Tidal restriction can lead to oxidation of sulfides and organics and result in the solubilization of metals into pore water. Exchange between pore water and surface water can then lead to restricted marshes becoming a source of previously immobilized metals. Acidification can accelerate this process by increasing metal solubility (Anisfeld and Benoit 1997).

Tidal restoration is expected to reverse this process and lead to renewed metal sequestration, especially since the restored marsh is often rapidly accumulating sediment (see earlier discussion). Surprisingly, there have been few, if any, measurements of metal accumulation rates following restoration. However, Andrews et al. (2008) measured metal accumulation rates at a natural marsh that they consider to be an analogue for managed realignment sites. They estimated that restoration of flow to 26 square kilometers (10 square miles) of land in the Humber Estuary (UK) could lead to removal of significant quantities of metals (e.g., 6 tons per year of zinc, 3 tons per year of lead). Andrews et al. (2006) used similar methods to estimate metal removal rates for the Humber Estuary under an "Extended Deep Green" scenario, in which tides are restored to approximately 85 square kilometers (33 square miles) of the estuary. They found that arsenic (As) removal would amount to approximately 85 percent of particulate As inputs to the estuary, while copper and lead removal would be approximately 9 percent of particulate inputs. They calculate the economic benefit of copper removal as approximately £1000 (US$1600) per year.

However, there is also the possibility that certain metals could be mobilized from marsh sediments during the process of tidal restoration, either because of increases in ionic strength or because of decreases in E_h. In their greenhouse experiments, Portnoy and Giblin (1997b) found that flooding of the restricted-drained sediment led to release of iron (II) and aluminum, leading to high concentrations in pore water. They attribute this to a combination of cation exchange and chemical reduction/solubilization.

Another element that might be susceptible to mobilization during restoration is cadmium, due to its relatively higher solubility (as chloride complexes) in saltwater compared to freshwater. At both the Orplands Farm (MacLeod et al. 1999) and Tollesbury (Chang et al. 2001) sites, sediment cadmium concentrations were lower after restoration than before, implying some loss of cadmium from marsh sediments to the estuary. At both sites, other metals mostly did not change significantly, though there were some relatively small increases in sediment concentrations of lead (Orplands Farm) and copper (Tollesbury), indicating additional sequestration.

Other Pollutants

Restricted tidal exchange can lead to poor flushing of tidal creeks and rivers, which—coupled with release of partly decomposed organic matter from oxidizing marshes—can cause water quality issues such as low dissolved oxygen and high levels of bacteria that could indicate human health problems (Soukup and Portnoy 1986; Portnoy 1991; Portnoy and Allen 2006). It is expected that restoration would significantly reduce these problems (Portnoy and Allen 2006; Maris et al. 2007), although it is also possible that increased loss of ammonium (NH_4^+) and dissolved organic carbon from the marsh sediments could lead to increased oxygen demand. In any case, there is little to no documentation in the marsh restoration literature of the response of oxygen (O_2) and indicator bacteria in tidal channels.

Restoration should also lead to indirect improvements in hypoxia due to nutrient removal. Marshes are hot spots for denitrification, and the tidal conditions in restored marshes (along with high primary productivity) should favor increased denitrification.

Biochemical Response Trajectories

There has been much discussion in the literature of the time course of marsh response to restoration, particularly the question of restoration trajectories. From the foregoing discussion and figure 3.1, it is clear that different biogeochemical parameters respond on different time frames. Three groups of parameters can be identified:

- *Fast responders* In most cases, salinity, redox, sulfide, and pH will reach levels roughly similar to reference marshes within two years of tidal restoration, although there may be some short-term deviations (e.g., a pH dip).
- *Variable responders* The length of time necessary for surface elevations to adjust to higher water levels can vary tremendously, depending on the depth of inundation, the availability of sediment, and other factors. As already discussed, restoring marshes can accrete extremely rapidly under the right circumstances, and in some cases can reach typical low-marsh elevations within a year. However, even with rapid accretion, it is also possible for restored sites to be wetter than reference (maybe even too wet to support marsh vegetation) for many years or decades. A third possibility is that a restored marsh may reach an alternate stable state through permanent conversion to intertidal mudflat or subtidal open water (drowning).
- *Slow responders* For marshes developing on substrate that is low in organic matter, there is good evidence that accumulation of organic matter and associated nutrients proceeds slowly over the course of several decades.

Overall, the evidence suggests that tidal restoration is often a biogeochemical success in two senses: achieving rough equivalence to reference in most biogeochemical parameters, and achieving a higher level of ecosystem functioning (e.g., higher metal sequestration rates). Still, there are risks associated with restoration. From the review in this chapter, the main risks appear to be the following:

- Too large an increase in water level may lead to marsh drowning in sub-sided marshes, especially given projected accelerations in RSLR.
- Release of nutrients, acidity, and metals (especially iron [II], aluminum, and cadmium) from marsh sediments may lead to a short-term deterioration in surface water quality.
- Especially in former agricultural lands, the properties of the relict marsh surface (high bulk density, poor drainage leading to anoxia) may impede colonization by marsh vegetation.

Research and Monitoring Opportunities

The lack of compelling and well-documented evidence for many of the changes discussed in this chapter demonstrates the need for more study of marsh restoration biogeochemistry. The biogeochemical aspects of tidal restoration are poorly studied relative to measures of hydrology and vegetation. Thus, for example, the comprehensive restoration monitoring protocol for the Gulf of Maine (Neckles et al. 2002) includes only one biogeochemical measure (pore water salinity) in its sixteen core variables. Six additional biogeochemical measures are included in the list of nineteen "additional variables to be monitored as warranted by the goals and resources of specific projects."

Table 3.1 illustrates a proposed framework for biogeochemical monitoring of marsh restoration. In order to accurately characterize the development of the variable and slow responders, monitoring needs to be carried out for twenty years or more, at least for some parameters. As more marshes are studied for longer periods of time, we will be able to understand more directly how marshes evolve in response to restoration, rather than relying on assumptions, mesocosm experiments, and space-for-time substitutions.

In addition to improved monitoring, there is a need for targeted research programs to address the following questions, among others:

- What controls the elevation response of marshes to restoration? Can we better predict the evolution of the marsh surface over time at a given site? How can we restore tidal flow while minimizing the risk of marsh drowning in an era of accelerated sea level rise?
- How do soil properties affect the success of marsh restoration? Is there anything that should be done differently in restoring marshes on sand substrates (e.g., in fill removal projects) versus peat substrates?

TABLE 3.1

Proposed framework for biogeochemical monitoring of marsh restoration

Category	Parameters[1]	Frequency[2]	Justification
Pore water	Water level Salinity pH Redox potential Sulfide (Nutrients) (Metals)	Prerestoration: monthly for two years (growing season only) Postrestoration: weekly for one month, monthly for two years (growing season only), then annually	Water level in wells, technically a hydrologic, rather than biogeochemical, parameter, is the driving force behind biogeochemical changes and can easily be monitored. Pore water salinity, pH, redox potential, and sulfide concentration are critical in determining ecosystem structure and function, including vegetation type and productivity. These parameters (especially redox and sulfide) are likely to vary greatly with depth, so measurement at two depths (near the top and bottom of the rooting zone, e.g., 10 and 30 centimeters) is recommended. Pore water nutrient (N, P) concentrations are also important in affecting rates of production and decomposition but tend to be highly variable temporally and spatially. Metal concentrations could be measured in polluted sites or where the fate of buried metals is of interest.
Surface water	Salinity pH Dissolved oxygen *Escherichia coli* (Nutrients) (Metals)	Prerestoration: monthly for two years (growing season only) Postrestoration: weekly for one month, monthly for two years (growing season only), then annually	Surface water salinity will respond quickly to restoration and will ultimately determine pore water salinity and the distribution of vegetation. Surface water pH should be monitored for excursions that could be harmful to aquatic life. Dissolved oxygen (DO) is critical for the health of aquatic organisms in tidal creeks, and *E. coli* is an important human pathogen indicator. DO and *E. coli* might be expected to respond to restoration, but these responses are poorly understood. Nutrient and metal concentrations do not need to be routinely measured, but targeted studies of sources and tidal exchange of these constituents would be useful.

Sediment properties	Bulk density Texture Organic content (Nutrients) (Metals) (Organic pollutants)	Prerestoration: once (within a year preceding restoration) Postrestoration: once in first year, then every two to three years	Monitoring sediment properties is important for (1) understanding how differences in soils under prerestoration conditions lead to differences in response to restoration, (2) identifying the changing conditions for benthic plants and animals, and (3) documenting changes in marsh function (e.g., metal sequestration). Sediment properties should be monitored through repeated collection of cores (e.g., 0 to 40 centimeters, sectioned into 5 centimeter intervals). Bulk density, texture, and organic content are all descriptive soil parameters that affect hydrology, biogeochemistry, and biology and would be expected to respond to restoration. Concentrations of nutrients, metals, and/or organic pollutants can be used, together with sedimentation data, to document the value of the marsh in removing these pollutants.
Sedimentation	Accretion (Elevation change)	Prerestoration: annually for at least three years Postrestoration: annually	Sedimentation is very responsive to restoration and is critical for the long-term survival of the marsh. Monitoring surface accretion (and/or accumulation) with marker horizons involves relatively low effort, although it requires preplanning to understand the sedimentation regime of the restricted marsh for at least three years prior to restoration. A fuller understanding of marsh elevation change can be gained through the use of sediment elevation tables, repeated surveying, or repeated Light Detection and Ranging (LIDAR) remote sensing. Although relatively expensive, elevation change measurements are highly recommended where marsh drowning is a potential concern or for projects involving placement of fill material to raise marsh elevations.

[1]Parameters in parentheses should be included where they are of particular concern or where additional resources are available. Other parameters should be included routinely in all monitoring programs.

[2]The suggested sampling frequency is a compromise between the potential rate of change in these parameters and the need for an affordable field sampling program. Both restoration and reference sites should be monitored. The number of sample locations needed per marsh depends on the size and complexity of the marsh, but a minimum of five is recommended.

• What is the rate of pollutant sequestration and removal (including denitrification) in restoring marshes? Can a defensible economic value be assigned to this marsh function?

Acknowledgments

This chapter was greatly improved by the comments, suggestions, and edits of David Burdick and two anonymous reviewers.

REFERENCES

Andrews, J. E., D. Burgess, R. R. Cave, E. G. Coombes, T. D. Jickells, D. J. Parkes, and R. K. Turner. 2006. "Biogeochemical Value of Managed Realignment, Humber Estuary, UK." *Science of the Total Environment* 371:19–30.

Andrews, J. E., G. Samways, and G. B. Shimmield. 2008. "Historical Storage Budgets of Organic Carbon, Nutrient and Contaminant Elements in Saltmarsh Sediments: Biogeochemical Context for Managed Realignment, Humber Estuary, UK." *Science of the Total Environment* 405:1–13.

Anisfeld, S. C., and G. Benoit. 1997. "Impacts of Flow Restrictions on Salt Marshes: An Instance of Acidification." *Environmental Science & Technology* 31:1650–57.

Anisfeld, S. C., M. J. Tobin, and G. Benoit. 1999. "Sedimentation Rates in Flow-Restricted and Restored Salt Marshes in Long Island Sound." *Estuaries* 22:231–44.

Bakker, J. P., P. Esselink, K. S. Dijkema, W. E. van Duin, and D. J. de Jong. 2002. "Restoration of Salt Marshes in the Netherlands." *Hydrobiologia* 478:29–51.

Blackwell, M. S. A., D. V. Hogan, and E. Maltby. 2004. "The Short-Term Impact of Managed Realignment on Soil Environmental Variables and Hydrology." *Estuarine Coastal and Shelf Science* 59:687–701.

Boumans, R. M. J., D. M. Burdick, and M. Dionne. 2002. "Modeling Habitat Change in Salt Marshes after Tidal Restoration." *Restoration Ecology* 10:543–55.

Burdick, D. M., M. Dionne, R. M. Boumans, and F. T. Short. 1997. "Ecological Responses to Tidal Restorations of Two Northern New England Salt Marshes." *Wetlands Ecology and Management* 4:129–44.

Cahoon, D. R., J. C. Lynch, B. C. Perez, B. Segura, R. D. Holland, C. Stelly, G. Stephenson, and P. Hensel. 2002. "High-Precision Measurements of Wetland Sediment Elevation, II: The Rod Surface Elevation Table." *Journal of Sedimentary Research* 72:734–39.

Chang, Y. H., M. D. Scrimshaw, C. L. Macleod, and J. N. Lester. 2001. "Flood Defence in the Blackwater Estuary, Essex, UK: The Impact of Sedimentological and Geochemical Changes on Salt Marsh Development in the Tollesbury Managed Realignment Site." *Marine Pollution Bulletin* 42:470–81.

Cornu, C. E., and S. Sadro. 2002. "Physical and Functional Responses to Experimental Marsh Surface Elevation Manipulation in Coos Bay's South Slough." *Restoration Ecology* 10:474–86.

Craft, C., J. Reader, J. N. Sacco, and S. W. Broome. 1999. "Twenty-five Years of Ecosystem Development of Constructed *Spartina alterniflora* (Loisel) Marshes." *Ecological Applications* 9:1405–19.

Craft, C., P. Megonigal, S. Broome, J. Stevenson, R. Freese, J. Cornell, L. Zheng, and J. Sacco. 2003. "The Pace of Ecosystem Development of Constructed *Spartina alterniflora* Marshes." *Ecological Applications* 13:1417–32.

Dent, D. L., E. J. B. Downing, and H. Rogaar. 1976. "Changes in Structure of Marsh Soils following Drainage and Arable Cultivation." *Journal of Soil Science* 27:250–65.

French, P. W. 2006. "Managed Realignment: The Developing Story of a Comparatively New Approach to Soft Engineering." *Estuarine Coastal and Shelf Science* 67:409–23.

Frenkel, R. E., and J. C. Morlan. 1991. "Can We Restore Our Salt Marshes? Lessons from the Salmon River, Oregon." *Northwest Environmental Journal* 7:119–35.

Hartig, E. K., V. Gornitz, A. Kolker, F. Mushacke, and D. Fallon. 2002. "Anthropogenic and Climate-Change Impacts on Salt Marshes of Jamaica Bay, New York City." *Wetlands* 22:71–89.

Intergovernmental Panel on Climate Change (IPCC). 2007. "Contribution of Working Groups I, II and III to the Fourth Assessment Report of the Intergovernmental Panel on Climate Change." *Climate Change 2007: Synthesis Report*. Geneva: IPCC. www.ipcc.ch/publications_and_data/ar4/syr/en/contents.html.

Konisky, R. A., D. M. Burdick, M. Dionne, and H. A. Neckles. 2006. "A Regional Assessment of Salt Marsh Restoration and Monitoring in the Gulf of Maine." *Restoration Ecology* 14:516–25.

Macleod, C. L., M. D. Scrimshaw, R. H. C. Emmerson, Y. H. Chang, and J. N. Lester. 1999. "Geochemical Changes in Metal and Nutrient Loading at Orplands Farm Managed Retreat Site, Essex, UK (April 1995–1997)." *Marine Pollution Bulletin* 38:1115–25.

Maris, T., T. Cox, S. Temmerman, P. De Vleeschauwer, S. Van Damme, T. De Mulder, E. Van den Bergh, and P. Meire. 2007. "Tuning the Tide: Creating Ecological Conditions for Tidal Marsh Development in a Flood Control Area." *Hydrobiologia* 588:31–43.

Morgan, P. A., and F. T. Short. 2002. "Using Functional Trajectories to Track Constructed Salt Marsh Development in the Great Bay Estuary, Maine/New Hampshire, USA." *Restoration Ecology* 10:461–73.

Neckles, H. A., M. Dionne, D. M. Burdick, C. T. Roman, R. Buchsbaum, and E. Hutchins. 2002. "A Monitoring Protocol to Assess Tidal Restoration of Salt Marshes on Local and Regional Scales." *Restoration Ecology* 10:556–63.

Paquette, C. H., K. L. Sundberg, R. M. J. Boumans, and G. L. Chmura. 2004. "Changes in Saltmarsh Surface Elevation due to Variability in Evapotranspiration and Tidal Flooding." *Estuaries* 27:82–89.

Pethick, J. 2002. "Estuarine and Tidal Wetland Restoration in the United Kingdom: Policy versus Practice." *Restoration Ecology* 10:431–37.

Portnoy, J. W. 1991. "Summer Oxygen Depletion in a Diked New England Estuary." *Estuaries* 14:122–29.

Portnoy, J. W. 1999. "Salt Marsh Diking and Restoration: Biogeochemical Implications of Altered Wetland Hydrology." *Environmental Management* 24:111–20.

Portnoy, J. W., and J. R. Allen. 2006. "Effects of Tidal Restrictions and Potential Benefits of Tidal Restoration on Fecal Coliform and Shellfish-Water Quality." *Journal of Shellfish Research* 25:609–17.

Portnoy, J. W., and A. E. Giblin. 1997a. "Effects of Historic Tidal Restrictions on Salt Marsh Sediment Chemistry." *Biogeochemistry* 36:275–303.

Portnoy, J. W., and A. E. Giblin. 1997b. "Biogeochemical Effects of Seawater Restoration to Diked Salt Marshes." *Ecological Applications* 7:1054–63.

Portnoy, J. W., and I. Valiela. 1997. "Short-Term Effects of Salinity Reduction and Drainage on Salt-Marsh Biogeochemical Cycling and *Spartina* (cordgrass) Production." *Estuaries* 20:569–78.

Roman, C. T., R. W. Garvine, and J. W. Portnoy. 1995. "Hydrologic Modeling as a Predictive Basis for Ecological Restoration of Salt Marshes." *Environmental Management* 19:559–66.

Rotman, R., L. Naylor, R. McDonnell, and C. MacNiocaill. 2008. "Sediment Transport on the Freiston Shore Managed Realignment Site: An Investigation Using Environmental Magnetism." *Geomorphology* 100:241–55.

Rozsa, R. 1995. "Tidal Wetland Restoration in Connecticut." Pp. 51–65 in *Tidal Marshes of Long Island Sound: Ecology, History and Restoration*, edited by G. D. Dreyer and W. A. Niering. New London, CT: Connecticut College Arboretum.

Short, F. T., D. M. Burdick, C. A. Short, R. C. Davis, and P. Morgan. 2000. "Developing Success Criteria for Restored Eelgrass, Salt Marsh and Mud Flat Habitats." *Ecological Engineering* 15:239–52.

Soukup, M. A., and J. W. Portnoy. 1986. "Impacts from Mosquito Control–Induced Sulphur Mobilization in a Cape Cod Estuary." *Environmental Conservation* 13:47–50.

Spencer, K. L., A. B. Cundy, S. Davies-Hearn, R. Hughes, S. Turner, and C. L. MacLeod. 2008. "Physicochemical Changes in Sediments at Orplands Farm, Essex, UK following 8 Years of Managed Realignment." *Estuarine Coastal and Shelf Science* 76:608–19.

Turner, R. E. 2004. "Coastal Wetland Subsidence Arising from Local Hydrologic Manipulations." *Estuaries* 27:265–72.

Warren, R. S., P. E. Fell, R. Rozsa, A. H. Brawley, A. C. Orsted, E. T. Olson, V. Swamy, and W. A. Niering. 2002. "Salt Marsh Restoration in Connecticut: 20 Years of Science and Management." *Restoration Ecology* 10:497–513.

Williams, P. B., and M. K. Orr. 2002. "Physical Evolution of Restored Breached Levee Salt Marshes in the San Francisco Bay Estuary." *Restoration Ecology* 10:527–42.

Zedler, J. B., and J. C. Callaway. 1999. "Tracking Wetland Restoration: Do Mitigation Sites Follow Desired Trajectories?" *Restoration Ecology* 7:69–73.

Chapter 4

Vegetation Responses to Tidal Restoration

STEPHEN M. SMITH AND R. SCOTT WARREN

Vegetation is perhaps the most conspicuous manifestation of hydrologic and physicochemical processes that have been transformed by both tidal restrictions and subsequent enhancements of tidal flow (restoration). The plants themselves, both living and dead, make up most of the above- and belowground physical structure of tidal marsh systems, while plant vigor, species composition, and phenology are indicators of ecosystem condition (Zhang et al. 1997; Tuxen et al. 2008). Thus monitoring and analysis of plant community change are vital for assessing the impacts of tidal restoration (Callaway et al. 2001).

Unfortunately, coastal marshes worldwide have been altered dramatically by the restriction of tidal exchange by various forms of human development. As a result, many tidal wetlands have undergone decades to centuries of degradation. Returning seawater flow to these systems started gaining popularity in New England in the 1980s (Warren et al. 2002), and these projects are now occurring with increasing frequency throughout the United States. However, the rate at which we are now returning tides to restricted salt marshes has greatly outpaced efforts to quantitatively monitor vegetation in ways that can support rigorous analyses of the impacts and effectiveness of this work. This has hampered our ability to evaluate progress and anticipate change. In reality, responses are quite variable. However, these variations are of great interest in that they enhance our understanding about the recovery process and the range of possible outcomes.

While there are some syntheses on vegetation responses to tidal restorations (Zedler 2001; Warren et al. 2002; Konisky et al. 2006), this chapter adds new information and further explores some of the ecological factors that may influence plant communities and trends of individual plant species during the restoration process. This chapter focuses on the New England region, but throughout there are multiple references to other regions of the United States and beyond.

Effects of Tidal Restrictions on Marsh Vegetation

It is fairly well understood how marsh vegetation responds to tidal restriction. In the northeastern United States and Atlantic Canada, salt marshes are dominated by perennial C_4 grasses *Spartina alterniflora* (smooth cordgrass), *Spartina patens* (salt meadow cordgrass), and *Distichlis spicata* (spike grass). With a reduction in seawater flow, salinities decline, marsh soils become drier and begin to oxidize, and these and other native halophytic graminoids and forbs are typically replaced with monocultures or mixtures of *Phragmites australis* (common reed), *Typha angustifolia* (narrowleaf cattail), *Typha latifolia* (cattail), and *Lythrum salicaria* (purple loosestrife) (Roman et al. 1984; Roman et al. 1995; Burdick et al. 1997). Where salinities fall below approximately 1 part per thousand, a variety of freshwater wetland taxa may become established (e.g., Roman et al. 1984). Severe drying from lack of tidal flooding can even allow upland forest communities to develop (Portnoy and Reynolds 1997). Irrespective of how marshes change in response to tidal restrictions, the resulting plant communities soon become dissimilar to the original community, and many of the ecosystem services provided by salt marshes are lost (Gedan et al. 2009). Thus the ultimate goal of tidal restoration is to regain to the maximum extent possible the ecological function of these wetlands, which encompasses many different processes, including the provision of suitable habitat for wildlife, nutrient transformations, water quality maintenance, primary and secondary productivity, carbon sequestration, flood abatement, and others. These functions can be evaluated in part through vegetation analysis.

Target Plant Communities

Restoration goals often relate to percent cover of native versus nonnative species. While target communities based on proportionally high cover of native halophytes are a reasonable objective, there is substantial variability in rates of change and the relative abundances of species in the eventual landscape that emerges. There are also problems with defining target communities. In some cases, there may be a lack of suitable reference marshes (Seigel et al. 2005). In New England, many nonrestricted marshes have been affected by ditching, facilitating the spread of high marsh species across elevations from which they normally would be excluded and reducing panne communities through enhanced drainage (Crain et al. 2009; Gedan and Bertness 2009; Gedan et al. 2009). Salt marsh haying has influenced marsh landscapes as well, mainly by promoting the growth of *S. patens* over *S. alterniflora* (Buchsbaum et al. 2008). Finally, reference marshes can show significant change at relatively short time scales, including annual (Dunton et al. 2001), multiyear (Donnelly and Bertness 2001), and decadal periods (Fell et al. 2000). Over the longer term, sea level rise may translate to continually changing targets (Christian et al. 2000). Thus it may not be realistic to expect restoring marshes to attain a high degree of similarity to reference sites, and adaptive management—the iterative process of making the best possible decisions within a spectrum of uncertainty based on system monitoring and assessment—may become an invaluable component of the restoration process.

An important point made by Warren et al. (2002) and others (Hackney 2000; Vasey and Holl 2007) provides some perspective on the reference marsh discussion. They emphasize the need to go beyond evaluating marsh restoration in terms of equivalence to reference communities, targeted end points, and time scales. Instead, they argue, the goal of tidal restoration lies in the process of change itself—from a degraded landscape to one that at least has the opportunity to develop similarities to the original ecosystem. Moreover, the value of achieving specific vegetation targets is limited anyway, since equivalent vegetation structure doesn't always translate to equivalent function (Zedler and Lindig-Cisneros 2001; Zedler 2007). Similarly, Hackney (2000) suggests that real progress toward restoration is best measured by positive trends rather than specific end points. Morgan and Short (2002) and Vasey and Holl (2007) emphasize the need to readjust our thinking about target communities and restoration success. Instead, they suggest that restoration targets need to be related to processes within a dynamic landscape in which a wider spectrum of ecosystem types are considered.

Examples of such trend analyses include Roman et al. (2002), who reported on short- term positive gains in native halophyte abundance coinciding with decreasing *Phragmites*. Smith et al. (2009) documented increasing halophyte cover corresponding with an expansion of salt marsh spatial area. Konisky et al. (2006) analyzed vegetation shifts from multiple restoration sites and generally observed initial decreases in *Phragmites* cover, with increasing halophyte cover over time. Buchsbaum et al. (2006) noted that, despite an initial lag phase, significant increases in *S. alterniflora* had occurred four years postrestoration, coincident with a decrease in *Typha* and *Phragmites*.

Vegetation Responses to Tidal Restoration

There is some debate in the literature about our ability to predict outcomes of tidal restoration (Simenstad and Thom 1996; Thom 2002). But while there certainly have been surprises, marshes in the New England region and elsewhere respond to tidal restoration in a somewhat predictable manner. Although individual species tolerances differ, freshwater wetland and upland taxa will succumb to seawater exposure rather quickly. In fact, a significant ancillary benefit of tidal restoration is that many tide-restricted floodplains and peripheral areas are infested with a variety of aquatic and terrestrial nonnative invasives and many of these species (e.g., *Rosa multiflora* [multiflora rose], *Lonicera* spp. [honeysuckles), *Celastrus orbiculata* [oriental bittersweet], and *Lythrum salicaria* [purple loosestrife]), are not salt or flood tolerant and are quickly eliminated by increased tidal flow.

The rapid decline of salt-intolerant species is followed by variable rates of recolonization by halophytes. Annual forbs, especially *Salicornia* spp. and *Suaeda* spp., tend to populate newly restored salt marsh areas first (Lindig-Cisneros and Zedler 2002; Wolters et al. 2005a; Fell et al. 2006; Smith 2007; Smith et al. 2009; Bowron et al. (2011) (fig. 4.1). This seems to be a widespread response (e.g., Atlantic Canada, New England, California, etc.) and is largely a consequence of small seed size and prolific seed production compared with perennial grasses (Ellison 1987). Colonization of new areas by *Spartina alterniflora* initially depends

FIGURE 4.1. *Spartina alterniflora*, *Salicornia virginica*, and *Limonium carolinianum* replacing *Phragmites australis* in response to increased tidal exchange. (Photo courtesy of Stephen Smith)

upon seeds, but thereafter vegetative growth becomes increasingly important for expansion (Metcalf et al. 1986). This has been shown to be the case for a closely related species, *Spartina densiflora* (denseflower cordgrass), which was introduced to California and after initial colonization rapidly proliferated through tiller production (Kittelson and Boyd 1997). In any event, rates of seed germination, recruitment, and vegetative growth are regulated by many factors, and there may be substantial lag time before *S. alterniflora* becomes dominant (Buchsbaum et al. 2006; Fell et al. 2006; Konisky et al. 2006). Overall, the relative importance of sexual versus asexual reproduction in the expansion of *S. alterniflora* is not yet well understood and may be unpredictable from site to site.

S. patens has smaller seeds than *S. alterniflora* and therefore disperses more easily. In this way, scattered, small populations can establish in areas of the marsh that are remote from source populations (Smith et al. 2009). If a significant amount of marsh subsidence has occurred, however, *S. alterniflora* may be initially favored over *S. patens* due to its ability to produce better developed aerenchyma to withstand longer periods of flooding (Naidoo et al. 1992). It may take some time for *S. patens* to become established, with cover increasing as accumulation of sediments and organic matter slowly increases marsh surface elevation. *Distichlis* responses are quite variable (Barrett and Niering 1993; Raposa 2008), perhaps because it frequently colonizes disturbance patches (bare ground) only to be outcompeted by *S. patens* (Bertness and Shumway 1993).

Brackish species can persist across a range of intermediate salinities and seem to be less predictable, although *Typha angustifolia* begins to die out at salinities of 10 parts per thousand, and *T. latifolia* is considered even less salt tolerant (Hutchinson 1988). *Phragmites* can thrive in areas with salinities between 10 and 25 parts per thousand, where interspecific competition from salt-intolerant taxa has been eliminated and osmotic stress is nonlethal, but significant decline occurs at salinities close to full-strength seawater (Vasquez et al. 2006; Smith et al. 2009), at which point it can be replaced by native halophytes (fig. 4.2). Meyerson et al. (2009) suggest that salinities greater than 18 parts per thousand are generally sufficient for keeping *Phragmites* populations from expanding, but stands can persist in restoration sites for years, even with salinities approaching 25 parts per thousand (Burdick et al. 2001; Warren et al. 2002). High sulfide concentrations tend to accelerate their demise (Hotes et al. 2005). Notwithstanding, clonal integration allows *Phragmites* to exploit suboptimal habitat, and mature clones can apparently withstand salinity up to 45 percent, sulfide concentrations up to 1.75 millimoles, and permanent inundation, at least for a season (Amesberry et al. 2000; Chambers et al. 1998, 2003). Purple loosestrife, another widespread exotic of restricted salt marshes in New England, generally does not tolerate salinities much above 8 parts per thousand (Hutchinson 1988; Smith et al. 2009).

FIGURE 4.2. Changes in a permanent vegetation monitoring plot (*Phragmites* to bare ground to *Spartina alterniflora*) during the course of four years in a restoring marsh, Hatches Harbor, Cape Cod National Seashore. (Photos courtesy of Stephen Smith)

Variability in Vegetation Responses to Tidal Restoration

There is some uncertainty as to exactly how tidal restoration will transform plant communities and over what time period it will happen. Warren et al. (2002) found that rates of vegetation recovery in Connecticut restoration projects differed by an order of magnitude. In Gulf of Maine marshes, Konisky et al. (2006) reported that the cover of halophyte species actually declined for the first two years following restoration but expanded thereafter. In New Hampshire, the return of salt marsh vegetation between a planned versus an unplanned hydrologic restoration occurred at vastly different rates and resulted in different taxonomic compositions (Burdick et al. 1997).

Some systems recover in less than a decade (Burdick et al.1997; Wolters et al. 2005b; Raposa 2008). In others, such as the Essex estuaries of southeast England, tidally restored salt marshes still differ in species richness, composition, and structure after 100 years (Garbutt and Wolters 2008). This uncertainty emphasizes the point that a multitude of variables can alter the trajectory of tidally driven vegetation restoration. As such, restoration responses are often quite site specific. Presented next (and in fig. 4.3) are a number of site factors that can contribute to inconsistencies in vegetation responses; it should be noted that these are often interrelated.

Duration and Magnitude of Tidal Restriction

The length of time under a regime of tidal restriction and the severity of the restriction will influence the degree to which vegetation, soil chemistry, soil subsidence, and other parameters have changed (Roman et al. 1984). These, in turn, can influence rates and patterns of vegetation recovery as discussed in the following subsections.

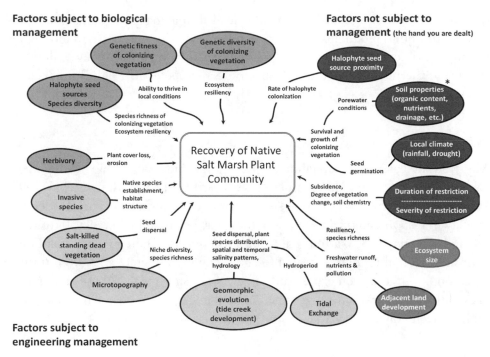

FIGURE 4.3. Factors that influence rates and patterns of salt marsh vegetation recovery following the reintroduction of tidal action.

Note: Those on the right reflect the history and landscape position of the restoration site; these must be considered in design of a restoration project but are otherwise outside the control of managers. Those on the bottom can be manipulated through project engineering, both in the design phase and, with adaptive management, throughout the follow-up and monitoring phase. Biological factors (top left) include source and nature of colonizing plants, in both naturally revegetating and planted systems, and are the subject of management decisions in the planning phase. Invasive plant species and herbivory are potential problems to be anticipated in the planning phase but are addressed through adaptive management during follow-up and monitoring. The asterisk on soil properties indicates that soil is sometimes managed/engineered in large-scale restoration efforts (e.g., spraying sediments on marsh surfaces, amending soils with organic matter, regulating sediment delivery through hydrologic management, etc.).

Hydrology

The importance of hydrology and the hydrologic analyses that must precede tidal restoration has long been understood (Coats et al. 1989; Roman et al. 1995). Although differences in the recovery of salt marsh taxa and reduction in undesirable taxa like *Phragmites* are related to differences in tidal flooding (Warren et al. 2002), more water is not always better. While facilitating maximum tidal exchange possible may seem desirable, restorations that result in areas with prolonged or permanent inundation can delay or inhibit vegetation establishment

(Rozas 1995). Such conditions may also lead to macroalgae blooms and/or increased mosquito production. Tidal restoration may also produce a much higher proportion of low marsh due to soil subsidence and a high frequency of flooding relative to the needs of high marsh species (Sinicrope et al. 1990; Burdick et al. 1997).

Topography

Marsh surface elevation in relation to tidal regime and topographic heterogeneity is a critical factor in the development of recovering marshes (Roman et al. 1995; Wolters et al. 2005b). For example, while salt-intolerant species generally decline quickly in response to tidal restoration, they may persist at higher elevations (Barrett and Niering 1993; Warren et al. 2002). A gentle elevation gradient, combined with incremental restoration, can allow *Phragmites* to migrate away from unsuitable conditions and even expand into new areas (Smith et al. 2009). Restoration sites with large elevation ranges and many topographic niches tend to support higher species diversity (Vivian-Smith 1997).

Geomorphology

Marsh geomorphology can affect a variety of physical and biological processes, including erosion, sedimentation, heterogeneity of habitat, and drainage (Torres et al. 2006). If the hydrologic network is lost as a result of tidal restriction, flooding and drainage dynamics may be dramatically altered from the original system. With restoration, it is difficult to predict how tidal creeks will redevelop, especially when subjected to disturbances like storm events (Teal and Weishar 2005; Zedler and West 2008). The emerging hydrologic patterns are important given that tidal creeks influence seed dispersal (Chang et al. 2007), gradients of plant species assemblages (Sanderson et al. 2000; Morzaria-Luna et al. 2004), plant vigor (Mendelssohn and Morris 2000), and marsh maturation (Tyler and Zieman 1999).

Pore Water Chemistry

Salinity and sulfide act individually and synergistically. Elevated salinity alone will eliminate or greatly reduce freshwater taxa, but well-drained soils with low levels of sulfide may not cause a significant decline in some brackish species like *Phragmites*. High salinity combined with high sulfide, the latter of which depends upon hydroperiod and soil properties, will result in more rapid losses (Chambers et al. 1998). However, tidal restoration can actually enhance *Phragmites* habitat by replacing freshwater with brackish conditions (Sinicrope et al. 1990; Smith et al. 2009). It should be noted that the duration of specific salinity levels over the course of the growing season is key to plant responses since transient exposures to high salinities may not be a significant stress (Whigham et al. 1989; Howard and Mendelssohn 1999).

Soil Properties

Organic matter content influences many soil properties, including water-holding capacity, porosity, nutrient storage, nutrient cycling, and the species composition and abundance of sediment-dwelling invertebrates (Broome et al. 2000). It is well known that organic matter amendments in created wetlands generally enhance the survivorship and growth of vegetation (O'Brien and Zedler 2006). Soil drainage influences the vigor of marsh plants, which is enhanced by oxygenation of the root zone (Pezeshki 1997; Mendelssohn and Morris 2000). Another important aspect of soils is the degree to which acid sulfates have developed. The reintroduction of seawater to acid sulfate soils could temporarily result in sulfide increases and nutrient release (Portnoy 1999; Johnston et al. 2009).

Nutrients

To a certain extent, nutrients are beneficial and will stimulate recovery of marsh vegetation. Yet an excess of nutrients may stimulate macroalgae blooms or alter species composition (Valiela et al. 1976; Hunter et al. 2008). It has been shown that *S. alterniflora* outcompetes *S. patens* under conditions of nitrogen enrichment (Levine et al. 1998; Bertness et al. 2002; Wigand et al. 2003). The abundance of *Phragmites* has also been positively correlated with nutrient inputs (Bertness et al. 2002; Minchinton and Bertness 2003). In addition, nutrient enrichment can have cascading trophic effects as vegetation is made more palatable for first-order consumers (Bertness et al. 2008).

Seeds

The recolonization of salt marsh vegetation is highly dependent upon proximity to existing seed sources, the size of source populations, extant seed banks, the dynamics of dispersal, and conditions for germination. At a Cape Cod National Seashore restoration site, lack of any existing or nearby source populations prompted multiple years of hand-seeding and plantings to inoculate the system with native halophytes. Restoration areas with some remnant halophytic vegetation or populations in close proximity to the restored areas with a hydrologic connection are likely to experience more rapid recolonization (Wolters et al. 2005a; Erfanzadeh et al. 2010).

Composition of extant halophyte populations is important too. Annual species, particularly *Salicornia* spp. and *Suaeda* spp., tend to colonize newly created salt marsh areas first (Lindig-Cisneros and Zedler 2002; Wolters et al. 2005a; Armitage et al. 2006; Smith et al. 2009). Seeds are typically dispersed by surface water currents (Dausse et al. 2008) and to a lesser extent by birds and mammals (Wolters et al. 2005a). Yet dispersal can be inhibited by large areas of standing dead salt-killed vegetation because many seeds are trapped in wrack material that becomes entangled in it or piled up along the downstream edge (Minchinton 2006; Smith 2007; Chang et al. 2007, 2008).

Numerous abiotic factors such as temperature, photoperiod, soil salinity, and soil moisture influence seed germination (Miller and Egler 1950; Zedler and Beare 1986; Bertness 1991; Noe and Zedler 2000, 2001). Salinity affects the physical buoyancy, and thus dispersal, of seeds (Elsey-Quirk et al. 2009). High levels of bioturbation can suppress seed germination, as has been found in subtidal eelgrass (Dumbald and Wyllie-Echeverria 2003). The development of algal mats can inhibit seed germination by smothering seeds and new seedlings (Jensen and Jefferies 1984; Callaway and Sullivan 2001).

Herbivory

A variety of herbivores, including muskrats, snow geese, insects, snails, and crabs can influence salt marsh vegetation (Lynch et al. 1947; Smith 1982; Holdredge et al. 2009). Their effects are variable, but at high enough levels grazing can result in an inhibition of vegetation development, declines in seed production, or, in some cases, a complete loss of vegetation (Bertness et al. 1987; Kuijper and Bakker 2003; Holdredge et al. 2009). Llewellyn and Shaffer (1993) advocated for the planting of herbivore-resistant species in Louisiana freshwater marsh restorations. Herbivore impacts on mangrove propagules have long been recognized as significant factors influencing the success of mangrove forest restoration (Kaly and Jones 1998).

Genotype

The genetic fitness of plants is perhaps an overlooked factor in marsh restoration (Proffitt et al. 2003, 2005). Seed germination and seedling growth are affected by genotype (Biber and Caldwell 2008). Genotype will also play a role in stress tolerance (Pezeshki and DeLaune 1995; Howard and Rafferty 2006), and the development of genetic diversity will offer a higher level of ecosystem resilience through stress and disturbance (Reusch and Hughes 2006).

Surrounding Land Use

Land development in the watershed can affect marsh hydrology, salinity, sediment dynamics, contaminant loadings, and vegetation (Greer and Stow 2003; Holland et al. 2004). Impervious surfaces increase freshwater runoff. Subsurface discharges of freshwater may be exploited by deep *Phragmites* roots and allow survival in places where it would otherwise be excluded by much higher surface water salinities (Burdick et al. 2001). A lack of freshwater flow can also impede restoration. Dams in particular can starve downstream marshes of sediment that would normally contribute to marsh building processes (Ravens et al. 2009).

Local Climate

An increase in the availability of freshwater due to high rainfall (corresponding with a reduction in salinity) can enhance seed germination of various species (Shumway and Bertness 1992; Allison 1992, 1996). Rates of vegetative growth and plant structure will also respond to annual precipitation (Callaway and Sabraw 1994; Dunton et al. 2001). Drought can negatively impact the growth of salt marsh plants primarily related to increased soil salinity (Visser et al. 2006; Alber et al. 2008). Alternatively, drought can ameliorate waterlogged conditions to make them more favorable for growth of *Spartina* grasses (Charles and Dukes 2009), while adversely impacting panne species (Gedan and Bertness 2009). Wind is not often modeled when studying the hydrology of tidal restoration, but it is a local factor that can have significant effects on flooding regime in microtidal systems (Nyman et al. 2009).

Size

There is minor evidence that large restoration sites recover more rapidly than small sites (Wolters et al. 2005b). In theory, larger ecosystems tend to be more stable and resilient to disturbance than smaller ones and are more likely to have habitat heterogeneity, which can support greater biodiversity (Crowley 1978; Brown and Dinsmore 1986; Cohen and Newman 1991; Nichols et al. 1998). Notwithstanding, more long-term data from tidal restoration projects with a range of aerial extents are needed to fully understand the relative importance of this variable.

The factors discussed are important but represent only a subset of those that can alter the trajectory of vegetation development. That said, response variability can be highly informative. Although it decreases our confidence in predicting outcomes, with rigorous monitoring and ancillary research it helps us to understand the restoration process and can guide us in developing better conceptual models and more realistic expectations.

Vegetation Management during Restoration

One objective of active vegetation management is to correct or reverse undesirable results or trends. Another is simply to enhance or accelerate the process of recovery. For example, the removal of salt-killed vegetation that presents a barrier to seed dispersal can be effective (Smith 2007). Hand-seeding and planting to inoculate areas distant from seed sources may be necessary as well. Selection of planting stock genetics may even be considered if certain attributes are needed, such as increased sulfide tolerance (Wang et al. 2003). In some cases it may be necessary to treat *Phragmites* or other exotics with herbicide or mechanical removal in order to prevent spreading during tidal restoration (Greenwood and McFarlane 2006). The re-creation or addition of tidal creeks has increased survivorship of desirable taxa (O'Brien and Zedler 2006). Constructed tidal channels can also enhance seawater and halophyte seed penetration upstream and may alleviate drainage problems. Pools have been created for fish enhancement and to enhance structural diversity of the restoring landscape (Larkin et al. 2009). In Australia, seagrass wrack has been used as nutrient amendments and for the creation of disturbance patches to promote species diversity (Chapman and Roberts 2004).

Restoration and Climate Change

It has been suggested that the case for restoring tidal wetlands is strengthened by the backdrop of climate change (Singh et al. 2007). The idea is that salt marshes that are free from significant stressors, like tidal restrictions, are likely to be more resilient to climate change than those that are degraded. Given the anticipated loss of tidal marshes to sea level rise, it is important that we restore as much acreage as possible to offset some of these eventual losses. Tidal marshes sequester huge amounts of carbon (Chmura et al. 2003; Brigham et al. 2006) and, if managed well, could be increasingly important in ameliorating rising atmospheric carbon dioxide (CO_2).

Tidal restoration will face numerous challenges and uncertainties with changing climate. Altered patterns of temperature, rainfall, and snowmelt could influence salinity gradients. Callaway et al. (2007) and others suggest that a rapid acceleration in sea level rise could result in a catastrophic loss of salt marshes, including restored systems. Certainly, the design of restoration projects (e.g., culvert, bridge, or reconfiguration of other tidal openings) must anticipate accelerated rates of sea level rise. Elevated CO_2 and temperature may differentially affect the growth of salt marsh species and, in doing so, could cause shifts in marsh structure, species composition (Arp et al. 1993; Gedan and Bertness 2009), marsh elevations through peat accumulation, and increased sediment deposition (Langley et al. 2009). Thus there are many questions to be answered with modeling and research to determine the long-term outcomes of restorations in relation to climate change.

Monitoring and Research Opportunities

Monitoring vegetation changes in a repeatable, quantitative way is essential to the future of tidal restoration (Roman et al. 2002). Neckles et al. (2002) provide comprehensive guidance on monitoring design and techniques. In addition, experimental field and laboratory research can provide specific knowledge regarding plant community response to specific abiotic and biotic environmental variables (Konisky and Burdick 2004; Callaway 2005). Callaway et al. (1997) suggest that mesocosms are an excellent way to assess experimental restoration techniques. Because the recolonization of restricted marshes by native halophytes is so dependent on seed dynamics, there is much research to be done on mechanisms that enhance dispersal and germination. In addition, various forms of vegetation management, both preceding and following restoration, are becoming increasingly important, in large part due to invasive species like *Phragmites*. Monitoring and research can provide information on (1) how prerestoration plant landscapes influence the subsequent trajectory of recovery, (2) how adaptive management techniques can facilitate desirable change, and (3) how climate change may be expected to influence the fate of tidal restorations.

Acknowledgments

We thank the coeditors and anonymous peers for their reviews of early drafts of this chapter.

REFERENCES

Alber, M., E. M. Swenson, S. C. Adamowicz, and I. A. Mendelssohn. 2008. "Salt Marsh Dieback: An Overview of Recent Events in the US." *Estuarine, Coastal and Shelf Science* 80:1–11.
Allison, S. K. 1992. "The Influence of Rainfall Variability on the Species Composition of a Northern California Salt-Marsh Plant Assemblage." *Vegetatio* 101:145–60.

Allison, S. K. 1996. "Recruitment and Establishment of Salt Marsh Plants following Disturbance by Flooding." *American Midland Naturalist* 136:232–47.

Amesberry, L., M. A. Baker, P. J. Ewanchuk, and M. D. Bertness. 2000. "Clonal Integration and the Expansion of *Phragmites australis*." *Ecological Applications* 10:1110–18.

Armitage, A. R., K. E. Boyer, R. R. Vance, and R. F. Ambrose. 2006. "Restoring Assemblages of Salt Marsh Halophytes in the Presence of a Rapidly Colonizing Dominant Species." *Wetlands* 26:667–76.

Arp, W. J., B. G. Drake, W. T. Pockman, P. S. Curtis, and D. F. Whigham. 1993. "Interactions between C₃ and C₄ Salt Marsh Plant Species during Four Years of Exposure to Elevated Atmospheric CO2." *Plant Ecology* 104:133–43.

Barrett, N. E., and W. A. Niering. 1993. "Tidal Marsh Restoration: Trends in Vegetation Change Using a Geographical Information System (GIS)." *Restoration Ecology* 1:18–28.

Bertness, M. D. 1991. "Zonation of *Spartina patens* and *Spartina alterniflora* in a New England Salt Marsh." *Ecology* 72:138–48.

Bertness, M. D., and S. W. Shumway. 1993. "Competition and Facilitation in Marsh Plants." *American Naturalist* 142:718–24.

Bertness, M. D., C. Wise, and A. M. Ellison. 1987. "Consumer Pressure and Seed Set in a Salt Marsh Perennial Plant Community." *Oecologia* 71:190–200.

Bertness, M. D., P. J. Ewanchuk, and B. R. Silliman. 2002. "Anthropogenic Modification of New England Salt Marsh Landscapes." *Proceedings of the National Academy of Sciences* 99:1395–98.

Bertness, M. D., C. M. Crain, C. Holdredge, and N. Sala. 2008. "Eutrophication Triggers Consumer Control of New England Salt Marsh Primary Production." *Conservation Biology* 22:131–39.

Biber, P. D., and J. D. Caldwell. 2008. "Seed Germination and Seedling Survival of *Spartina alterniflora* Loisel." *American Journal of Agricultural and Biological Science* 3:633–38.

Bowron, T., N. Neatt, D. van Proosdij, J. Lundholm, and J. Graham. 2011. "Macro-tidal Salt Marsh Ecosystem Response to Culvert Expansion." *Restoration Ecology* 19:307–22.

Brigham, S. D., J. P. Megonigal, J. K. Keller, N. P. Bliss, and C. Trettin. 2006. "The Carbon Balance of North American Wetlands." *Wetlands* 26:889–916.

Broome, S. W., C. B. Craft, and W. A. Toomey Jr. 2000. "Soil Organic Matter (SOM) Effects on Infaunal Community Structure in Restored and Created Tidal Marshes." Pp. 737–47 in *Concepts and Controversies in Tidal Marsh Ecology*, edited by M. P. Weinstein and D. A. Kreeger. Dordrecht, Netherlands: Kluwer Academic.

Brown, M., and J. Dinsmore. 1986. "Implications of Marsh Size and Isolation for Marsh Bird Management." *Journal of Wildlife Management* 50:392–97.

Buchsbaum, R. N., J. Catena, E. Hutchins, and M. J. James-Pirri. 2006. "Changes in Salt Marsh Vegetation, *Phragmites australis*, and Nekton in Response to Increased Tidal Flushing in a New England Salt Marsh." *Wetlands* 26:544–57.

Buchsbaum, R. N., L. A. Deegan, J. Horowitz, R. H. Garritt, A. E. Giblin, J. P. Ludlam, and D. H. Shull. 2008. "Effects of Regular Salt Marsh Haying on Marsh Plants, Algae, Invertebrates and Birds at Plum Island Sound, Massachusetts." *Wetlands Ecology and Management* 17:469–87.

Burdick, D. M., M. Dionne, R. M. Boumans, and F. T. Short. 1997. "Ecological Responses to Tidal Restorations in Two Northern New England Salt Marshes." *Wetlands Ecology and Management* 4:129–44.

Burdick, D. M., R. Buchsbaum, and E. Holt. 2001. "Variation in Soil Salinity Associated with Expansion of *Phragmites australis* in Salt Marshes." *Environmental and Experimental Botany* 46:247–61.

Callaway, J. C. 2005. "The Challenge of Restoring Functioning Salt Marsh Ecosystems." *Journal of Coastal Research* 40:24–36.

Callaway, R. M., and C. S. Sabraw. 1994. "Effects of Variable Precipitation on the Structure and Diversity of a California Salt Marsh Community." *Journal of Vegetation Science* 5:433–438.

Callaway, J. C., and G. Sullivan. 2001. "Sustaining Restored Wetlands: Identifying and Solving Management Problems." Pp. 337–59 in *Handbook for Restoring Tidal Wetlands*, edited by J. B. Zedler. New York: CRC Press.

Callaway, J. C., J. B. Zedler, and D. L. Ross. 1997. "Using Tidal Salt Marsh Mesocosms to Aid Wetland Restoration." *Restoration Ecology* 5:135–46.

Callaway, J. C., G. Sullivan, J. S. Desmond, G. D. Williams, and J. B. Zedler. 2001. "Assessment and Monitoring." Pp. 271–324 in *Handbook for Restoring Tidal Wetlands*, edited by J. B. Zedler. New York: CRC Press.

Callaway, J. C., V. T. Parker, M. C. Vasey, and L. M. Schile. 2007. "Emerging Issues for the Restoration of Tidal Marsh Ecosystems in the Context of Predicted Climate Change." *Madroño* 54:234–48.

Chambers, R. M., T. J. Mozdzer, and J. C. Ambrose. 1998. "Effects of Salinity and Sulfide on the Distribution of *Phragmites australis* and *Spartina alterniflora* in a Tidal Saltmarsh." *Aquatic Botany* 62:161–69.

Chambers, R. M., D. T. Osgood, D. J. Bart, and F. Montalto. 2003. "*Phragmites australis* Invasion and Expansion in Tidal Wetlands: Interactions among Salinity, Sulfide, and Hydrology." *Estuaries* 26:398–406.

Chang, E. R., R. M. Veeneklaas, and J. P. Bakker. 2007. "Seed Dynamics Linked to Variability in Movement of Tidal Water." *Journal of Vegetation Science* 18:253–62.

Chang, E. R., R. M. Veeneklaas, R. Buitenwerf, J. P. Bakker, and T. J. Bouma. 2008. "To Move or Not to Move: Determinants of Seed Retention in a Tidal Marsh." *Functional Ecology* 22:4720–27.

Chapman, M. G., and D. E. Roberts. 2004. "Use of Seagrass Wrack in Restoring Disturbed Australian Salt Marshes." *Ecological Management and Restoration* 5:183–90.

Charles, H., and J. S. Dukes. 2009. "Effects of Warming and Altered Precipitation on Plant and Nutrient Dynamics of a New England Salt Marsh." *Ecological Applications* 19:1758–73.

Chmura, G. L., S. C. Anisfeld, D. R. Cahoon, and J. C. Lynch. 2003. "Global Carbon Sequestration in Tidal, Saline Wetland Soils." *Global Biogeochemical Cycles* 17:1111.

Christian, R. R., L. E. Stasavich, C. R. Thomas, and M. M. Brinson. 2000. "Reference Is a Moving Target in Sea-Level Controlled Wetlands." Pp. 805–26 in *Concepts and Controversies in Tidal Marsh Ecology*, edited by M. P. Weinstein and D. A. Kreeger. Dordrecht, Netherlands: Kluwer Academic.

Coats, R., M. Swanson, and P. Williams. 1989. "Hydrologic Analysis for Coastal Wetland Restoration." *Environmental Management* 13:715–27.

Cohen, J. E., and C. M. Newman. 1991. "Community Area and Food-Chain Length: Theoretical Predictions." *American Naturalist* 138:1542–54.

Crain, C. M., K. B. Gedham, and M. Dionne. 2009. "Tidal Restrictions and Mosquito Ditching in New England Marshes; Case Studies of the Biotic Evidence, Physical Extent, and Potential for Restoration of Altered Hydrology." Pp. 149–70 in *Human Impacts on Salt Marshes. A Global Perspective*, edited by B. R. Silliman, E. D. Grosholz, and M. D. Bertness. Berkeley: University of California Press.

Crowley, P. H. 1978. "Effective Size and the Persistence of Ecosystems." *Oecologia* 35:185–95.

Dausse, A., A. Bonis, J. B. Bouzillé, and J. C. Lefeuvre. 2008. "Seed Dispersal in a Polder after Partial Tidal Restoration: Implications for Salt-Marsh Restoration." *Applied Vegetation Science* 11:3–12.

Donnelly, J. P., and M. D. Bertness. 2001. "Rapid Shoreward Encroachment of Salt Marsh Vegetation in Response to Accelerated Sea-Level Rise." *Proceedings of the National Academy of Sciences* 98:14218–23.

Dumbald, B., and S. Wyllie-Echeverria. 2003. "The Influence of Burrowing Thalassinid Shrimp on the Distribution of Intertidal Seagrasses in Willapa Bay, Washington." *Aquatic Botany* 77:27–42.

Dunton, K. H., B. Hardegree, and T. E. Whitledge. 2001. "Response of Estuarine Marsh Vegetation to Interannual Variations in Precipitation." *Estuaries and Coasts* 24:851–61.

Ellison, A. M. 1987. "Effects of Competition, Disturbance, and Herbivory on *Salicornia europaea*." *Ecology* 68:576–86.

Elsey-Quirk, T., B. A. Middleton, and C. E. Proffitt. 2009. "Seed Flotation and Germination of Salt Marsh Plants: The Effects of Stratification, Salinity, and/or Inundation Regime." *Aquatic Botany* 91:40–46.

Erfanzadeh, R., A. Garbutt, J. Pétillon, J. P. Maelfait, and M. Hoffmann. 2010. "Factors Affecting the Success of Early Salt-Marsh Colonizers: Seed Availability Rather Than Site Suitability and Dispersal Traits." *Plant Ecology* 206:335–47.

Fell, P. E., R. S. Warren, and W. A. Niering. 2000. "Restoration of Salt and Brackish Tidelands in Southern New England." Pp. 845–58 in *Concepts and Controversies in Tidal Marsh Ecology*, edited by M. P. Weinstein and D. A. Kreeger. Dordrecht, Netherlands: Kluwer Academic.

Fell, P. E., R. S. Warren, A. E. Curtis, and E. M. Steiner. 2006. "Short-Term Effects of Herbiciding and Mowing *Phragmites australis*–Dominated Tidal Marsh on Macroinvertebrates and Fishes." *Northeastern Naturalist* 13:191–212.

Garbutt, A., and M. Wolters. 2008. "The Natural Regeneration of Salt Marsh on Formerly Reclaimed Land." *Vegetation Science* 11:335–44.

Gedan, K. B., and M. D. Bertness. 2009. "Experimental Warming Causes Rapid Loss of Plant Diversity in New England Salt Marshes." *Ecology Letters* 12:842–48.

Gedan, K. B., B. R. Silliman, and M. D. Bertness. 2009. "Centuries of Human-Driven Change in Salt Marsh Ecosystems." *Annual Review of Marine Science* 1:117–41.

Greenwood, M. E., and G. R. MacFarlane. 2006. "Effects of Salinity and Temperature on the Germination of *Phragmites australis, Juncus kraussii*, and *Juncus acutus*: Implications for Estuarine Restoration Initiatives." *Wetlands* 14:854–61.

Greer, K., and D. Stow. 2003. "Vegetation Type Conversion in Los Peñasquitos Lagoon, California: An Examination of the Role of Watershed Urbanization." *Environmental Management* 31:489–503.

Hackney, C.T. 2000. "Restoration of Coastal Habitats: Expectation and Reality." *Ecological Engineering* 15:165–70.

Holdredge, C., M. D. Bertness, and A. Altieri. 2009. "Role of Crab Herbivory in Die-off of New England Salt Marshes." *Conservation Biology* 23:672–79.

Holland, A. F., D. M. Sanger, C. P. Gawle, S. B. Lerberg, M. S. Santiago, G. H. M. Riekerk, L. E. Zimmerman, and G. I. Scott. 2004. "Linkages between Tidal Creek Ecosystems and the Landscape and Demographic Attributes of Their Watersheds." *Journal of Experimental Marine Biology and Ecology* 298:151–78.

Hotes, S., E. B. Adema, A. P. Grootjans, T. Inoue, and P. Poschlod. 2005. "Reed Die-back Related to Increased Sulfide Concentration in a Coastal Mire in Eastern Hokkaido, Japan." *Wetlands Ecology and Management* 13:83–91.

Howard, R. J., and I. A. Mendelssohn. 1999. "Salinity as a Constraint on Growth of Oligohaline Marsh Macrophytes, II: Salt Pulses and Recovery Potential." *American Journal of Botany* 86:795–806.

Howard, R. J., and P. S. Rafferty. 2006. "Clonal Variation in Response to Salinity and Flooding Stress in Four Marsh Macrophytes of the Northern Gulf of Mexico, USA." *Environmental and Experimental Botany* 56:301–13.

Hunter, A., N. M. B. Morris, and J. Cebrian. 2008. "Effects of Nutrient Enrichment on *Distichlis spicata* and *Salicornia bigelovii* in a Salt Marsh Pan." *Wetlands* 28:760–775.

Hutchinson, I. 1988. *Salinity Tolerance of Plants of Estuarine Wetlands and Associated Uplands*. Vancouver, BC: Simon Fraser University, Washington State Shorelands and Coastal Zone Management Program: Wetlands Section.

Jensen, A., and R. L. Jefferies. 1984. "Fecundity and Mortality in Populations of *Salicornia europaea* agg. at Skallingen, Denmark." *Holarctic Ecology* 7:399–412.

Johnston S. G., R. T. Bush, L. A. Sullivan, E. D. Burton, D. Smith, M. A. Martens, A. E. McElnea, C. R. Ahern, B. Powell, L. P. Stephens, S. T. Wilbraham, and S. van Heel. 2009. "Changes in Water Quality following Tidal Inundation of Coastal Lowland Acid Sulfate Soil Landscapes." *Estuarine, Coastal and Shelf Science* 81:257–66.

Kaly, U. L., and G. P. Jones. 1998. "Mangrove Restoration: A Potential Tool for Coastal Management in Tropical Developing Countries. *Ambio* 27:656–66.

Kittelson, P. M., and M. J. Boyd. 1997. "Mechanisms of expansion for an introduced

Species of Cordgrass, *Spartina densiflora*, in Humboldt Bay, California." *Estuaries* 2:770–78.

Konisky, R. A., and D. M. Burdick. 2004. "Effects of Stressors on Invasive and Halophytic Plants of New England Salt Marshes: A Framework for Predicting Response to Tidal Restoration." *Wetlands* 24:434–47.

Konisky, R. A., D. M. Burdick, M. Dionne, and H. A. Neckles. 2006. "A Regional Assessment of Salt Marsh Restoration and Monitoring in the Gulf of Maine." *Restoration Ecology* 14:516–25.

Kuijper, D. P. J., and J. P. Bakker. 2003. "Large-Scale Effects of a Small Herbivore on Salt-Marsh Vegetation Succession: A Comparative Study on Three Wadden Sea Islands." *Journal of Coastal Conservation* 9:179–88.

Langley, J. A., K. L. McKee, D. R. Cahoon, J. A. Cherry, and J. P. Megonigal. 2009. "Elevated CO_2 Stimulates Marsh Elevation Gain, Counterbalancing Sea-Level Rise." *Proceedings of the National Academy of Sciences* 106:6182–86.

Larkin, D. J., J. M. West, and J. B. Zedler. 2009. "Created Pools and Food Availability for Fishes in a Restored Salt Marsh." *Ecological Engineering* 35:65–74.

Levine, J., S. J. Brewer, and M. D. Bertness. 1998. "Nutrient Availability and the Zonation of Marsh Plant Communities." *Journal of Ecology* 86:285–92.

Lindig-Cisneros, R., and J. B. Zedler. 2002. "Halophyte Recruitment in a Salt Marsh Restoration Site." *Estuaries* 25:1174–83.

Llewellyn, D. W., and G. P. Shaffer. 1993. "Marsh Restoration in the Presence of Intense Herbivory: The Role of *Justicia lanceolata* (Chapm.) Small." *Wetlands* 13:176–84.

Lynch, J. J., T. O'Neil, and D. W. Lay. 1947. "Management Significance of Damage by Geese and Muskrats to Gulf Coast Marshes." *Journal of Wildlife Management* 11:50–76.

Mendelssohn, I. A., and J. T. Morris. 2000. "Eco-physiological Controls on the Productivity of *Spartina alterniflora* Loisel." Pp. 59–80 in *Concepts and Controversies in Tidal Marsh Ecology*, edited by M. P. Weinstein and D. A. Kreeger. Dordrecht, Netherlands: Kluwer Academic.

Metcalfe, W. S., A. M. Ellison, and M. D. Bertness. 1986. "Survivorship and Spatial Development of *Spartina alterniflora* Loisel. (Gramineae) seedlings in a New England Salt Marsh." *Annals of Botany* 58:249–58.

Meyerson, L. A., K. Saltonstall, and R. M. Chambers. 2009. "*Phragmites australis* in Eastern North America: A Historical and Ecological Perspective." Pp. 57–82 in *Salt Marshes under Global Siege*, edited by B. R. Silliman, E. Grosholz, and M. D. Bertness. Berkeley: University of California Press.

Miller, W. B., and F. E. Egler. 1950. "Vegetation of the Wequetequock-Pawcatuck Tidal Marshes, Connecticut." *Ecological Monographs* 20:143–72.

Minchinton, T. E. 2006. "Rafting on Wrack as a Mode of Dispersal for Plants in Coastal Marshes." *Aquatic Botany* 84:372–76.

Minchinton, T. E., and M. D. Bertness. 2003. "Disturbance-Mediated Competition and the Spread of *Phragmites australis* in a Coastal Marsh." *Ecological Applications* 13:1400–16.

Morgan, P., and F. T. Short. 2002. "Using Functional Trajectories to Model Constructed Salt Marsh Development in the Great Bay Estuary, Maine/New Hampshire." *Restoration Ecology* 10:461–73.

Morzaria-Luna, H., J. C. Callaway, G. Sullivan, and J. B. Zedler. 2004. "Relationship between Topographic Heterogeneity and Vegetation Patterns in a Californian Salt-Marsh." *Journal of Vegetation Science* 15:523–30.

Naidoo, G., K. L. McKee, and I. A. Mendelssohn. 1992. "Anatomical and Metabolic Responses to Waterlogging and Salinity in *Spartina alterniflora* and *S. patens*." *American Journal of Botany* 79:765–70.

Neckles, H. A., M. Dionne, D. M. Burdick, C. T. Roman, R. Buchsbaum, and E. Hutchins. 2002. "A Monitoring Protocol to Assess Tidal Restoration of Salt Marshes on Local and Regional Scales." *Restoration Ecology* 10:556–63.

Nichols, W. F., K. T. Killingbeck, and P. V. August. 1998. "The Influence of Geomorphological Heterogeneity on Biodiversity: A Landscape Perspective." *Conservation Biology* 2:371–79.

Noe, G. B., and J. B. Zedler. 2000. "Differential Effects of Four Abiotic Factors on the Germination of Salt Marsh Annuals." *American Journal of Botany* 87:1679–92.

Noe, G. B., and J. B. Zedler. 2001. "Spatio-temporal Variation of Salt Marsh Seedling Establishment in Relation to the Abiotic and Biotic Environment." *Journal of Vegetation Science* 12:61–74.

Nyman, J. A., M. K. La Peyre, A. Caldwell, S. Piazza, C. Thom, and C. Winslow. 2009. "Defining Restoration Targets for Water Depth and Salinity in Wind-Dominated *Spartina patens* (Ait.) Muhl. Coastal Marshes." *Journal of Hydrology* 376:327–36.

O'Brien, E., and J. B. Zedler. 2006. "Accelerating the Restoration of Vegetation in a Southern California Salt Marsh." *Wetlands Ecology Management* 14:269–86.

Pezeshki, S. R. 1997. "Photosynthesis and Root Growth in *Spartina alterniflora* in Relation to Root Zone Aeration." *Photosynthetica* 34:107–14.

Pezeshki, S. R., and R. D. DeLaune. 1995. "Variation in Response of Two U.S. Gulf Coast Populations of *Spartina alterniflora* to Hypersalinity." *Journal of Coastal Research* 11:89–95.

Portnoy, J. W. 1999. "Salt Marsh Diking and Restoration: Biogeochemical Implications of Altered Wetland Hydrology." *Environmental Management* 24:111–20.

Portnoy, J., and M. Reynolds. 1997. "Wellfleet's Herring River: The Case for Habitat Restoration." *Environment Cape Cod* 1:35–43.

Proffitt, C. E., S. E. Travis, and K. R. Edwards. 2003. "Genotype and Elevation Influence *Spartina alterniflora* Colonization and Growth in a Created Salt Marsh." *Ecological Applications* 13:180–92.

Proffitt, C. E., R. L. Chiasson, A. B. Owens, K. R. Edwards, and S. E. Travis. 2005. "*Spartina alterniflora* Genotype Influences Facilitation and Suppression of High Marsh Species Colonizing an Early Successional Salt Marsh." *Journal of Ecology* 93:404–16.

Raposa, K. B. 2008. "Early Ecological Responses to Hydrologic Restoration of a Tidal Pond and Salt Marsh Complex in Narragansett Bay, Rhode Island." *Journal of Coastal Research* 55:180–92.

Ravens, T. M., R. C. Thomas, K. A. Roberts, and P. H. Santschi. 2009. "Causes of Salt Marsh Erosion in Galveston Bay, Texas." *Journal of Coastal Research* 25:265–72.

Reusch, T. B. H., and A. R. Hughes. 2006. "The Emerging Role Of Genetic Diversity for Ecosystem Functioning: Estuarine Macrophytes as Models." *Estuaries and Coasts* 29:159–64.

Roman, C. T., W. A. Niering, and R. S. Warren. 1984. "Salt Marsh Vegetation Change in Response to Tidal Restriction." *Environmental Management* 8:141–50.

Roman, C. T., R. W. Garvine, and J. W. Portnoy. 1995. "Hydrologic Modeling as a Predictive Basis for Ecological Restoration of Salt Marshes." *Environmental Management* 19:559–66.

Roman, C. T., K. B. Raposa, S. C. Adamowicz, M. J. James-Pirri, and J. G. Catena. 2002. "Quantifying Vegetation and Nekton Response to Tidal Restoration of a New England Salt Marsh." *Restoration Ecology* 10:450–60.

Rozas, L. P. 1995. "Hydroperiod and Its Influence on Nekton of the Salt Marsh: A Pulsing Ecosystem." *Estuaries* 18:579–90.

Sanderson, E. W., S. L. Ustin, and T. C. Foin. 2000. "The Influence of Tidal Channels on the Distribution of Salt Marsh Plant Species in Petaluma Marsh, CA, USA." *Plant Ecology* 146:29–41.

Seigel, A., C. Hatfield, and J. M. Hartman. 2005. "Avian Response to Restoration of Urban Tidal Marshes in the Hackensack Meadowlands, New Jersey." *Urban Habitats* 3:87–116.

Shumway, S. W., and M. D. Bertness. 1992. "Salt Stress Limitation of Seedling Recruitment in a Salt Marsh Plant Community." *Oecologia* 92:490–97.

Simenstad, C. A., and R. M. Thom. 1996. "Functional Equivalency Trajectories of the Restored Gog-Le-Hi-Te Estuarine Wetland." *Ecological Applications* 6:38–56.

Singh, K., B. B. Walters, and J. Ollerhead. 2007. "Climate Change, Sea-Level Rise and the Case for Salt Marsh Restoration in the Bay of Fundy, Canada." *Environments* 35:71.

Sinicrope, T. L., P. G. Hine, R. S. Warren, and W. A. Niering. 1990. "Restoration of an Impounded Salt Marsh in New England." *Estuaries* 13:25–30.

Smith, S. M. 2007. "Removal of Salt-Killed Vegetation during Tidal Restoration of a New England Salt Marsh: Effects on Wrack Movement and the Establishment of Native Halophytes." *Ecological Restoration* 25:268–73.

Smith, S. M., C. T. Roman, M. J. James-Pirri, K. Chapman, J. Portnoy, and E. Gwilliam. 2009. "Responses of Plant Communities to Incremental Hydrologic Restoration of a Tide-Restricted Salt Marsh in Southern New England (Massachusetts, U.S.A.)." *Restoration Ecology* 17:606–18.

Smith, T. 1982. "Alteration of Salt Marsh Plant Community Composition by Grazing Snow Geese." *Ecography* 6:204–10.

Teal, J. M., and L. Weishar. 2005. "Ecological Engineering, Adaptive Management, and Restoration Management in Delaware Bay Salt Marsh Restoration." *Ecological Engineering* 25:304–14.

Thom, R. M., R. Zeigler, and A. B. Borde. 2002. "Floristic Development Patterns in a Restored Elk River Estuarine Marsh, Grays Harbor, Washington." *Restoration Ecology* 10:487–96.

Torres, R., S. Fagherazzib, D. van Proosdijc, and C. Hopkinson. 2006. "Salt Marsh Geomorphology: Physical and Ecological Effects on Landform." *Estuarine, Coastal and Shelf Science* 69:309–10.

Tuxen, K., L. Schile, M. Kelly, and S. Siegel. 2008. "Vegetation Colonization in a Restoring Tidal Marsh: A Remote Sensing Approach." *Restoration Ecology* 16:313–23.

Tyler, A. C., and J. C. Zieman. 1999. "Patterns of Development in the Creekbank Region of a Barrier Island *Spartina alterniflora* Marsh." *Marine Ecology Progress Series* 180:161–77.

Valiela, I., J. M. Teal, and N. Y. Persson. 1976. "Production and Dynamics of Experimentally Enriched Salt Marsh Vegetation: Belowground Biomass." *Limnology and Oceanography* 21:245–52.

Vasey, M. C., and K. D. Holl. 2007. "Ecological Restoration in California: Challenges and Prospects." *Madroño* 54:215–24.

Vasquez, E. A., E. P. Glenn, G. R. Guntenspergen, J. J. Brown, and S. G. Nelson. 2006. "Salt Tolerance and Osmotic Adjustment of *Spartina alterniflora* (Poaceae) and the Invasive M Haplotype of *Phragmites australis* (Poaceae) along a Salinity Gradient." *American Journal of Botany* 93:1784–90.

Visser, J. M., C. E. Sasser, and B. S. Cade. 2006. "The Effect of Multiple Stressors on Salt Marsh End-of-Season Biomass." *Estuaries and Coasts* 29:331–42.

Vivian-Smith, G. 1997. "Microtopographic Heterogeneity and Floristic Diversity in Experimental Wetland Communities." *Journal of Ecology* 85:71–82.

Wang, J. B., D. M. Seliskar, and J. L. Gallagher. 2003. "Tissue Culture and Plant Regeneration of *Spartina alterniflora*: Implication for Wetland Restoration." *Wetlands* 23:386–93.

Warren, R. S., P. E. Fell, R. Rozsa, A. H. Brawley, A. C. Orsted, E. T. Olson, V. Swamy, and W. A. Niering. 2002. "Salt Marsh Restoration in Connecticut: 20 Years of Science and Management." *Restoration Ecology* 10:497–513.

Whigham, D. F., T. E. Jordan, and J. Miklas. 1989. "Biomass and Resource Allocation of *Typha angustifolia* L. (Typhaceae): The Effects of Within and Between Year Variations in Salinity." *Bulletin of the Torrey Botanical Club* 116:364–70.

Wigand, C., R. McKinney, M. Chintala, M. Charpentier, and G. Thursby. 2003. "Relationships of Nitrogen Loadings, Residential Development, and Physical Characteristics with Plant Structure in New England Salt Marshes." *Estuaries* 26: 1494–1504.

Wolters, M., A. Garbutt, and J. P. Bakker. 2005a. "Plant Colonization after Managed Realignment: The Relative Importance of Diaspore Dispersal." *Journal of Applied Ecology* 42:770–77.

Wolters, M., A. Garbutt, and J. P. Bakker. 2005b. "Salt-Marsh Restoration: Evaluating the Success of De-embankments in North West Europe." *Biological Conservation* 123:249–68.

Zedler, J. B. 2001. *Handbook for Restoring Tidal Wetlands*. Boca Raton, FL: CRC Press.

Zedler, J. B. 2007. "Success: An Unclear, Subjective Descriptor of Restoration Out-comes." *Ecological Restoration* 25:162–68.

Zedler, J. B., and P. A. Beare. 1986. "Temporal Variability of Salt Marsh Vegetation: The Role of Low Salinity Gaps and Environmental Stress. Pp. 295–306 in *Estuarine Variability*, edited by D. Wolfe. New York: Academic.

Zedler, J. B., and R. Lindig-Cisneros. 2001. "Functional Equivalency of Restored and Natural Salt Marshes." Pp. 565–82 in *Concepts and Controversies in Tidal Marsh Ecology*, edited by M. Weinstein and D. Kreeger. Dordrecht, Netherlands: Kluwer Academic.

Zedler, J. B., and J. M. West. 2008. "Declining Diversity in Natural and Restored Salt Marshes: A 30-Year Study of Tijuana Estuary." *Restoration Ecology* 16:249–62.

Zhang, M., S. L. Ustin, E. Rejmankova, and E. W. Sanderson. 1997. "Monitoring Pacific Coast Salt Marshes Using Remote Sensing." *Ecological Applications* 7:1039–53.

Chapter 5

Ecology of Phragmites australis *and Responses to Tidal Restoration*

RANDOLPH M. CHAMBERS, LAURA A. MEYERSON,
AND KIMBERLY L. DIBBLE

Tidal wetland restoration typically has as one of its primary goals the reestablishment of ecosystem-level functions and services to marsh habitats degraded by reductions in tidal flow. On a fundamental level, reduction or restriction of tidal flooding alters the wetland environment so dramatically that soils, hydrology, and vegetation are all impacted, so that wetland function de facto is changed. Luckily, restoration of tidal flows in many wetlands can reverse some of the functional changes caused by tidal restriction.

Smith and Warren (chap. 4, this volume) considered the more general topic of vegetative responses to tidal restoration. This chapter focuses on one notable plant species—*Phragmites australis* (common reed)—that has a checkered past with respect to its invasion, spread, and impacts in tidal wetlands. Although a number of *Phragmites* haplotypes may be native to North American wetlands, a putative, nonnative haplotype introduced sometime in the nineteenth century rapidly expanded into tidal wetlands of New England. The nonnative haplotype has since expanded across the entire continent of North America, wreaked havoc on wetland plant diversity, altered animal communities, and changed soil and hydrologic features of invaded wetlands, to the point where many wetland management programs specifically target the removal of nonnative *Phragmites*. *Phragmites* is fairly salt tolerant, but the species appears to be better adapted to high marsh and lower salinity conditions. Whether *Phragmites* invasion and expansion are causes or consequences of wetland alteration, restoration of tidal flows is often an effective tool for replacing the nonnative *Phragmites* with native vegetation adapted to more extensive flooding and elevated salinity. Within the broader context of wetland management in a time of coastal eutrophication and rising sea level, however, *Phragmites*-dominated wetlands can contribute valuable ecosystem functions and services that contribute to human and estuarine welfare. This chapter reviews the ecology of *Phragmites australis*, its legacy borne out of past alteration of tidal wetland habitats, and its future in a coastal landscape transformed by anthropogenic and other forces.

Expansion of *Phragmites* into Tidal Wetlands

A number of prior reviews have summarized the history of *Phragmites* expansion into tidal wetlands in North America (Marks et al. 1994; Chambers et al. 1999; Meyerson et al. 2009). For probably thousands of years prior to the industrial revolution, *Phragmites* was part of the mixed-plant community in some high marshes (Orson 1999). In the past two hundred years, however, the species has become more broadly distributed at local, regional, and national scales, forming extensive monocultures often extending into lower tidal elevations, especially in oligohaline and mesohaline marshes. In the coastal environment, *Phragmites* tends to grow densely around urban population centers (New York, NY; Boston, MA; Philadelphia, PA; Wilmington, DE; Baltimore, MD; and New Orleans, LA), suggesting the initial invasion and subsequent spread of a nonnative form of *Phragmites* was facilitated by shipping and boat traffic, or that disturbance of intertidal habitats (e.g., mosquito ditching, shoreline alteration) has been greatest in these developed regions. *Phragmites* is now extensive in coastal wetlands throughout New England, the focal region for this book.

Lelong et al. (2007) suggest the "sleeper-weed" phenomenon to describe the extended period of acclimatization of the introduced form of *Phragmites* prior to its recent and sudden appearance and spread into wetland habitats. Plants establish in a new location via either seedling or rhizome dispersal, then clonal growth via root and rhizome extension allows *Phragmites* to displace other species and inhibit the growth of competitors.

The first comprehensive work on genetic diversity of *Phragmites* in tidal wetlands of North America was completed by Saltonstall, who used variation in both chloroplast and nuclear DNA to identify thirteen haplotypes either native to North America or introduced in the recent past (Saltonstall 2002; Saltonstall 2003; Meadows and Saltonstall 2007). These two closely related lineages of *Phragmites australis* have been designated as subspecies by Saltonstall et al. (2004). Furthermore, based on past work, multiple introductions of the introduced haplotype M to North America seem likely (Saltonstall et al. 2010). If this proves indeed to be the case, genetic diversity of introduced *Phragmites* in North America may be higher in its introduced range, as is the case for other introduced grasses such as *Phalaris arundinaceae* (reed canarygrass) (Lavergne and Molofsky

2007), potentially contributing to the highly invasive behavior of the introduced lineage.

Because the native and introduced subspecies are so closely related, it seems likely that they should be able to interbreed. However, to date no conclusive evidence for this has been detected in wild populations. Various researchers have speculated that this was due to a phenological barrier, but Meyerson and colleagues (2010) have demonstrated overlap of anthesis between multiple native and introduced populations both in the field and in a common garden experiment. In addition, they have also successfully produced hybrids of the native and introduced subspecies, although so far all viable offspring have introduced pollen parents and native seed parents suggesting unidirectional gene flow. Speculation persists as to why wild hybrids have not thus far been detected in tidal marshes and include reasons such as undersampling and outbreeding depression (Meyerson et al. 2010) and salinity constraints on germination and seedling growth (Bart and Hartman 2003). More recently, dozens of new native populations have been identified that are within range to interbreed with nonnative populations (Blossey and Hazelton, pers. comm.). These discoveries increase the possibility that wild hybrids will be found.

Introduction of the nonnative strain of *Phragmites* is considered a prerequisite for the invasion and spread of this species into tidal wetlands of New England and Atlantic Canada (Saltonstall 2002). *Phragmites* exhibits many characteristics of successful invasive species, summarized by Meyerson et al. (2009). As an early colonizer of disturbed environments, *Phragmites* typically establishes in wetlands that have been recently altered by human activities (Bart et al. 2006; Peterson and Partyka 2006). Physiologically, the nonnative haplotype exhibits a number of adaptations to tidal wetland habitats, including effective osmoregulation of rhizome-started plants in brackish water (Vasquez et al. 2005), tolerance to flooding and to toxic sulfide relative to freshwater species (Chambers et al. 2003), greater nutrient use efficiency (Saltonstall and Stevenson 2007), greater rates of photosynthesis and stomatal conductance (Mozdzer and Zieman 2010), greater rhizome growth (League et al. 2006), and decreased susceptibility to herbivory relative to native haplotypes (Park and Blossey 2008).

This is the enigma for tidal wetland restoration, in that many wetlands altered by human intervention—and thus targeted for restoration—have provided the optimum conditions to encourage *Phragmites* invasion and spread (Roman et al. 1984). Indeed, Niering and Warren (1980) considered the presence of *Phragmites* a signature of wetland alteration in New England tidal marshes. Even though *Phragmites* may become established in apparently pristine wetland sites, it is human facilitation of *Phragmites* introduction and spread that is most commonly observed (Bart et al. 2006). For much of the twentieth century, *Phragmites* was observed invading and spreading into tidal wetlands where some form of physical disturbance of wetland hydrology, soils, or plant community structure had recently occurred. Much like an early successional species, *Phragmites* is capable of exploiting small-scale "safe sites" commensurate with physical habitat disturbance (Bart and Hartman 2003). Once established, the plant expands primarily via clonal growth (Amsberry et al. 2000) and quickly becomes a dominant species.

   Phragmites is found growing in different wetland types, but this summary focuses on the incursions of *Phragmites* into tidal marshes where alterations of tidal flow have occurred. Implicit in the discussion of the science and management of tidal flow restoration is the science and management of *Phragmites*, simply because the history of *Phragmites* invasion and spread is tied so closely with wetland alteration.

Impacts of *Phragmites* in Tidal Wetlands

Once established, dense stands of *Phragmites* may grow to heights exceeding 3 to 4 meters and physically prevent or displace native marsh vegetation, including *Spartina alterniflora* (smooth cordgrass), *S. patens* (salt meadow cordgrass), *Distichlis spicata* (spikegrass), *Juncus romerianus* (black needlerush), and other high marsh species, via competitive dominance for light and nutrients (Windham and Meyerson 2003; Meyerson et al. 2009) and perhaps allelopathy (Bains et al. 2009). Some of these changes in plant community structure are driven both by the presence of *Phragmites* and by the initial alteration of the wetland via restricted tidal flows. *Phragmites* modifies the biotic environment both at the soil surface and aboveground, effectively excluding potential competitors (Minchinton et al. 2006) and reducing species diversity (Lambert and Casagrande 2007; Lelong et al. 2007). Beyond the obvious changes in vegetation, however, *Phragmites*-dominated wetlands are significantly different from other tidal marshes in other ways.

Soil Structure

Phragmites invests a tremendous amount of carbon storage belowground, both as roots and as rhizomes. Bulk soil organic matter is tied up in both live and dead *Phragmites* tissues that effectively form a mat 5 to 20 centimeters thick. The *Phragmites* mat and accumulated litter sit atop the wetland soils; combined with enhanced sediment trapping, the wetland surface in a *Phragmites*-dominated tidal marsh tends to grow higher in elevation (Rooth et al. 2003). Further, the root and rhizome mat creates a more uniform surface elevation of the marsh, yielding less variation in microtopography and reducing the density of incipient channel formation (Lathrop et al. 2003). Interestingly, *Phragmites* is also capable of extending deep roots through shallow, saltier marsh pore water to plumb deeper freshwater lenses (Burdick et al. 2001).

In many wetlands where tidal flows have been reduced, however, decreased tidal flooding and, as a consequence, decreased salinity tend to encourage the introduction and expansion of *Phragmites*. These same, reduced flow conditions also tend to oxidize previously deposited wetland peat, so that soil elevation decreases relative to sea level. These wetlands with restricted tidal flows experience both subsidence associated with peat oxidation (Anisfeld, chap. 3, this volume) and increasing surface elevation associated with *Phragmites* root and rhizome deposition.

Hydrology

Especially in tidal wetlands where the flow of saline water has been inhibited, *Phragmites* is capable of establishing and expanding (Burdick and Konisky 2003). Beyond this initial human facilitation, however, *Phragmites* "engineers" wetland hydrology in a number of ways. First, because *Phragmites* grows in dense stands, movement of tidal water across wetland surfaces is slowed, and hydroperiods are decreased as a consequence. Not only is the time of inundation typically reduced in wetlands where *Phragmites* has been introduced; the depth of flooding is also shallower. As a result, the *Phragmites* rhizosphere remains oxidized in even saline wetlands for extended periods of time. During neap tidal phases, *Phragmites*-dominated tidal wetlands may not flood at all (Chambers et al. 2003). These long periods of rhizosphere exposure are apparently sufficient to offset oxygen or carbon stress during more extensive periods of inundation during spring tidal phases.

In many tidal wetlands *Phragmites* competes with *Spartina* spp. that are C_4 plants adapted to reducing water loss in a hypersaline environment. As a C_3 plant, evapotranspirative fluxes by *Phragmites* pull more water from the soils than C_4 plants and tend to draw down the water table. Water evaporated from the leaf surfaces is replaced by root uptake of soil water below the soil surface; in turn, soil water is replaced by infiltration. Relative to C_4-dominated wetlands, the turnover of pore water is much faster in *Phragmites*-dominated wetlands, which may facilitate both the flushing of toxins and the delivery of nutrient- and oxygen-rich water into the rhizosphere. Collectively, these hydrologic characteristics appear to create positive feedbacks of decreased flooding stress and increased plant growth that allow *Phragmites* to establish and thrive in wetlands with reduced or restricted tidal flooding.

Animals

Numerous studies have examined the possible change in habitat function and value as introduced *Phragmites* expands through tidal-marsh ecosystems. Some have reported declines in juvenile and larval habitat as the *Phragmites* invasion progresses, especially for *Fundulus heteroclitus* (mummichog; Able et al. 2003; Able and Hagan 2003; Osgood et al. 2003; Hunter et al. 2006). *F. heteroclitus* is the most abundant species in these habitats and is considered a major conduit for transfer of marsh, epibenthic, and water column production to higher trophic levels (Kneib 1986). In *Phragmites*-dominated tidal marshes, fish species and crustaceans may not be reduced in terms of diversity and total abundance (Warren et al. 2001). Fell et al. (2003) suggested *Phragmites* might be a better nursery habitat for fish and macroinvertebrates relative to oligohaline marshes vegetated by other plant species. Similarities in nekton abundance have also been shown for *Phragmites* and non-*Phragmites* marshes at similar elevation and flooding frequency (Osgood et al. 2006), suggesting the change in faunal community structure in tidally restricted wetlands is not driven by *Phragmites* per se but more by reduction in tidal flooding. Carbon, nitrogen, and sulfur from *Phragmites* wetlands is detected in estuarine food webs (Wainright et al. 2000; Wozniak et al. 2006) and contributes to the production of marine resident and transient species (Weinstein et al. 2000), but the relative importance of *Phragmites* plant detritus in supporting secondary production in adjacent estuaries has not been quantified.

Phragmites-dominated tidal wetlands in Connecticut have been shown to exhibit fewer birds overall and fewer state-listed species relative to short-grass marshes dominated by *Spartina* and other species (Benoit and Askins 1999). In a recent study in Rhode Island, foraging egrets were never observed in *Phragmites* stands (Trocki and Paton 2006). Another study found that neither blue herons nor egrets nested in *Phragmites* stands, but that *Phragmites* patches were critical nesting habitat for some wading birds and also provided a buffer from human disturbance (Parsons 2003). As one component of a wetland matrix including other vegetation, mudflats, and tidal creeks, *Phragmites* stands appear to serve as valuable avian habitat for numerous species, including some marsh specialists. In many degraded wetlands with restricted tidal flow, however, *Phragmites* monocultures are not utilized as extensively, an outcome related both to the loss of other habitat types in *Phragmites* monocultures and perhaps to the absence of tidal exchange.

Biogeochemistry/Nutrient Cycling

Phragmites is a colonizing, early successional species, but it exhibits some characteristics typical of a mature, climax species. So, for example, nutrients tend to be recycled fairly tightly in *Phragmites* stands, and most nutrients are stored in organic form in the mass of live and standing dead material (Windham and Meyerson 2003). The total standing stock of nitrogen in a *Phragmites*-dominated wetland is very large; thus the nitrogen requirements for the plant are large (Meyerson et al. 2000). Silliman and Bertness (2004) were first to associate local nitrogen enrichment from shoreline development with *Phragmites* invasion. Since then, a number of studies have verified the association of *Phragmites* establishment and spread with local nitrogen enrichment from adjacent upland environments. Wigand and McKinney (2007) used stable isotopic analysis of plant tissue to show that *Phragmites* incorporates nitrogen derived from different types of shoreline development. King et al. (2007) found positive correlations between coastal urbanization and nitrogen content of *Phragmites*, a proxy for nitrogen availability/uptake. Chambers et al. (2008) found that *Phragmites* occurrence along the shoreline of Chesapeake Bay was correlated with agriculture in the adjacent uplands. Although other mechanisms are plausible, collectively these studies indicate that nutrients derived from upland sources may supplement the limiting pool of available nitrogen in wetlands. Especially if these high marsh locations have experienced localized disturbance such as peat disturbance and rhizome burial (Bart et al. 2006), the combination with nutrient enrichment may facilitate the invasion and spread of *Phragmites*.

Phragmites appears to exploit forms of nitrogen that are not as readily available to other wetland species. Mozdzer et al. (2010) found that both *Phragmites* and *Spartina alterniflora* were able to incorporate dissolved organic nitrogen species, but that the introduced *Phragmites* had significantly greater urea assimilation rates than either native *Phragmites* or *S. alterniflora*. Ecophysiological differences including larger nitrogen demands and overall greater photosynthetic rates may contribute to the success of the introduced haplotype in North American tidal wetlands, especially in areas of nitrogen enrichment (Mozdzer and Zieman 2010). To this end, watershed-derived nitrogen from freshwater runoff, groundwater discharge, and atmospheric deposition all may contribute to satisfying nitrogen demand and facilitate *Phragmites* establishment and spread in coastal wetlands.

The combination of restricted tidal flows and *Phragmites* growth tends to draw down the water table; thus the soils are relatively oxidized. Oxidation of reduced sulfur compounds (iron monosulfide and pyrite) in *Phragmites*-dominated marsh soils can create slight to moderate acidity in the rhizosphere. Metal oxides and manganese and iron plaques may form on the roots and rhizomes of *Phragmites* where phosphorus may be immobilized. These plaques do not seem to limit plant growth (Batty et al. 2002), although Packett and Chambers (2006) measured high foliar N:P molar ratios of over 50:1 in *Phragmites* marshes, suggestive of phosphorus limitation. In *Phragmites* marshes with restricted tidal flows, the physical growth form of *Phragmites* (Windham and Meyerson 2003; Minchinton et al. 2006) and apparent sequestration of a large nutrient pool together are sufficient to exclude most other plant species.

Responses of *Phragmites* to Tidal Restoration

Implicit in tidal marsh restoration is a relative increase in the exposure of marsh soils to flooding by saline water, accomplished via larger culverts, breaching of dikes, or ditch-plugging (Konisky et al. 2006). Simenstad et al. (2006) ask what we are restoring to, noting that shifts in both the restoring landscape and external forcing functions shaping that landscape have to be considered along with the goals of restoration. With respect to *Phragmites*, the object of most tidal restoration efforts is to eliminate the nonnative species while encouraging reestablishment of a tidal wetland dominated by native marsh vegetation. Unfortunately, restoration pathways are not simply the reverse of prior degradation pathways (Zedler and Kercher 2005); that is, a hysteretic effect exists for which the wetland may fail to return to its original state once the external force (in this case, the tidal restriction or invasive species *Phragmites*) is removed (Valega et al. 2008). To some extent, tidal wetlands will self-organize when tidal flows are restored and *Phragmites* dies back, but some constraints may exist associated with respect to factors both within and external to the wetland. So, for example, wetland soil structure may be altered so dramatically that the "restored" vegetation community is different from that prior to *Phragmites* invasion. The plant community outcome after reflooding may be a unique result of species interactions involving the relative stress tolerance to physical factors and competitive strength (Konisky and Burdick 2004). In this study, for example, *Phragmites* continued growing even in higher-salinity zones with regular flooding. Finally, the wetland ecosystem may exhibit sufficient heterogeneity that only some portions successfully "restore" (Callaway 2005). The surrounding upland watershed and regional environment may have changed to the extent that the "before *Phragmites*" wetland type cannot be a realistic end point for restoration (Warren et al. 2002).

Nevertheless, success stories documenting restoration of tidal flows leading to reduced or eliminated growth of introduced *Phragmites* are common. Despite its remarkable success in degraded polyhaline and even some euhaline marshes, *Phragmites* is a brackish marsh species typically occurring in the high marsh. *Phragmites* owes its success in degraded marshes to the reduction in tidal flows that converts a flooded salt marsh to a more exposed brackish marsh. In this sense, restoration of tidal flows does offer opportunity to allow tidal water, salt, and sulfide to stress the invader and encourage native halophytic vegetation (Chambers et al. 2003; Konisky and Burdick 2004). More saltwater tends to eliminate many of the advantages *Phragmites* has in wetlands with restricted tidal flow, and the species moves to higher elevations (Smith et al. 2009). Warren et al. (2002) summarized twenty years of science and management of marsh restoration in Connecticut, noting that "recovery rates"—assessed in part as reduction in various plant parameters related to *Phragmites* vigor and coverage—varied more with restoration hydroperiod than with the salinity of the tidal water. *Phragmites* reduction was more rapid in wetlands with lower elevation, longer hydroperiod, and higher water table. Low marsh habitats invaded by *Phragmites* tended to restore to *S. alterniflora* wetlands in just a few years after reintroduction of tidal flows, whereas reduction in *Phragmites* expanse in higher marsh habitats took one to two decades. Return of marsh invertebrates, fish, and avian fauna was variable and did not always parallel the rates of recovery of native vegetation (Brawley et al. 1998).

Konisky et al. (2006) reported monitoring results from thirty-six tidal salt marsh restoration projects in Maine and found that replacement of brackish species including *Phragmites* and recovery of halophytes was typically observed three or more years following restoration. Restoration of tidal flushing in a salt marsh in Ipswich, Massachusetts, led to variable decreases in *Phragmites* vigor after four years (Buchsbaum et al. 2006). In Rhode Island, Roman et al. (2002) found that restoration of tidal flow in a restricted marsh significantly decreased *Phragmites* abundance and height after just one year, and that the entire vegetation community was converging toward that of an adjacent, unrestricted tidal marsh. Tidal exchange has been enhanced at over sixty-five wetland sites along the Connecticut coast, with reduction or elimination of *Phragmites* one of the major outcomes (Warren et al. 2002; Rozsa, chap. 8, this volume). Based on these studies, restoration of full wetland functionality for higher trophic levels including fish and birds may take decades (Fell et al. 2000). Further, restoration of tidal flows and elimination of *Phragmites* may yield other unintended consequences, as found in a Rhode Island marsh where greater nest failure by sharp-tailed sparrows initially was observed owing to flooding of the nests (DeQuinzio et al. 2002).

Restoration success of tidal marshes invaded by *Phragmites* has primarily been measured by changes in both plant and nekton communities. Marshes with restricted tidal flow may support large numbers of nekton, but the restrictions limit contributions of those nekton to estuarine productivity (Eberhardt et al. 2011). Reducing restrictions and enhancing tidal flooding typically increase the abundance and diversity of fishes and crustaceans (Roman et al. 2002; Jivoff and Able 2003; Able et al. 2004), and increase the support of the estuarine food web (Wozniak et al. 2006). In fact, nekton use of the restored marsh surface may occur before *Phragmites* is replaced (James-Pirri et al. 2001), with restoration of the most restricted wetland sites exhibiting the most dramatic shift in nekton assemblages (Raposa and Roman 2003). In other studies, however, changes in nekton could not be demonstrated as a result of tidal restoration and reduction in growth of *Phragmites* (Buchsbaum et al. 2006; Raposa 2008), suggesting that nekton abundance and diversity are not directly tied to the presence or vigor of *Phragmites*.

Most successful tidal restoration programs in polyhaline wetlands have also been successful *Phragmites* removal efforts, but many tidally restricted salt marshes have been opened to tidal flow already; that is, fewer opportunities for restoration of restricted salt marshes remain in New England. For oligohaline and mesohaline wetlands, the timescale for *Phragmites* removal postrestoration of tidal flow is variable (Buchsbaum et al. 2006) and may be on the order of decades, if at all (Warren et al. 2002). Nationally, more opportunities exist for restoring tidal flows to wetlands in which *Phragmites* has not yet fully invaded (e.g., the extensive diked tidelands along the southeastern US coast). In these marshes, tidal restoration can perhaps prevent, rather than cure, invasion of nonnative *Phragmites*.

Future of *Phragmites*

As most scientists and managers realize, the objective of tidal wetland restoration is not necessarily to eliminate *Phragmites* but to control its dominance. The general tidal marsh restoration goal is to restore specific functions and services, some of which may be provided by *Phragmites*. Invasion and spread of *Phragmites* historically have been viewed in a negative sense because the functions and ecosystem services of "pre-*Phragmites*" wetlands are deemed more valuable (Hershner and Havens 2008), yet some *Phragmites* wetlands do provide important ecosystem services, including shoreline stabilization in a time of rising sea level, energy dissipation during storm surges, and rapid nutrient uptake during a time of coastal eutrophication (Ludwig et al. 2003). None of these services are provided by native *Phragmites* that historically has been a minor component of tidal wetland communities; beyond the argument for maintaining native diversity, its future in New England tidal marshes is in doubt. Given the ongoing transformation of the coastal landscape driven by an expanding human population, global climate change, and other environmental pressures, the nonnative haplotype of *Phragmites* may become more valued in New England and Atlantic Canada tidal marshes during the twenty-first century.

Acknowledgments

RMC acknowledges funding from USDA-CSREES grant no. 2003-35320-13395; LAM from USDA Agricultural Experiment Station grant no. 0208537 and the Michael R. Paine Conservation Trust; KLD from the NSF Coastal Institute IGERT Project, NOAA National Estuarine Research Reserve System, EPA STAR graduate fellowship program, P.E.O. International, Rhode Island Natural History Survey, The Nature Conservancy of Rhode Island, and the Northeast Aquatic Plant Management Society.

REFERENCES

Able, K. W., and S. M. Hagan. 2003. "Impact of Common Reed, *Phragmites australis*, on Essential Fish Habitat: Influence on Reproduction, Embryological Development, and Larval Abundance of Mummichog (*Fundulus heteroclitus*)." *Estuaries* 26:40–50.

Able, K. W., S. M. Hagan, and S. A. Brown. 2003. "Mechanisms of Marsh Habitat Alteration due to *Phragmites*: Response of Young-of-the-Year Mummichog (*Fundulus heteroclitus*) to Treatment for *Phragmites* Removal." *Estuaries* 26:484–94.

Able, K. W., D. M. Nemerson, and T. M. Grothues. 2004. "Evaluating Salt Marsh Restoration in Delaware Bay: Analysis of Fish Response at Former Salt Hay Farms." *Estuaries* 27:58–69.

Amsberry, L., M. A. Baker, P. J. Ewanchuk, and M. D. Bertness. 2000. "Clonal Integration and the Expansion of *Phragmites australis*." *Ecological Applications* 10:1110–18.

Bains, G., A. S. Kumar, T. Rudrappa, E. Alff, T. E. Hanson, and H. P. Bais. 2009. "Native Plant and Microbial Contributions to a Negative Plant–Plant Interaction." *Plant Physiology* 151:2145–51.

Bart, D., and J. M. Hartman. 2003. "The Role of Large Rhizome Dispersal and Low Salinity Windows in the Establishment of Common Reed, *Phragmites australis*, in Salt Marshes: New Links to Human Activities." *Estuaries* 26:436–43.

Bart, D. M., D. Burdick, R. Chambers, and J. Hartman. 2006. "Human Facilitation of *Phragmites australis* Invasions in Tidal Marshes: A Review and Synthesis." *Wetland Ecology and Management* 14:53–65.

Batty, L. C., A. J. M. Baker, and B. D. Wheeler. 2002. "Aluminum and Phosphate Uptake by *Phragmites australis*: The Role of Fe, Mn and Al Root Plaques." *Aquatic Botany* 89:443–49.

Benoit, L. K., and R. A. Askins. 1999. "Impact of the Spread of *Phragmites* on the Distribution of Birds in Connecticut Marshes." *Wetlands* 19:194–208.

Brawley, A. H., R. S. Warren, and R. A. Askins. 1998. "Bird Use of Restoration and Reference Marshes within the Barn Island Wildlife Management Area, Stonington, Connecticut, USA." *Environmental Management* 22:625–33.

Buchsbaum, R. N., J. Catena, E. Hutchins, and M. J. James-Pirri. 2006. "Changes in Salt Marsh Vegetation, *Phragmites australis*, and Nekton Response to Increased Tidal Flushing in a New England Salt Marsh." *Wetlands* 26:544–57.

Burdick, D. M., and R. A. Konisky. 2003. "Determinants of Expansion for *Phragmites australis*, Common Reed, in Natural and Impacted Coastal Marshes." *Estuaries* 26:407–16.

Burdick, D. M., R. Buchsbaum, and E. Holt. 2001. "Variation in Soil Salinity Associated with Expansion of *Phragmites australis* in Salt Marshes." *Environmental and Experimental Botany* 46:247–61.

Callaway, J. C. 2005. "The Challenge of Restoring Functioning Salt Marsh Ecosystems." *Journal of Coastal Research* 40:24–36.

Chambers, R. M., L. A. Meyerson, and K. Saltonstall. 1999. "Expansion of *Phragmites australis* into Tidal Wetlands of North America." *Aquatic Botany* 64:261–73.

Chambers, R. M., D. T. Osgood, D. J. Bart, and F. Montalto. 2003. "*Phragmites australis* Invasion and Expansion in Tidal Wetlands: Interactions among Salinity, Sulfide, and Hydrology." *Estuaries* 26:398–406.

Chambers, R. M., K. J. Havens, S. Killeen, and M. Berman. 2008. "Common Reed *Phragmites australis* Occurrence and Adjacent Land Use along Estuarine Shoreline in Chesapeake Bay." *Wetlands* 28:1097–1103.

DiQuinzio, D. A., P. W. C. Paton, and W. R. Eddleman. 2002. "Nesting Ecology of Salt-marsh Sharp-Tailed Sparrows in a Tidally Restricted Salt Marsh." *Wetlands* 22:179–85.

Eberhardt, A. L., D. M. Burdick, and M. Dionne. 2011. "The Effects of Road Culverts on Nekton in New England Salt Marshes: Implications for Tidal Restoration." *Restoration Ecology* 9:776–85.

Fell, P. E., R. S. Warren, and W. A. Niering. 2000. "Restoration of Salt and Brackish Tidelands in Southern New England." Pp. 845–58 in *Concepts and Controversies in Tidal Marsh Ecology*, edited by M. P. Weinstein and D. A. Kreeger. Boston, MA: Kluwer Academic.

Fell, P. E., R. S. Warren, J. K. Light, R. L. Lawson Jr., and S. M. Fairley. 2003. "Comparison of Fish and Macroinvertebrate Use of *Typha angustifolia*, *Phragmites australis*, and Treated *Phragmites* Marshes along the Lower Connecticut River." *Estuaries* 26:534–51.

Hershner, C., and K. J. Havens. 2008. "Managing Invasive Aquatic Plants in a Changing System: Strategic Consideration of Ecosystem Services." *Conservation Biology* 22:544–50.

Hunter, K. L., D. A. Fox, L. M. Brown, and K. W. Able. 2006. "Responses of Resident Marsh Fishes to Stages of *Phragmites australis* Invasion in Three Mid Atlantic Estuaries." *Estuaries and Coasts* 29:487–98.

James-Pirri, M. J., K. B. Raposa, and J. G. Catena. 2001. "Diet Composition of Mummichogs, *Fundulus heteroclitus*, from Restoring and Unrestricted Regions of a New England (USA) Salt Marsh." *Estuarine, Coastal and Shelf Science* 53:205–13.

Jivoff, P. R., and K. W. Able. 2003. "Blue Crab, *Callinectes sapidus*, Response to the Invasive Common Reed, *Phragmites australis*: Abundance, Size, Sex Ratio, and Molting Frequency." *Estuaries and Coasts* 26:587–95.

King, R. S., W. V. Deluca, D. F. Whigham, and P. P. Marra. 2007. "Threshold Effects of Coastal Urbanization on *Phragmites australis* (Common Reed) Abundance and Foliar Nitrogen in Chesapeake Bay." *Estuaries and Coasts* 30:469–81.

Kneib, R. T. 1986. "The Role of *Fundulus heteroclitus* in Salt Marsh Trophic Dynamics." *American Zoologist* 26:259–69.

Konisky, R. A., and D. M. Burdick. 2004. "Effects of Stressors on Invasive and Halophytic Plants of New England Salt Marshes: A Framework for Predicting Response to Tidal Restoration." *Wetlands* 24:434–47.

Konisky, R. A., D. M. Burdick, M. Dionne, and H. A. Neckles. 2006. "A Regional Assessment of Salt Marsh Restoration and Monitoring in the Gulf of Maine." *Restoration Ecology* 14:516–25.

Lambert, A. M., and R. A. Casagrande. 2007. "Characteristics of a Successful Estuarine Invader: Evidence of Self-Compatibility in Native and Non-native Lineages of *Phragmites australis*." *Marine Ecology Progress Series* 337:299–301.

Lathrop, R. G., L. Windham, and P. Montesano. 2003. "Does *Phragmites* Expansion Alter the Structure and Function of Marsh Landscapes? Patterns and Processes Revisited." *Estuaries* 26:423–35.

Lavergne, S., and J. Molofsky. 2007. "Increased Genetic Variation and Evolutionary Potential Drive the Success of an Invasive Grass." *Proceedings of the National Academy of Sciences* 104:3883–88.

League, M. T., E. P. Colbert, D. M. Seliskar, and J. L. Gallagher. 2006. "Rhizome Growth Dynamics of Native and Exotic Haplotypes of *Phragmites australis* (Common Reed)." *Estuaries and Coasts* 29:269–76.

Lelong, B., C. Lavoie, Y. Jodoin, and F. Belzile. 2007. "Expansion Pathways of the Exotic Common Reed (*Phragmites australis*): A Historical and Genetic Analysis." *Diversity and Distributions* 13:430–37.

Ludwig, D. F., T. J. Iannuzzi, and A. N. Esposito. 2003. "*Phragmites* and Environmental Management: A Question of Values." *Estuaries and Coasts* 26:624–30.

Marks, M. B., B. Lapin, and J. Randall. 1994. "*Phragmites australis (P. communis)*: Threats, Management, and Monitoring." *Natural Areas Journal* 14:285–94.

Meadows, R. F., and K. Saltonstall. 2007. "Distribution of Native and Non-native Populations of *Phragmites australis* in Oligohaline Marshes of the Delmarva Peninsula and Southern New Jersey." *Journal of the Torrey Botanical Society* 134:99–107.

Meyerson, L. A., K. Saltonstall, L. Windham, E. Kiviat, and S. Findlay. 2000. "A Comparison of *Phragmites australis* in Freshwater and Brackish Marsh Environments in North America." *Wetlands Ecology and Management* 8:89–103.

Meyerson, L. A., K. Saltonstall, and R. M. Chambers. 2009. "*Phragmites australis* in Coastal Marshes of North America: A Historical and Ecological Perspective." Pp. 57–82 in *Human Impacts on Salt Marshes: A Global Perspective*, edited by B. R. Silliman, E. Grosholz, and M. D. Bertness. Berkeley: University of California Press.

Meyerson, L. A., D. V. Viola, and R. N. Brown. 2010. "Hybridization of Invasive *Phragmites australis* with a Native Subspecies in North America." *Biological Invasions* 12:103–11.

Minchinton, T. E., J. C. Simpson, and M. D. Bertness. 2006. "Mechanisms of Exclusion of Native Coastal Marsh Plants by an Invasive Grass." *Journal of Ecology* 94:342–54.

Mozdzer, T. J., and J. C. Zieman. 2010. "Ecophysiological Differences between Genetic Lineages Facilitate the Invasion of Non-native *Phragmites australis* in North American Atlantic Coast Wetlands." *Journal of Ecology* 98:451–58.

Mozdzer, T. J., J. C. Zieman, and K. J. McGlathery. 2010. "Nitrogen Uptake by Native and Invasive Temperate Coastal Macrophytes: Importance of Dissolved Organic Nitrogen." *Estuaries and Coasts* 33:784–97.

Niering, W. A., and R. S. Warren. 1980. "Vegetation Patterns and Processes in New England Salt Marshes." *BioScience* 30:301–7.

Orson, R. 1999. "A Paleoecological Assessment of *Phragmites australis* in New England Tidal Marshes: Changes in Plant Community Structure during the Last Millennium." *Biological Invasions* 1:149–58.

Osgood, D. T., D. J. Yozzo, R. M. Chambers, D. Jacobson, T. Hoffman, and J. Wnek. 2003. "Tidal Hydrology and Habitat Utilization by Resident Nekton in *Phragmites* and Non-*Phragmites* Marshes." *Estuaries* 26:522–33.

Osgood, D. T., D. J. Yozzo, R. M. Chambers, S. Pianka, C. LePage, and J. Lewis. 2006. "Patterns of Habitat Utilization by Resident Nekton in *Phragmites* and *Typha* Marshes of the Hudson River Estuary." *American Fisheries Society Symposium* 51:151–73.

Packett, C. R., and R. M. Chambers. 2006. "Distribution and Nutrient Status of Haplotypes of the Marsh Grass *Phragmites australis* along the Rappahannock River in Virginia." *Estuaries and Coasts* 29:1222–25.

Park, M. G., and B. Blossey. 2008. "Importance of Plant Traits and Herbivory for Invasiveness of *Phragmites australis* (Poaceae)." *American Journal of Botany* 95:1557–68.

Parsons, K. C. 2003. "Reproductive Success of Wading Birds Using *Phragmites* Marsh and Upland Nesting Habitats." *Estuaries* 26:596–601.

Peterson, M. S., and M. L. Partyka. 2006. "Baseline Mapping of *Phragmites australis* (Common Reed) in Three Coastal Mississippi Estuarine Basins." *Southeastern Naturalist* 5:747–56.

Raposa, K. B. 2008. "Early Ecological Responses to Hydrologic Restoration of a Tidal Pond and Salt Marsh Complex in Narragansett Bay, Rhode Island." *Journal of Coastal Research* 55:180–92.

Raposa, K. B., and C. T. Roman. 2003. "Using Gradients in Tidal Restriction to Evaluate Nekton Community Responses to Salt Marsh Restoration." *Estuaries* 26:98–105.

Roman, C. T., W. A. Niering, and R. S. Warren. 1984. "Salt Marsh Vegetation Changes in Response to Tidal Restriction." *Environmental Management* 8:141–50.

Roman, C. T., K. B. Raposa, S. C. Adamowicz, M. J. James-Pirri, and J. G. Catena. 2002. "Quantifying Vegetation and Nekton Response to Tidal Restoration of a New England Salt Marsh." *Restoration Ecology* 10:450–60.

Rooth, J. E., J. C. Stevenson, and J. C. Cornwell. 2003. "Increased Sediment Accretion Rates following Invasion by *Phragmites australis*: The Role of Litter." *Estuaries* 26:475–83.

Saltonstall, K. 2002. "Cryptic Invasion by a Non-native Genotype of the Common Reed, *Phragmites australis*, into North America." *Proceedings of the National Academy of Sciences* 99:2445–49.

Saltonstall, K. 2003. "Genetic Variation among North American Populations of *Phragmites australis*: Implications for Management." *Estuaries* 26:444–51.

Saltonstall, K., and J. C. Stevenson. 2007. "The Effect of Nutrients on Seedling Growth of Native and Introduced *Phragmites australis*." *Aquatic Botany* 86:331–36.

Saltonstall, K., P. M. Peterson, and R. Soreng. 2004. "Recognition of *Phragmites australis* Subsp. *americanus* (Poaceae: Arundinaceae) in North America: Evidence from Morphological and Genetic Analyses." *Sida* 21:683–92.

Saltonstall, K., A. Lambert, and L. A. Meyerson. 2010. "Genetics and Reproduction of Common (*Phragmites australis*) and Giant Reed (*Arundo donax*). *Invasive Plant Science and Management* 3:495–505.

Silliman, B. R., and M. D. Bertness. 2004. "Shoreline Development Drives Invasion of *Phragmites australis* and the Loss of Plant Diversity on New England Salt Marshes." *Conservation Biology* 18:1424–34.

Simenstad, C., D. Reed, and M. Ford. 2006. "When Is Restoration Not? Incorporating Landscape-Scale Processes to Restore Self-Sustaining Ecosystems in Coastal Restoration." *Ecological Engineering* 26:27–39.

Smith, S. M., C. T. Roman, M. J. James-Pirri, K. Chapman, J. Portnoy, and E. Gwilliam. 2009. "Responses of Plant Communities to Incremental Hydrologic Restoration of a Tide-Restricted Salt Marsh in Southern New England (Massachusetts, U.S.A.)." *Restoration Ecology* 17:606–18.

Trocki, C. L., and P. W. C. Paton. 2006. "Assessing Habitat Selection by Foraging Egrets in Salt Marshes at Multiple Spatial Scales." *Wetlands* 26:307–12.

Valega, M., A. I. Lillebo, M. E. Pereira, A. C. Duarte, and M. A. Pardal. 2008. "Long-Term Effects of Mercury in a Salt Marsh: Hysteresis in the Distribution of Vegetation following Recovery from Contamination." *Chemosphere* 71:765–72.

Vasquez, E. A., E. P. Glenn, J. J. Brown, G. R. Guntenspergen, and S. G. Nelson. 2005. "Salt Tolerance Underlies the Cryptic Invasion of North American Salt Marshes by an Introduced Haplotype of the Common Reed *Phragmites australis* (Poaceae)." *Marine Ecology Progress Series* 298:1–8.

Wainright, S. C., M. P. Weinstein, K. W. Able, and C. A. Currin. 2000. "Relative Importance of Benthic Microalgae, Phytoplankton, and the Detritus of Smooth Cordgrass *Spartina alterniflora* and the Common Reed *Phragmites australis* to Brackish-Marsh Food Webs." *Marine Ecology Progress Series* 200:77–91.

Warren, R. S., P. E. Fell, J. L. Grimsby, E. L. Buck, G. C. Rilling, and R. A. Fertik. 2001. "Rates, Patterns, and Impacts of *Phragmites australis* Expansion and Effects of Experimental *Phragmites* Control on Vegetation, Macroinvertebrates, and Fish within Tidelands of the Lower Connecticut River." *Estuaries* 24:90–107.

Warren, R. S., P. E. Fell, R. Rozsa, A. H. Brawley, A. C. Orsted, E. T. Olson, V. Swamy, and W. A. Niering. 2002. "Salt Marsh Restoration in Connecticut: 20 Years of Science and Management." *Restoration Ecology* 10:497–513.

Weinstein, M. P., S. Y. Litvin, K. L. Bosley, C. M. Fuller, and S. C. Wainright. 2000. "The Role of Tidal Salt Marsh as an Energy Source for Marine Transient and Resident Finfishes: A Stable Isotope Approach." *Transactions of the American Fisheries Society* 129:797–810.

Wigand, C., and R. A. McKinney. 2007. "Varying Stable Nitrogen Isotope Ratios of Different Coastal Marsh Plants and Their Relationship with Wastewater Nitrogen and Land Use in New England, USA." *Environmental Monitoring and Assessment* 131:71–81.

Windham, L., and L. A. Meyerson. 2003. "Effects of Common Reed (*Phragmites australis*) Expansion on Nitrogen Dynamics of Tidal Marshes of the Northeastern U.S." *Estuaries* 26:452–64.

Wozniak, A. S., C. T. Roman, S. C. Wainright, R. A. McKinney, and M. J. James-Pirri. 2006. "Monitoring Food Web Changes in Tide-Restored Salt Marshes: A Carbon Stable Isotope Approach." *Estuaries and Coasts* 29:568–78.

Zedler, J. B., and S. Kercher. 2005. "Wetland Resources: Status, Trends, Ecosystem Services, and Restorability." *Annual Review of Environment and Resources* 30:39–74.

Chapter 6

A Meta-analysis of Nekton Responses to Restoration of Tide-Restricted New England Salt Marshes

KENNETH B. RAPOSA AND DREW M. TALLEY

Salt marsh landscapes provide a mosaic of valuable habitats for a diverse array of estuarine nekton (defined here as free-swimming fish, shrimp, and crabs). Different species and life-history stages of nekton use salt marshes for foraging (Allen et al. 1994; James-Pirri et al. 2001; McMahon et al. 2005; Nemerson and Able 2005; Shervette and Gelwick 2007), overwintering (Smith and Able 1994; Raposa 2003), spawning (Taylor et al. 1977; Kneib 1997), as nurseries (Talbot and Able 1984; Rountree and Able 1992; Able and Fahay 1998; Minello et al. 2003), and as a refuge from predation (Kneib 1987; Minello 1993; Kneib 1997). Nektonic marsh species in turn play an important ecological role in adjacent estuaries by physically transporting energy from salt marshes into deeper estuarine habitats (Cicchetti and Diaz 2000; Deegan et al. 2000). Marsh nekton are also the primary prey items of economically valuable fishery species (e.g., Grant 1962; Nelson et al. 2003; Gartland et al. 2006) and charismatic estuarine birds such as herons and egrets (e.g., Parsons and Master 2000; McCrimmon et al. 2001). The importance of salt marshes to nekton is well documented, and the link that nekton forges between marshes and estuaries is clear. Unfortunately, human activities have long compromised the direct, physical connections between salt marshes and estuaries that are essential for nekton transport.

In the northeastern United States, humans have altered salt marshes for a variety of purposes dating back to at least colonial times (Rozsa 1995; Philipp 2005). A common type of alteration was the construction of a tide-restricting barrier across a marsh. These barriers led to changes in marsh hydrology and sediments (Portnoy 1991; Portnoy and Giblin 1997), vegetation (Roman et al. 1984), and marsh elevation (Portnoy and Giblin 1997), all of which can act in concert to change and often negatively impact nekton. Further, these effects on nekton seem to be related to the degree or severity of tidal restrictions (Raposa and Roman 2003; Wozniak et al. 2006).

In an effort to remediate the negative ecological impacts of tidal restrictions, projects to restore natural tidal regimes to tide-restricted marshes are under way in New England and elsewhere (e.g., Warren et al. 2002; Teal and Peterson 2005; Konisky et al. 2006). Successful tidal restoration projects should lead to more natural marsh conditions and improve nekton habitat function and assemblage structure. The only way to validate this is to compare nekton monitoring data before and after restoration from restoration and reference/control sites. This is occurring throughout coastal New England, and a number of case studies have been published (Burdick et al. 1997; Dionne et al. 1999; Raposa 2002; Roman et al. 2002; Buchsbaum et al. 2006; Raposa 2008).

This chapter presents a meta-analysis of tidal restriction impacts on nekton communities in New England and explores how these communities in turn respond to tidal restoration efforts. The chapter's focus includes the five New England coastal states of Maine, New Hampshire, Massachusetts, Rhode Island, and Connecticut. Datasets reviewed in this chapter were obtained from the peer-reviewed scientific literature and by soliciting unpublished reports, posters, and complete and summary datasets from restoration practitioners and scientists throughout New England. This chapter also includes a broader discussion of nekton's role in marsh–estuarine connectivity with an emphasis on how connectivity is affected when humans alter tidal flow.

Assessing Nekton in New England Marshes

Nekton data for this review were primarily obtained from monitoring programs in Rhode Island and Massachusetts. All of these data were collected with 1-square-meter (11-square-foot) throw traps following the same standardized protocol (Raposa and Roman 2001), which requires that samples be taken from shallow unvegetated habitats such as creeks and pools during the lower stages of the tide when water is not on the vegetated marsh surface. The use of the same sampling gear and monitoring protocol ensures that all of these data are directly comparable.

We used throw trap data collected from a total of twenty sites in this review (table 6.1; fig. 6.1). In some cases, a site contained two marshes (e.g., the upstream tide-restricted marsh and downstream reference marsh at Galilee, RI). We considered each marsh (total equal to twenty-five), including pairs of marshes within the same site, independent for statistical purposes. In most cases data were collected from an individual marsh for more than one year. Because variability was often high among years within the same marsh, and because data were sometimes collected by different personnel, annualized means of structural nekton variables (e.g., density and richness) were calculated and considered as independent datasets. Thus from the twenty sites/twenty-five marshes, sixty-nine total annualized throw trap datasets were used in this analysis. Of these sixty-nine datasets, thirty-three were from reference marshes, twenty-five from tide-restricted marshes, and eleven from restoring marshes (table 6.1).

Throw trap datasets were augmented with data from seven additional sites throughout New England, including one each from Connecticut, Massachusetts, and Maine, and four from New Hampshire (table 6.1; fig. 6.1). These data were only available in summary form and were collected with seines, lift nets, fyke nets, and minnow traps. These data therefore were not directly compatible with the throw trap datasets. Instead they were used to examine relative changes in nekton between pre- and postrestoration conditions and to place the findings from Rhode Island and Massachusetts into a broader perspective. While this is surely not an exhaustive collection of nekton data associated with tidal restrictions and restoration in New England, these datasets provide a quantitative aggregate dataset from which general patterns and trends can be examined. In total, this review of nekton in tide-restricted, restoring, and reference marshes in New England is based on datasets from twenty-seven sites throughout the entire region (fig. 6.1).

Data Analyses

The throw trap datasets were used for examining trends in nekton density, richness, and community composition. Datasets were placed into three groups (reference, tide-restricted, and restoring marshes), and statistical comparisons were made among these three groups of marshes. For each individual dataset, mean densities of each species or species group were calculated and the number of species was totaled (and considered an indicator of species richness). Densities of individual species, all fishes combined, all decapods combined, and all nekton combined were analyzed among reference, tide-restricted, and restoring marshes using one-way analysis of variance (ANOVA). To address the assumptions of normality and equal variance, all data were log $(x + 1)$ transformed prior to analysis. If significant differences in density were detected, pairwise comparisons among treatment groups were then conducted using Tukey's Honestly-Significant-Difference (HSD) Test. The same technique was used to compare richness of total decapods, total fish, and total nekton among the three groups of marshes. All density and richness statistical analyses were conducted using SYSTAT version 12 (SYSTAT Software, Inc.).

Comparisons of overall nekton community structure were made among marsh groups (reference, tide-restricted, and restoring) using analysis of similarity

TABLE 6.1

Nekton dataset characteristics from reference, tide-restricted, and restoring salt marshes in New England

Marsh name	Marsh type	Sampling gear	Years sampled	Months sampled	Sample size per year and (total)	Dataset type
Barn Island, CT (upstream)	Restoring	Minnow trap	1999	Feb–Nov	Unspecified	Summary
Barn Island, CT (downstream)[1]	Reference	Minnow trap	1999	Feb–Nov	Unspecified	Summary
Argilla, MA (downstream)	Reference	Seine	1997; 1999	Jul–Oct	8, 8 (16)	Summary
Argilla, MA (upstream)	Restoring	Seine	1999	Jul–Oct	8 (8)	Summary
Argilla, MA (upstream)	Tide-restricted	Seine	1997	Jul–Oct	8 (8)	Summary
East Harbor, MA	Tide-restricted	Throw trap	2003–2006	Jul–Oct	20, 30, 53, 56 (159)	Full
Hatches Harbor, MA (downstream)	Reference	Throw trap	1997; 1999; 2003; 2005	Jun–Sep	120, 120, 21, 31, (292)	Full
Hatches Harbor, MA (upstream)	Restoring	Throw trap	1999; 2003; 2005	Jun–Sep	90, 13, 74 (177)	Full
Hatches Harbor, MA (upstream)	Tide-restricted	Throw trap	1997	Jun–Sep	90 (90)	Full
Herring River, MA (downstream)	Reference	Throw trap	1998; 2005	Jul–Oct	40, 40 (80)	Full
Herring River, MA (upstream)	Tide-restricted	Throw trap	1998; 2005	Jul–Oct	80, 34 (114)	Full
Moon Pond, MA	Tide-restricted	Throw trap	2003–2006	Jul–Oct	25, 31, 45, 24 (125)	Full
Nauset, MA	Reference	Throw trap	1998; 2004–2006	Jun–Oct	200, 68, 42, 27 (337)	Full
Nonquitt, MA	Tide-restricted	Throw trap	2005–2006	Aug–Sep	20, 22 (42)	Full
Round Hill, MA	Reference	Throw trap	2005–2006	Aug–Sep	15, 14 (29)	Full
Drake's Island, ME	Restoring	Fyke net	1 year (unspecified)	unspecified	2 (2)	Summary
Drake's Island, ME	Tide-restricted	Fyke net	1 year (unspecified)	unspecified	2 (2)	Summary
Bass Beach, NH	Restoring	Lift net	1 year (unspecified)	unspecified	21 (21)	Summary
Brown's River, NH	Restoring	Fyke net	1 year (unspecified)	unspecified	2 (2)	Summary
Brown's River, NH	Tide-restricted	Fyke net	1 year (unspecified)	unspecified	2 (2)	Summary

Site	Type	Gear	Years	Months	Samples	Dataset
Little River, NH	Restoring	Lift net	3 years (unspecified)	unspecified	18, 24, 20 (62)	Summary
Parson's Creek, NH	Restoring	Fyke net	1 year (unspecified)	unspecified	2 (2)	Summary
Parson's Creek, NH	Tide-restricted	Fyke net	1 year (unspecified)	unspecified	2 (2)	Summary
Coggeshall, RI	Reference	Throw trap	2000; 2003–2005	Jul–Sep	50, 50, 50, 20 (170)	Full
Fox Hill, RI	Reference	Throw trap	2005	Aug	20 (20)	Full
Galilee, RI (downstream)	Reference	Throw trap	1997–1999	Jun–Sep	40, 40, 40 (120)	Full
Galilee, RI (upstream)	Restoring	Throw trap	1998–1999	Jun–Sep	64, 64 (128)	Full
Galilee, RI (upstream)	Tide-restricted	Throw trap	1997	Jun–Sep	64 (64)	Full
Gooseneck Cove, RI	Tide-restricted	Throw trap	2006; 2008	Jul, Aug	30, 31 (61)	Full
Jacob's Point (downstream)	Tide-restricted	Throw trap	2005; 2007–2008	Jul, Sep	10, 10, 10 (30)	Full
Jacob's Point, RI (upstream)	Reference	Throw trap	2005; 2007–2008	Jul, Sep	12, 12, 12 (36)	Full
Potter Pond, RI	Restoring	Throw trap	2003–2004	Jul, Sep	50, 50 (100)	Full
Potter Pond, RI	Tide-restricted	Throw trap	2000	Jul, Sep	50 (50)	Full
Round Marsh, RI	Reference	Throw trap	2005; 2008	Jul–Aug	20, 34 (54)	Full
Sachuest Point, RI (downstream)	Reference	Throw trap	1997–1999; 2004	Jul–Oct	30, 60, 60, 66 (216)	Full
Sachuest Point, RI (upstream)	Restoring	Throw trap	1998–1999	Aug–Oct	60, 60 (120)	Full
Sachuest Point, RI (upstream)	Tide-restricted	Throw trap	1997	Aug–Oct	30 (30)	Full
Sapowet, RI	Reference	Throw trap	2005	Aug	20 (20)	Full
Silver Creek, RI	Tide-restricted	Throw trap	2005; 2007–2008	Jul, Sep	30, 30, 29 (89)	Full
Succotash, RI	Reference	Throw trap	2008	Jul, Aug	39 (39)	Full
Thatch, RI	Reference	Throw trap	2006–2007	Aug	15, 12 (27)	Full
Walker Farm, RI	Restoring	Throw trap	2006–2007	Jul, Sep	22, 22 (44)	Full
Walker Farm, RI	Tide-restricted	Throw trap	2002	Sep	15 (15)	Full

Note: Full dataset types are those with structural nekton data from each individual sample; summary datasets contained only a summary of a few basic parameters indicative of nekton structure. Refer to figure 6.1 for locations of each marsh.

[1]The Barn Island (downstream) marsh is referred to as Headquarters Marsh in Warren et al. (2002).

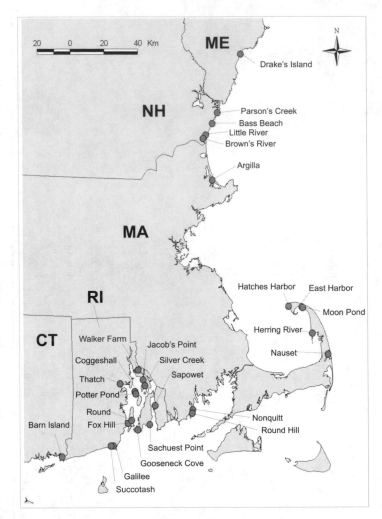

FIGURE 6.1. New England salt marshes where structural nekton data have been collected and are included in this review. These twenty-seven sites include tide-restricted, restoring, and reference salt marshes.

(ANOSIM). All data were square-root transformed prior to analysis to lessen the overall effects of the most abundant species on community structure. To create the resemblance matrix prior to analysis, Bray-Curtis similarity was calculated among samples. To identify the relative contributions of individual species to any significant differences in community structure between paired marsh groups, one-way similarity percentages (SIMPER) was used. ANOSIM and SIMPER analyses were conducted using PRIMER (Clarke and Warwick 2001; Clarke and Gorley 2006).

To assess nekton responses to restoration over time, the relative change (expressed as a percentage) in density and richness was compared between pre- and postrestoration conditions within each marsh where data were available (e.g., between tide-restricted and restoring conditions within the upstream Sachuest Point marsh), and then averaged across all marshes. The relative changes were broken into the first year postrestoration and then into an aggregate group of two or more years after restoration. This is the same approach used by Konisky et al. (2006) and was necessary because replication was limited beyond the second year following restoration.

Results

Forty-two nektonic taxa were collected in the three marsh groups (table 6.2). As is typical of salt marshes in general, the southern New England nekton community was dominated by a small number of highly abundant species, and the majority of species were relatively rare. Based on density, over 90 percent of the nekton assemblage consisted of *Palaemonetes* spp. (grass shrimp; 69 percent), *Fundulus heteroclitus* (mummichog; 18 percent), *Cyprinodon variegatus* (sheepshead minnow; 2 percent), and *Crangon septemspinosa* (sand shrimp; 2 percent). In contrast, thirty-three of the forty-two species each made up less than 1 percent of the overall community, and fourteen species were found in only one or two of the datasets. When considering only fishes, the families Fundulidae (killifish; 77 percent of all fish based on density) and Atherinopsidae (silversides; 10 percent) were overwhelmingly dominant; the family Palaemonidae (grass shrimp) represented 95 percent of all decapods.

Based on ANOSIM, nekton community composition did not differ between reference and restoring marshes (ANOSIM, $R = -0.04$, $p = 0.77$), nor did it differ between restoring and tide-restricted marshes (ANOSIM, $R = -0.04$, $p = 0.72$). A significant difference was detected, however, between reference and tide-restricted marshes (ANOSIM, $R = 0.18$, $p = 0.001$). The species that were most responsible for the significant difference in community structure between reference and tide-restricted marshes include *Palaemonetes* spp. (responsible for 29 percent of the dissimilarity), *F. heteroclitus* (10 percent), *Fundulus majalis* (striped killifish; 7 percent), and *C. septemspinosa*, *C. variegatus*, and *Menidia menidia* (Atlantic silverside; 6 percent each).

The mean number of species per marsh was relatively low in each marsh group, with reference marshes supporting an average of 10.5 nekton species and tide-restricted and restoring marshes supporting 9.2 and 10.0 nekton species, respectively. Neither the total number of fish species nor all nekton species combined differed among the three groups (ANOVA, $p = 0.97$ and 0.35, respectively;

TABLE 6.2.

Mean nekton density (number m^{-2} ± SE) in reference, tide-restricted, and restoring salt marshes in Rhode Island and Massachusetts

Species	Common name	Reference (n = 33)	Restricted (n = 25)	Restoring (n = 11)	Overall (n = 69)
Palaemonetes spp.[1]	Grass shrimp	69.57 (22.30)	29.79 (9.48)	50.82 (35.32)	52.17 (12.55)
Fundulus heteroclitus	Mummichog	13.83 (1.46)[ab]	10.96 (1.82)[b]	20.09 (3.41)[a]	13.79 (1.15)
Cyprinodon variegatus	Sheepshead minnow	1.13 (0.33)	1.85 (0.74)	1.99 (0.72)	1.53 (0.33)
Crangon septemspinosa	Sand shrimp	1.85 (0.60)	0.75 (0.39)	0.60 (0.49)	1.25 (0.33)
Lucania parva	Rainwater killifish	0.31 (0.16)	2.96 (2.06)	0.13 (0.11)	1.24 (0.76)
Menidia menidia	Atlantic silverside	1.64 (0.81)	0.96 (0.26)	0.68 (0.34)	1.24 (0.40)
Fundulus majalis	Striped killifish	1.74 (0.44)[a]	0.06 (0.02)[b]	0.97 (0.34)[a]	1.01 (0.23)
Pagurus longicarpus	Long-armed hermit crab	1.95 (0.69)[a]	0.01 (0.00)[b]	0.01 (0.01)[b]	0.94 (0.35)
Menidia beryllina	Inland silverside	0.12 (0.05)[b]	1.04 (0.43)[a]	0.34 (0.10)[ab]	0.49 (0.16)
Apeltes quadracus	Fourspine stickleback	0.21 (0.10)[b]	0.91 (0.26)[a]	0.01 (0.01)[b]	0.43 (0.12)
Brevoortia tyrannus	Atlantic menhaden	0.70 (0.44)	0.23 (0.22)	0.03 (0.02)	0.42 (0.23)
Carcinus maenas	Green crab	0.64 (0.14)[a]	0.17 (0.06)[b]	0.29 (0.14)[ab]	0.41 (0.08)
Menidia spp.	Silversides	0.16 (0.10)	0.51 (0.51)	0.00	0.26 (0.19)
Pungitius pungitius	Ninespine stickleback	0.06 (0.04)	0.28 (0.20)	<0.01	0.13 (0.07)
Anguilla rostrata	American eel	0.05 (0.02)[b]	0.18 (0.05)[a]	0.05 (0.02)[ab]	0.10 (0.02)
Callinectes sapidus	Blue crab	0.10 (0.05)	0.03 (0.02)	0.15 (0.07)	0.08 (0.03)
Gasterosteus aculeatus	Threespine stickleback	0.16 (0.15)	<0.01	0.01 (0.01)	0.08 (0.07)
Panopeidae[2]	Mud crabs	0.12 (0.06)	0.02 (0.02)	<0.01	0.07 (0.03)
Morone americana	White perch	0.00	0.10 (0.06)	<0.01	0.04 (0.02)
Mugil curema	White mullet	0.07 (0.05)	<0.01	<0.01	0.03 (0.02)
Gobiosoma spp.[3]	Gobies	0.05 (0.04)	<0.01	0.02 (0.02)	0.03 (0.02)
Syngnathus fuscus	Northern pipefish	0.03 (0.01)	0.04 (0.02)	<0.01	0.03 (0.01)
Limulus polyphemus	Atlantic horseshoe crab	0.02 (0.02)	<0.01	0.09 (0.09)	0.03 (0.02)

Species	Common name				
Pseudopleuronectes americanus	Winter flounder	0.01 (0.01)	0.02 (0.01)	<0.01	0.02 (0.01)
Lepomis macrochirus	Bluegill	0.00	0.03 (0.03)	0.00	0.01 (0.01)
Centropristis striata	Black sea bass	<0.01	0.00	0.03 (0.03)	<0.01
Ovalipes ocellatus	Lady crab	<0.01	0.00	<0.01	<0.01
Tautogolabrus adspersus	Cunner	<0.01	0.00	0.00	<0.01
Unknown crab	Unknown crab	<0.01	<0.01	0.00	<0.01
Alosa pseudoharengus	Alewife	<0.01	<0.01	0.00	<0.01
Tautoga onitis	Tautog	<0.01	0.00	0.00	<0.01
Notropis spp.	Shiners	0.00	<0.01	<0.01	<0.01
Trinectes maculatus	Hogchoker	0.00	0.00	0.00	<0.01
Opsanus tau	Oyster toadfish	<0.01	0.00	0.00	<0.01
Libinia spp.[4]	Spider crabs	0.00	<0.01	0.00	<0.01
Unknown crayfish	Unknown crayfish	0.00	<0.01	0.00	<0.01
Ammodytes americanus	American sand lance	<0.01	0.00	0.00	<0.01
Uca pugnax	Atlantic marsh fiddler	<0.01	0.00	0.00	<0.01
Clupea harengus	Atlantic herring	<0.01	0.00	0.00	<0.01
Alosa aestivalis	Blueback herring	0.00	0.00	<0.01	<0.01
Hemigrapsus sanguineus	Asian shore crab	0.00	0.00	<0.01	<0.01

Note: For each marsh group, all data were averaged across available annualized datasets. For each species, significant differences in density among the three groups of marshes are indicated by different superscripts (ANOVA, Tukey, $p < 0.05$).

[1] Predominantly *Palaemonetes pugio*, but all individuals were not identified to species.

[2] Identified species include *Dyspanopeus sayi*, *Panopeus herbstii*, and unidentified mud crab species.

[3] Includes *Gobiosoma ginsburgi* and *Gobiosoma bosc*.

[4] Includes *Libinia dubia* and unidentified *Libinia* species.

fig. 6.2). Tide-restricted marshes, however, supported a significantly smaller number of decapod species than did reference marshes (ANOVA/Tukey's HSD, $p <$ 0.001; fig. 6.2).

Of the forty-two species or species groups collected with throw traps, the densities of only seven were found to differ among the three marsh groups (table 6.2). Reference marshes supported significantly higher densities of *F. majalis* and *Carcinus maenas* (green crab) than tide-restricted marshes, and more *Pagurus longicarpus* (long-armed hermit crab) than either tide-restricted or restoring marshes. Tide-restricted marshes supported higher densities of *Menidia beryllina*

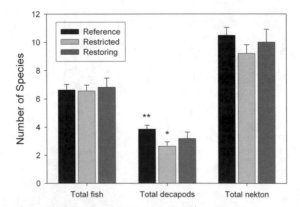

FIGURE 6.2. Mean nekton density and richness in reference, tide-restricted, and restoring salt marshes in Rhode Island and Massachusetts. All data were collected with 1 m² throw traps. N = 33 nekton datasets for reference marshes, 25 for tide-restricted marshes, and 11 for restoring marshes. The asterisks denote a significant difference in decapod richness between reference and restricted marshes (ANOVA, Tukey, $p < 0.05$). Error bars are ±1 SE.

(inland silverside) and *Anguilla rostrata* (American eel) relative to reference marshes and more *Apeltes quadracus* (fourspine stickleback) than did either reference or restoring marshes. Finally, restoring marshes supported higher densities of *F. heteroclitus* and *F. majalis* than tide-restricted marshes. Densities of all other species, all fish combined, all decapods combined, and all nekton combined did not differ among marsh groups (ANOVA, $p > 0.05$ in all cases; table 6.2; fig. 6.2).

When compared directly to tide-restricted conditions on a site-specific basis, densities of total fish, total decapods, and total nekton all increased during the first year postrestoration and continued to increase in subsequent years (fig. 6.3). Rich-

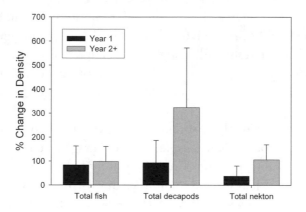

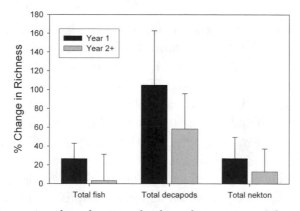

Figure 6.3. Changes in nekton density and richness between pre- and postrestoration conditions. The relative changes, expressed as a percentage of the level before restoration, are given for the first year postrestoration and then combined for years 2 and beyond. For density of each nekton group and richness of total nekton, $n = 8$ marshes for year 1 and $n = 6$ for year 2. For fish and decapod richness, $n = 6$ marshes for year 1 and $n = 5$ for year 2. Error bars are +1 SE.

ness of each of the three groups of nekton also increased on a percentage basis during the first year of restoration, though this increase was lower in subsequent postrestoration years (fig. 6.3).

Overall Trends and Patterns

Based on the foregoing results, few significant differences in nekton community structure were detected among reference, tide-restricted, and restoring salt marshes in New England at the regional scale. When considering nekton density and richness, differences between any two groups of marshes were found for only 17 percent of all species and for decapod richness (on average, one fewer decapod species was found in tide-restricted marshes compared to reference marshes). The most notable finding is that nekton community composition differed significantly between reference and tide-restricted salt marshes. On a regionwide scale, these findings suggest that tide-restricted, restoring, and reference salt marshes may differ less than previously thought in nekton structure (or nekton habitat value); instead they may simply tend to support different assemblages of marsh nekton species.

At the scale of an individual marsh, however, nekton tends to respond favorably to tidal restoration efforts. For example, when averaged across eight marshes where data were available, total nekton density and richness both increased on a percentage basis over tide-restricted conditions during the first year after restoration. However, even this is a simplification since nekton density was decreased at four of the eight sites where data were available. In fact, the percent change in nekton density during the first year of restoration ranged from negative 68 percent to 225 percent.

Based on the findings presented here, it is difficult to generalize nekton responses to tidal restoration, which appear to depend on environmental and hydrologic conditions inherent to each site and on the scale at which analyses are conducted. At the regional scale, variability in nekton structure among marshes within the same group (i.e., reference, tide-restricted, restoring) can be introduced by factors that include latitudinal species shifts, estuarine geomorphology, and levels of anthropogenic impacts. This large-scale variability may mask patterns of nekton responses to restoration that would be apparent if examined on a smaller scale or within more homogeneous groupings of marsh types.

Although the effects of restoration can be quantified at any one site (if the proper experimental design is used), these results may not broadly apply to many other sites, thus limiting the ability to generalize based on data from a small number of studies. In light of this, the best ways to better understand the effects of restoration on nekton at multiple scales are to (1) continue to compile case studies that include before and after monitoring data at restoring and control/reference marshes, and (2) ensure that it is possible to classify and group each marsh prior to analysis by better defining the specific ecological, geomorphic, and hydrologic conditions at each restoration site.

Diked/Drained versus Diked/Impounded

One way to better understand nekton responses to restoration is to group tide-restricted marshes into subgroups of diked/drained or diked/impounded marshes as described by Warren et al. (2002). Raposa and Roman (2003) showed an inverse relationship between nekton density and the degree of tidal restriction among three diked/drained marshes in southern New England. At each of these sites, nekton density increased immediately after restoration. Conversely, nekton density decreased after restoration at Potter Pond in Narragansett Bay, Rhode Island (Raposa 2008) and Argilla Marsh in Massachusetts (Buchsbaum et al. 2006), both of which were diked/impounded marshes prior to restoration.

Opposite responses of nekton to restoration of these two different types of tidal restrictions make ecological sense. By definition, diked/drained marshes hold little water over the tidal cycle and therefore cannot support large nekton populations. Drained conditions also dampen high tides, thereby limiting nekton access to foraging, spawning, and refuge areas on the vegetated marsh surface. These factors can worsen already physiologically demanding conditions of tidal marshes and eliminate less tolerant species (i.e., a reduction in richness) and reduce densities of those species that remain.

In contrast, diked/impounded marshes provide marsh nekton with a relatively stable and deep body of water. In some cases, storm and spring tides can lead to extended periods when the marsh surface is flooded and is therefore accessible to nekton. Along the Gulf Coast of the United States, extended periods of marsh surface flooding from meteorological forcing can be beneficial for nekton (Rozas 1995), and it is possible that the same benefit is conveyed to nekton in impounded marshes in New England. Impounded conditions can also lead to the proliferation of subtidal macroalgal mats. At a site in Narragansett Bay, Rhode Island, these mats provided some benefit to nekton by providing a vegetated refuge from wading bird predation (Raposa 2008). These and other factors may help explain why nekton abundance is artificially high at some diked/impounded marshes in New England and why decreases in abundance can be observed after tidal restoration (e.g., Buchsbaum et al. 2006; Raposa 2008).

Species-Specific Assessments

A shift in upstream nekton community composition appears to be a fundamental effect of tidal restrictions in New England. The statistical difference in composition between tide-restricted and reference marshes, but not between restoring and reference or between restoring and restricted marshes, suggests that compositional shifts generally take longer than two years to develop, and most monitoring programs have not been in place long enough to detect such changes.

Species that are indicative of areas upstream of tidal restrictions are generally tolerant of variable salinity, temperature, and dissolved oxygen levels and are often associated with macroalgae. In southern New England, these species include *F. heteroclitus*, *C. variegatus*, *Lucania parva* (rainwater killifish), and *M. beryllina*. These species are less abundant in reference and restoring marshes than in tide-restricted marshes. In contrast, tidal restoration generally improves conditions for species less tolerant of extreme environmental conditions and those species that favor sandy, high-energy environments, including *F. majalis*, *M. menidia*, *P. longicarpus*, and *C. septemspinosa*. Except under moderate to severe diked/drained conditions, when density and richness both decline, the difference between tide-restricted and restoring conditions is a shift between these two groups of marsh species.

Restoration Trajectories and Reference Conditions

Postrestoration changes in nekton occur rapidly and reflect the high mobility and reproductive capacity of many species. Although the trajectory of structural variables may differ depending on prerestoration hydrology, the response of nekton to restoration is generally toward a community that is more similar to unrestricted or reference marshes. The lack of a significant difference in nekton community structure between reference and restoring and between tide-restricted and restoring marshes suggests that marshes in the early stages of restoration represent an intermediate phase between these two end points (restricted and reference/restored). Whether or not nekton structure in restoring marshes continues to trend toward reference marshes is generally unknown due to a lack of long-term monitoring data. In Connecticut, the nekton community from a restoring marsh was similar to that in a nearby reference marsh twenty-one years after restoration (Swamy et al. 2002), but this is only one example. In addition, this site was not continuously monitored following restoration, making it impossible to know how nekton changed over time. Monitoring nekton by collecting quantitative data at intervals over the long term is critically needed for salt marsh restoration projects in New England.

The selection of a reference marsh is an important consideration when designing monitoring plans for restoration projects (White and Walker 1997). Often, the area downstream of a tidal restriction is chosen as the reference marsh, which helps control for differences in geographic location, setting within the estuary, and other factors. However, downstream marshes are not entirely independent of upstream tide-restricted marshes since free-swimming nekton can move between them after tidal restoration. For example, if a decrease in nekton density is observed after restoring an impounded marsh, a concurrent density increase could theoretically occur in the downstream reference marsh as nekton travel downstream with ebbing tides. When possible, it may be more appropriate to choose a nearby, though separate, marsh to ensure independence, and since nekton populations vary greatly among marshes, it may be best to choose multiple reference marshes as suggested by Underwood (1992). Alternatively, if an ecologically similar marsh cannot be found to establish reference conditions, other marshes should be monitored anyway to serve as an experimental control for the restoration marsh. Using a reference allows one to distinguish changes in nekton actually due to restoration from large-scale changes occurring throughout the estuary.

Confounding Factors to Consider when Monitoring Nekton

Actual nekton responses to restoration could be superimposed and obscured by the chronic effects of eutrophication on marsh nekton. For example, true reference conditions are difficult to obtain in Narragansett Bay, Rhode Island, because many marshes in this urbanized estuary are impacted by eutrophication, and they support altered nekton communities (specifically, high densities of *Palaemonetes* spp., James-Pirri unpublished data; Raposa unpublished data). In this case, it is difficult to define the structure of a natural nekton community and, subsequently, to define restoration targets based on nekton. Thus the degree of eutrophication of a prospective restoration site (or a proposed reference site) should be considered when one is setting reasonable targets for postrestoration nekton assemblage structure.

Choice of sampling gear can also influence the interpretation of nekton responses to restoration. For example, two separate nekton monitoring programs at the Galilee salt marsh in Rhode Island, one using throw traps (Raposa 2002) and the other using minnow traps (Golet et al., chap. 20, this volume), yielded different results; these differences can be solely attributed to the use of these two sampling gears that differ drastically in terms of capture efficiencies, biases, and the sizes and species of captured nekton (Rozas and Minello 1997; Kneib and Craig 2001; Steele et al. 2006). In some cases it is possible to carefully combine datasets collected with different sampling gear. For example, Konisky et al. (2006) were able to combine data obtained from a variety of methods to examine patterns across the Gulf of Maine. In the analysis presented here, data collected with seines, lift nets, and throw traps were all normalized to look for relative changes in nekton over time expressed as a percentage. Density data from each gear type were available but were not directly comparable on an absolute basis due to substantial differences in capture efficiencies among the gear (Rozas and Minello 1997; Steele et al. 2006). Because of data compatibility issues, the most powerful analyses are derived from programs that consistently use the same gear and protocol. It is also important to compute the statistical power of each gear and sampling program. Without knowing the statistical power (and therefore the number of samples necessary for statistical analysis) a finding of no significant difference between treatments may simply be due to too few samples being collected. For throw traps, approximately twenty to twenty-five samples should be collected to accurately assess nekton within a marsh (Raposa et al. 2003), but there is a need to determine statistical power for other sampling gear as well.

Moving Beyond Structural Assessments

Most nekton monitoring of salt marsh restoration projects in New England has focused on collecting structural data (e.g., density and community composition). This focus is likely a reflection of the relative ease with which these kinds of data can be collected. Enough structural data have been collected to support the regional analysis presented here, and our understanding should improve as more data from additional case studies emerge. There is a pressing need, however, to move beyond examining structure only and into investigations of the effects of marsh restoration on the functional roles of estuarine nekton.

A few studies of nekton function from New England salt marsh restoration projects already exist. Notable examples include studies on nekton foraging (Allen et al. 1994; James-Pirri et al. 2001) and a carbon stable isotope food web analysis (Wozniak et al. 2006). While these studies are illustrative and provide a solid starting point, the difficulties encountered when trying to identify trends in simple structural metrics across the region despite a relatively large number of datasets indicate that much more functional research is needed. An excellent precedent has been set in Delaware Bay, where an impressive number of publications on the effects of salt marsh restoration on both nekton structure and function are emerging (see brief review in Teal and Peterson 2005).

Tidal Restrictions in the Context of Habitat Connectivity

Recently, ecologists have begun to recognize a multitude of connections between seemingly distinct and independent habitat types (e.g., Polis and Strong 1996). These connections take many forms, including trophic, demographic, and physical, with powerful ecological effects that can ramify through entire ecosystems (see Talley et al. 2006 and references therein). Wetlands form a nexus between the aquatic and terrestrial realms, and often between freshwater and saltwater habitats; thus habitat connectivity is a prominent feature of this zone. These functional linkages are not restricted to the interface itself. Habitat linkages have effects that extend beyond the habitats of the coastal zone, ranging from the continental shelf and seaward to hundreds of kilometers inland (e.g., Willson et al. 2004; Talley et al. 2006).

Tidal restrictions can impact these connections, both directly and indirectly. Access to the intertidal and secondary production in estuarine nekton are strongly linked (e.g., Weisberg and Lotrich 1982; Irlandi and Crawford 1997; Kneib 1997), but the potential effects of altering flow can extend over a much broader spatial and taxonomic scale. Altering the exchange of oceanic water may not only interrupt the movement of adult or larval forms of nekton (e.g., Nordby and Zedler 1991; Roman et al. 2002; Ritter et al. 2008) but also may affect species interactions. For example, limiting their movement not only directly affects anadromous fish populations (e.g., Nehlsen et al. 1991) but also may reduce the input of marine nutrients to terrestrial plants (Nagasaka et al. 2006) and animals (Gende et al. 2002; Willson et al. 2004) far upstream. This terrestrial reliance on wetland input is not an isolated phenomenon restricted to anadromous fishes. More than eighty fish and brachyuran species are prey for maritime mammals (Carlton and Hodder 2003), and that number is undoubtedly an underestimate when indirect transfer of nutrients as described by Gende et al. (2002) is included. Thus water control structures can affect taxa and ecosystems hundreds of kilometers from a project site.

Similarly, hydrologic changes that degrade nursery habitat or alter the rate of movement of organisms or materials from a marsh can have profound effects on distant marine habitats. In addition to the potential alteration of populations of migratory species that facultatively (e.g., *Paralichthys californicus* [California halibut]; Fodrie and Mendoza 2006) or obligately (e.g., *Brevoortia tyrannus* [Atlantic menhaden]; Kroger and Guthrie 1973) use wetlands, these effects can ramify through offshore ecosystems. For example, the English sole, *Pleuronectes vetutus*, disproportionately uses wetlands as nursery habitat (Gillanders et al. 2003) and yet can potentially affect offshore ecosystems, where it plays a role both as an abundant predator (down to 250 meters [820 feet] deep) and as prey for marine mammals (Lassuy 1989). Similarly, marsh resident species, such as killifish (*Fundulus* spp.), are prey for many taxa and thus can act as net exporters of energy from impounded areas (Kneib 1997; Talley 2008).

Changes to flow in wetland systems have extensive, nonlinear, and site- and taxa-specific effects, most of which are poorly understood at present. Teasing apart these linkages will require a sustained research effort across a variety of hydrologic conditions, wetland types, and regions to better understand the consequences of these modifications and the best methods of mitigating them.

Summary of Responses and Research Opportunities

Nekton responds rapidly to hydrologic restoration of tide-restricted salt marshes in New England, but trends in nekton responses across the region are not clear. Although nekton density and richness both tend to increase after tidal restoration, the response by nekton depends on prerestoration hydrologic conditions upstream of the restriction. Limited evidence suggests that moderate to severe diked/drained marshes support degraded nekton communities, which then respond favorably to salt marsh restoration. In contrast, tide-restricted marshes that are diked/impounded can support viable nekton assemblages, and nekton density may decrease at these sites after restoration. Based on the available structural data, the only clear trend across the region is that restoration of tide-restricted salt marshes in New England elicits a change in overall nekton density and a shift in upstream nekton community composition.

Numerous opportunities and gaps remain with regard to understanding the effects of tidal restrictions and restorations on estuarine nekton. Specific needs include (1) long-term monitoring of nekton responses to restoration from multiple sites, (2) additional published case studies on restorations of different types of tidal restrictions to improve the ability to identify large-scale patterns in nekton responses, and (3) more emphasis on examining functional responses of nekton to tidal restrictions and hydrologic restorations.

Acknowledgments

We thank Elizabeth DeCelles, Michele Dionne, Mary-Jane James-Pirri, and Ray Konisky for providing full and summary nekton datasets for use in this meta-analysis.

REFERENCES

Able, K. W., and M. P. Fahay. 1998. *The First Year in the Life of Estuarine Fishes in the Middle Atlantic Bight*. New Brunswick, NJ: Rutgers University Press.

Allen, E. A., P. E. Fell, M. A. Peck, J. A. Gieg, C. R. Guthke, and M. D. Newkirk. 1994. "Gut Contents of Common Mummichogs, *Fundulus heteroclitus* L., in a Restored Impounded Marsh and in Natural Reference Marshes." *Estuaries* 17:462–71.

Buchsbaum, R. N., J. Catena, E. Hutchins, and M. J. James-Pirri. 2006. "Changes in Salt Marsh Vegetation, *Phragmites australis*, and Nekton in Response to Increased Tidal Flushing in a New England Salt Marsh." *Wetlands* 26:544–57.

Burdick, D. M., M. Dionne, R. M. Boumans, and F. T. Short. 1997. "Ecological Responses to Tidal Restorations of Two Northern New England Salt Marshes." *Wetlands Ecology and Management* 4:129–44.

Carlton, J. T., and J. Hodder. 2003. "Marine Mammals: Terrestrial Mammals as Consumers in Marine Intertidal Communities." *Marine Ecology Progress Series* 256:271–86.

Cicchetti, G., and R. J. Diaz. 2000. "Types of Salt Marsh Edge and Export of Trophic Energy from Marshes to Deeper Habitats." Pp. 515–41 in *Concepts and Controversies in Tidal Marsh Ecology*, edited by M. Weinstein and D. Kreeger. Dordrecht, Netherlands: Kluwer.

Clarke, K. R., and R. N. Gorley. 2006. *PRIMER (Plymouth Routines in Multivariate Ecological Research) v6.12: User Manual/Tutorial*. Plymouth, UK: PRIMER-E.

Clarke, K. R., and R. M. Warwick. 2001. *Change in Marine Communities: An Approach to Statistical Analysis and Interpretation*. 2nd ed. Plymouth, UK: PRIMER-E.

Deegan, L. A., J. E. Hughes, and R. A. Rountree. 2000. "Salt Marsh Ecosystem Support of Marine Transient Species." Pp. 333–68 in *Concepts and Controversies in Tidal Marsh Ecology*, edited by M. Weinstein and D. Kreeger. Dordrecht, Netherlands: Kluwer.

Dionne, M., F. T. Short, and D. M. Burdick. 1999. "Fish Utilization of Restored, Created, and Reference Salt-Marsh Habitat in the Gulf of Maine." Pp. 384–404 in *Fish Habitat: Essential Fish Habitat and Rehabilitation*, edited by L. Benaka. American Fisheries Society Symposium 22. Bethesda, MD: American Fisheries Society.

Fodrie, F. J., and G. Mendoza. 2006. "Availability, Usage and Expected Contribution of Potential Nursery Habitats for the California Halibut." *Estuarine, Coastal and Shelf Science* 68:149–64.

Gartland, J., R. J. Latour, A. D. Halvorson, and H. M. Austin. 2006. "Diet Composition of Young-of-the-Year Bluefish in the Lower Chesapeake Bay and Coastal Ocean of Virginia." *Transactions of the American Fisheries Society* 135:371–78.

Gende, S. M., R. T. Edwards, M. F. Willson, and M. S. Wipfli. 2002. "Pacific Salmon in Aquatic and Terrestrial Ecosystems." *BioScience* 52:917–28.

Gillanders, B. M., K. W. Able, J. A. Brown, D. B. Eggleston, and P. F. Sheridan. 2003. "Evidence of Connectivity between Juvenile and Adult Habitats for Mobile Marine Fauna: An Important Component of Nurseries." *Marine Ecology Progress Series* 247:281–95.

Grant, G. C. 1962. "Predation of Bluefish on Young Atlantic Menhaden in Indian River, Delaware." *Chesapeake Science* 3:45–47.

Irlandi, E. A., and M. K. Crawford. 1997. "Habitat Linkages: The Effect of Intertidal Saltmarshes and Adjacent Subtidal Habitats on Abundance, Movement and Growth of an Estuarine Fish." *Oecologia* 110:222–30.

James-Pirri, M. J., K. B. Raposa, and J. G. Catena. 2001. "Diet Composition of Mummichogs, *Fundulus heteroclitus*, from Restoring and Unrestricted Regions of a New England (U.S.A.) Salt Marsh." *Estuarine, Coastal and Shelf Science* 53:205–13.

Kneib, R. T. 1987. "Predation Risk and Use of Intertidal Habitats by Young Fishes and Shrimp." *Ecology* 68:379–86.

Kneib, R. T. 1997. "The Role of Tidal Marshes in the Ecology of Estuarine Nekton." *Oceanography and Marine Biology: An Annual Review* 35:163–220.

Kneib, R. T., and A. H. Craig. 2001. "Efficacy of Minnow Traps for Sampling Mummichogs in Tidal Marshes." *Estuaries* 24:884–93.

Konisky, R. A., D. M. Burdick, M. Dionne, and H. A. Neckles. 2006. "A Regional Assessment of Salt Marsh Restoration and Monitoring in the Gulf of Maine." *Restoration Ecology* 14:516–25.

Kroger, R. L., and J. F. Guthrie. 1973. "Migrations of Tagged Juvenile Atlantic Menhaden." *Transactions of the American Fisheries Society* 2:417–22.

Lassuy, D. R. 1989. "Species Profiles: Life Histories and Environmental Requirements of Coastal Fishes and Invertebrates (Pacific Northwest)—English Sole." U.S. *Fish and Wildlife Service Biological Report* 82 TR EL-82-4.

McCrimmon, D. A., Jr., J. C. Ogden, and G. T. Bancroft. 2001. "Great Egret (*Ardea alba*)." No. 570 in *The Birds of North America*, edited by A. Poole and F. Gill. Philadelphia, PA: Birds of North America.

McMahon, K. T., B. J. Johnson, and W. G. Ambrose Jr. 2005. "Diet and Movement of the Killifish, *Fundulus heteroclitus*, in a Maine Salt Marsh Assessed Using Gut Contents and Stable Isotope Analyses." *Estuaries* 28:966–73.

Minello, T. J. 1993. "Chronographic Tethering: A Technique for Measuring Prey Survival Time and Testing Predation Pressure in Aquatic Habitats." *Marine Ecology Progress Series* 101:99–104.

Minello, T. J., K. W. Able, M. P. Weinstein, and C. G. Hays. 2003. "Salt Marshes as Nurseries for Nekton: Testing Hypotheses on Density, Growth and Survival through Meta-analysis." *Marine Ecology Progress Series* 246:39–59.

Nagasaka, A., Y. Nagasaka, K. Ito, T. Mano, M. Yamanaka, A. Katayama, Y. Sato, A. L. Grankin, A. I. Zdorikov, and G. A. Boronov. 2006. "Contributions of Salmon-Derived Nitrogen to Riparian Vegetation in the Northwest Pacific Region." *Journal of Forest Resources* 11:377–82.

Nehlsen, W., J. E. Williams, and J. A. Jicharowich. 1991. "Pacific Salmon at the Crossroads: Stocks at Risk from California, Oregon, Idaho, and Washington." *Fisheries* 16:4–21.

Nelson, G. A., B. C. Chase, and J. Stockwell. 2003. "Food Habits of Striped Bass (*Morone saxatilis*) in Coastal Waters of Massachusetts." *Journal of Northwest Atlantic Fishery Science* 32:1–25.

Nemerson, D. M., and K. W. Able. 2005. "Juvenile Sciaenid Fishes Respond Favorably to Delaware Bay Marsh Restoration." *Ecological Engineering* 25:260–74.

Nordby, C. S., and J. B. Zedler. 1991. "Responses of Fish and Macrobenthic Assemblages to Hydrologic Disturbances in Tijuana Estuary and Los Peñasquitos Lagoon, California." *Estuaries* 14:80–93.

Parsons, K. C., and T. L. Master. 2000. "Snowy Egret (*Egretta thula*)." No. 489 in *The Birds of North America*, edited by A. Poole and F. Gill. Philadelphia, PA: Birds of North America.

Philipp, K. R. 2005. "History of Delaware and New Jersey Salt Marsh Restoration Sites." *Ecological Engineering* 25:214–30.

Polis, G. A., and D. R. Strong. 1996. "Food Web Complexity and Community Dynamics." *American Naturalist* 147:813–46.

Portnoy, J. W. 1991. "Summer Oxygen Depletion in a Diked New England Estuary." *Estuaries* 14:122–29.

Portnoy, J. W., and A. E. Giblin. 1997. "Effects of Historic Tidal Restrictions on Salt Marsh Sediment Chemistry." *Biogeochemistry* 36:275–303.

Raposa, K. B. 2002. "Early Responses of Fishes and Crustaceans to Restoration of a Tidally Restricted New England Salt Marsh." *Restoration Ecology* 10:665–76.

Raposa, K. B. 2003. "Overwintering Habitat Selection by the Mummichog, *Fundulus heteroclitus*, in a Cape Cod (USA) Salt Marsh." *Wetlands Ecology and Management* 11:175–82.

Raposa, K. B. 2008. "Early Ecological Responses to Hydrologic Restoration of a Tidal Pond and Salt Marsh Complex in Narragansett Bay, Rhode Island." *Journal of Coastal Research* SI (55):180–92.

Raposa, K. B., and C. T. Roman. 2001. *Monitoring Nekton in Shallow Estuarine Habitats: A Protocol for the Long-Term Coastal Ecosystem Monitoring Program at Cape Cod National Seashore.* Final Report for the Long-Term Ecosystem Monitoring Program, Cape Cod National Seashore. http://www.nature.nps.gov/im/monitor/protocoldb.cfm.

Raposa, K. B., and C. T. Roman. 2003. "Using Gradients in Tidal Restriction to Evaluate Nekton Community Responses to Salt Marsh Restoration." *Estuaries* 26:98–105.

Raposa, K. B., C. T. Roman, and J. F. Heltshe. 2003. "Monitoring Nekton as a Bioindicator in Shallow Estuarine Habitats." *Environmental Monitoring and Assessment* 81:239–55.

Ritter, A. F., K. Wasson, S. I. Lonhart, R. K. Preisler, A. Woolfolk, K. A. Griffith, S. Connors, and K. W. Heiman. 2008. "Ecological Signatures of Anthropogenically Altered Tidal Exchange in Estuarine Ecosystems." *Estuaries and Coasts* 31:554–71.

Roman, C. T., W. A. Niering, and R. S. Warren. 1984. "Salt Marsh Vegetation Change in Response to Tidal Restriction." *Environmental Management* 8:141–50.

Roman, C. T., K. B. Raposa, S. C. Adamowicz, M. J. James-Pirri, and J. G. Catena. 2002. "Quantifying Vegetation and Nekton Response to Tidal Restoration of a New England Salt Marsh." *Restoration Ecology* 10:450–60.

Rountree, R. A., and K. W. Able. 1992. "Fauna of Subtidal Polyhaline Marsh Creeks in Southern New Jersey: Composition, Abundance, and Biomass." *Estuaries* 15:171–85.

Rozas, L. P. 1995. "Hydroperiod and Its Influence on Nekton Use of the Salt Marsh: A Pulsing Ecosystem." *Estuaries* 18:579–90.

Rozas, L. P., and T. J. Minello. 1997. "Estimating Densities of Small Fishes and Decapod Crustaceans in Shallow Estuarine Habitats: A Review of Sampling Design with Focus on Gear Selection." *Estuaries* 20:199–213.

Rozsa, R. 1995. "Tidal Wetland Restoration in Connecticut." Pp. 51–65 in *Tidal Marshes of Long Island Sound: Ecology, History and Restoration*, edited by G. D. Dreyer and W. A. Niering. Connecticut College Arboretum Bulletin 34. New London: Connecticut College Arboretum.

Shervette, V. R., and F. Gelwick. 2007. "Habitat-Specific Growth in Juvenile Pinfish." *Transactions of the American Fisheries Society* 136:445–51.

Smith, K. J., and K. W. Able. 1994. "Salt-Marsh Tide Pools as Winter Refuges for the Mummichog, *Fundulus heteroclitus*, in New Jersey." *Estuaries* 17:226–34.

Steele, M. A., S. C. Schroeter, and H. M. Page. 2006. "Sampling Characteristics and Biases of Enclosure Traps for Sampling Fishes in Estuaries." *Estuaries and Coasts* 29:630–38.

Swamy, V., P. E. Fell, M. Body, M. B. Keaney, M. K. Nyaku, E. C. McIlvan, and A. L. Keen. 2002. "Macroinvertebrate and Fish Populations in a Restored Impounded Salt Marsh 21 Years after the Re-establishment of Tidal Flooding." *Environmental Management* 29:516–30.

Talbot, C. W., and K. W. Able. 1984. "Composition and Distribution of Larval Fishes in New Jersey High Marshes." *Estuaries* 7:434–43.

Talley, D. M. 2008. "Spatial Subsidy." Pp. 3325–31 in *Encyclopedia of Ecology*, edited by S. E. Jorgensen and B. D. Fath. Oxford: Elsevier.

Talley, D. M., G. R. Huxel, and M. Holyoak. 2006. "Habitat Connectivity at the Land–Water Interface." Pp. 97–129 in *Connectivity Conservation*, edited by K. Crooks and M. Sanjayan. Cambridge: Cambridge University Press.

Taylor, M. H., L. DiMichele, and G. J. Leach. 1977. "Egg Stranding in the Life Cycle of the Mummichog, *Fundulus heteroclitus*." *Copeia* 1977:397–99.

Teal, J. M., and S. B. Peterson. 2005. "Introduction to the Delaware Bay Salt Marsh Restoration." *Ecological Engineering* 25:199–203.

Underwood, A. J. 1992. "Beyond BACI: The Detection of Environmental Impacts on Populations in the Real, but Variable, World." *Journal of Experimental Marine Biology and Ecology* 161:145–78.

Warren, R. S., P. E. Fell, R. Rozsa, A. H. Brawley, A. C. Orsted, E. T. Olson, V. Swamy, and W. A. Niering. 2002. "Salt Marsh Restoration in Connecticut: 20 Years of Science and Management." *Restoration Ecology* 10:497–515.

Weisberg, S. B., and V. A. Lotrich. 1982. "The Importance of an Infrequently Flooded Intertidal Marsh Surface as an Energy Source for the Mummichog *Fundulus heteroclitus*: An Experimental Approach." *Marine Biology* 66:307–10.

White, P. S., and J. L. Walker. 1997. "Approximating Nature's Variation: Selecting and Using Reference Information in Restoration Ecology." *Restoration Ecology* 5:338–49.

Willson, M. F., S. M. Gende, and P. A. Bisson. 2004. "Anadromous Fishes as Ecological Links between Ocean, Fresh Water, and Land." Pp. 284–300 in *Food Webs at the Landscape Level*, edited by G. A. Polis, M. E. Power, and G. R. Huxel. Chicago: University of Chicago Press.

Wozniak, A. S., C. T. Roman, S. C. Wainright, and R. A. McKinney. 2006. "Monitoring Food Web Changes in Tide-Restored Salt Marshes: A Carbon Stable Isotope Approach." *Estuaries and Coasts* 29:568–78.

Chapter 7

Avian Community Responses to Tidal Restoration along the North Atlantic Coast of North America

W. Gregory Shriver and Russell Greenberg

Salt marshes in the New England and Atlantic Canada region are widely affected by humanmade structures such as dikes, tide gates, bridges, and culverts, and from other impacts like dredge-spoil fill and mosquito ditches (Hansen and Shriver 2006), all of which alter the volume, velocity, and spatial pattern of tidal flow. In response to concerns about the impact of such management, coastal managers are now actively engaged in salt marsh restoration practices focused on the return of natural hydrology to impacted marshes (Konisky et al. 2006). Federal, state, and provincial agencies have initiated more than a hundred salt marsh restoration projects in the Gulf of Maine region since 1990 (Cornelisen 1998), and these practices continue to be a major management emphasis.

Despite the magnitude of habitat change associated with human alterations to salt marshes, only recently have agencies concerned with wildlife conservation begun to systematically survey salt marsh avifauna in northeastern North America (Hansen and Shriver 2006) or monitor the response of the avian community to tidal restoration actions (DiQuinzio et al. 2002; Seigel et al. 2005; Raposa 2008). Quantitative information about species occurrence, relative abundance, and density of several wetland species within the Northeast is unknown or only recently gathered (Benoit and Askins 1999; Shriver et al. 2004; Hansen and Shriver 2006); thus there is little information regarding the effects of tidal restoration on these species (Konisky et al. 2006). This chapter reviews the existing literature on the avian response to tidal restoration and summarizes the habitat requirements for avian species that could be affected by tidal marsh restoration projects in New England and Atlantic Canada.

Avian Response to Tidal Restoration in the North Atlantic

Avian responses to salt marsh habitat alterations are poorly understood. This is due in part to the relatively small spatial scale of many of the tidal restoration projects, the temporal component (10–15 years) necessary for vegetation to return to some state of equilibrium, and the inherent variability in avian sampling. Most studies that included avian sampling in the assessment of tidal marsh restoration have been conducted over very short time periods at only a few sites, making it difficult to detect a difference before and after a restoration at the same site, or between a restored area and a reference site. Also, because avian guilds will respond in different ways (positively, negatively, or no change) and over different times since restoration, analyses that do not specify guilds or species can be confounded. Systematic studies of the response of bird assemblages to tidal marsh restoration vary in their focus, from general studies of avifaunal change to those more focused on specialized species, restricted to the tidal marsh ecosystem.

Changes in Overall Abundance, Composition, and Species Richness

Studying highly altered *Phragmites australis* (common reed) marshes, Seigel et al. (2005) surveyed the bird community response to tidal restoration at Harrier Marsh, Hackensack, New Jersey (table 7.1). These marshes have suffered from reduced freshwater flow and tidal influence as well as their near complete domination by *Phragmites*. Avian surveys were conducted one year prior and three years after restoration of 22 hectares of a 32 hectare marsh. Metrics used to characterize the avian communities included avian abundance, species richness, diversity, evenness, and similarity. The authors also categorized species into six foraging guilds (generalist, aerial, upland, *Phragmites*, open-water, and mudflat foragers). Avian species richness and average avian abundance were greater two years postrestoration compared to prerestoration estimates at Harrier Marsh (Seigel et al. 2005). The major avian change, primarily in response to an increase in open water, included a transition from an avian community dominated by passerines to one dominated by waterbirds.

Konisky et al. (2006) compiled the results from thirty-six salt marsh restoration projects in the Gulf of Maine to provide a regional assessment of indicators used to monitor salt marsh restoration projects (table 7.1). Avian species richness was summarized for 25 percent of the projects and avian density for 53 percent. Based on the analyses of datasets that used somewhat different methodologies to estimate avian species richness and density, there were no clear patterns. The authors did find that both avian species richness and density were lowest at ditch plug sites but attributed these differences to methodological, not ecological, issues.

TABLE 7.1

Review of projects that have measured the response of birds to tidal marsh restoration in the North Atlantic

Reference	State	Restoration	Names and sizes of surveyed areas (hectares)	Timing of survey in relation to restoration event	Avian survey methods
Brawley et al. 1998	CT	Restored to tidal inundation following breaching of the dikes	Wequetequock Cove (1) Headquarters (8) Impoundment One (21) Impoundment Three (12) Davis Marsh (19)	14 years post	25 × 100 m plots Thirty-minute surveys conducted four per year May–August
Warren et al. 2002	CT	Restored to tidal inundation following breaching of the dikes		5 yrs post 15 yrs post	25 × 100 m plots Thirty-minute surveys conducted four per year May–August all birds seen and heard
DeQuinzo et al. 2002	RI	Tidal restoration	Galilee Marsh (52) 1 year post	5 years pre	Mist nets/banding Nest searching
Seigel et al. 2005	NJ	Tidal restoration	Harrier Marsh (22) Mill Creek (38)		Unlimited radius five-minute surveys Five surveys in each of three seasons (spring, summer, fall)
Konisky et al. 2006	ME	Tidal restoration, ditch plug, excavation	NA	1 year post 2 years post	Review of thirty-six separate studies
Raposa 2008	RI	Tidal restoration	Upper Impoundment (0.24) Lower Impoundment (2)	2 years pre 1 and 2 years post	One visual point count at each impoundment Ten-minute counts Seven visits per year (biweekly, June–Sept)

Raposa (2008) conducted a comparison of the bird community response to re-moval of a tidal restriction on Prudence Island, Rhode Island. New culverts were installed to reconnect two hydrologically connected impoundments (2.28 hec-tares total) to Narragansett Bay (table 7.1). Two avian sampling locations (one at each impoundment) were sampled at the restoration site, and four were sampled at a nearby reference site. Multiple years postrestoration were sampled, but the first year after restoration resulted in the greatest change in the avian community with a shift from open-water foragers, such as the *Sterna albifrons* (little tern), *Megaceryle alcyon* (belted kingfisher), and *Phalacrocorax auritus* (double-crested cormorant), to shorebirds, including the *Charadrius semipalmatus* (semipalmated plover), *Calidris pusilla* (semipalmated sandpiper), *Egreta thula* (snowy egret), *Limnodromus griseus* (short-billed dowitcher), and *Charadrius vociferous* (kill-deer). The bird community did not change in the *Phragmites*-dominated upper impoundment or the reference site.

Studies That Focus on Salt Marsh Specialists

Brawley et al. (1998) studied avian species use of a 21-hectare impoundment from 1993 to 1994 at Barn Island (Stonington, CT) where tidal flow had been restored for fourteen years prior to the avian sampling event. Other distinct areas ranging in size from 1.2 hectares to 12 hectares within Barn Island served as reference areas and were sampled for bird use during two breeding seasons, 1993–1994 (table 7.1). Bird species were divided into guilds for analysis (marsh specialists, long-legged waders, shorebirds, and marsh generalists) with the mean number of individuals in each guild used as the response variable. Fourteen years after the tidal regime was restored at Impoundment One, the dominant vegetation changed from *Typha angustifolia* (narrowleaf cattail) and *Phragmites* to *Spartina alterniflora* (smooth cordgrass) and other more typical salt marsh vegetation (Brawley et al. 1998). In conjunction with this change in vegetation, the restored marsh (Impoundment One) at Barn Island provided adequate breeding habitat for *Ammodramus caudacutus* and A. *maritimus* (saltmarsh and seaside sparrows, respectively). The restoration of tidal flow may initially increase the amount of surface water on a marsh and eliminate breeding habitat for birds that nest on the marsh surface. However, the reestablishment of S. *alterniflora* on Impoundment One demonstrated that tidal restoration can eventually create suitable conditions for these specialized species. This study clearly showed the need for vegetation to become established before tidal marsh obligate species will colonize the site and the importance of time since restoration in this process.

Warren et al. (2002) conducted the most comprehensive and long-term study of the effects of tidal restoration on the ecological trajectory of tidal wetlands in the North Atlantic. In this study, birds were surveyed on four separate marsh sites, many the same as those sampled by Brawley et al. (1998). Generally, salt marsh specialist bird species tended to recover in relative abundance and were comparable with reference sites fifteen years after restoration. This pattern was especially true for saltmarsh sparrows but not as apparent for seaside sparrow or *Tringa semipalmatus* (willet). Seaside sparrows were detected on two of the restoration sites in 1994–1995 surveys but then only on the reference marsh during the 1999 surveys. Willets were detected only on the reference and the one restoration site during the 1994–1995 surveys and not detected during the 1999 survey on any sites. This could be due to differences in the number of visits to each marsh between survey years (eight visits 1994–1995 and three visits 1999) or other unknown factors. Warren et al. (2002) also showed that marsh generalists—egrets, herons, and migratory shorebirds—tended to use the restoration sites quickly and soon after a restoration event and then declined over time as the relative abundance of marsh specialists increased. The authors suggested that it may take as many as fifteen years for these marshes to be comparable to reference sites in avian use, or even decades for rare species like the willet to return.

DiQuinzio et al. (2002) may be one of the only studies to assess the effects of tidal restoration on the reproductive ecology of breeding marsh birds; most others use counts and estimates of relative abundance. They investigated nest success, population size, and nest site selection of saltmarsh sparrows in response to the installation of self-regulating tide gates at a salt marsh near Galilee, Rhode Island. DiQuinzio et al. (2002) used banding to estimate sparrow population size, nest search to estimate nest survival, and vegetation measures to estimate nest site selection. They found a significant decline in the number of saltmarsh sparrows (48 percent decline for juveniles, 25 percent decline for females, and 17 percent decline for males), nesting period survival rates declined from 83 percent (1993) to 5 percent (1998), and the cause of nest failure changed from predation to flooding after the installation of the tide gates. The researchers suggest that more time is required for the vegetation to recover before tidal marsh obligate nesters like saltmarsh sparrows return.

Research on *Rallus longirostris* (clapper rail; Schwarzbach et al. 2006) and tidal marsh sparrows (Greenberg et al. 2006) suggest that reproductive success may be influenced by a precarious balance between nest exposure to potential tidal flooding and predation. Nesting in higher vegetation or higher in the tidal gradient may allow birds to avoid the most likely flooding events, whereas nesting at lower marsh levels may reduce the exposure to generalist predators concentrated at the marsh edge. Food availability and nesting requirements also appear to be decoupled in populations leading to the concentration of nests at higher tidal elevations while foraging at productive areas of low marsh, mudflats, and slough edges. Therefore, the restoration of natural tidal regimes in already altered and fragmented marshes may eliminate important tidal refugia that are not located close to marsh edge development.

In reviewing other tidal marsh restoration projects from other US regions, habitat management in California provides some important insights as these marshes are home to *Rallus longirostris livepes* (light-footed clapper rail) and *Rallus longirostris obsoletus* (California clapper rail), the only salt marsh birds that are listed as endangered or threatened by the US Fish and Wildlife Service. The light-footed clapper rail is restricted to a few relatively small marshes that persist in Southern California and adjacent Baja California, where they depend heavily upon luxuriant growth of the *Spartina* foliosa. Loss and fragmentation of habitat and possibly high contaminant loads have led to severe declines in this sub-species, particularly in highly populated areas of Southern California, where a few hundred individuals persist. In addition to habitat loss, marshes have been de-graded—particularly by the blockage of tidal access and reductions of freshwater flow, resulting in vegetation changes (Zedler 2001). The 2007 population assess-ment indicated that tidal restoration and increases in *Spartina* coverage probably contributed to an overall doubling of the California population from approxi-mately two hundred in the 1980s to over four hundred in recent years.

The California clapper rail was formerly abundant in the San Francisco Bay marshes, with smaller populations on other marshes of the Central California coast. With the loss of over 85 percent of the original salt marshes, it is not surpris-ing that by the 1980s the population was restricted to the San Francisco and Suisun Bays and may have consisted of as few as two hundred birds. But recently, the reestablishment of tidal flow and salt marsh vegetation may have contributed substantially to the population increase to approximately eighteen hundred birds (http://www.werc.usgs.gov/dixon/rails/index.html).

Avian Species of Tidal Marshes in the Northeast

Salt marshes provide habitat for a variety of bird species during all stages of the an-nual cycle (breeding, molting, migration, and wintering). Because of the contin-uum of conditions from grassland to wetland, the heterogeneous distribution of microscale habitat features, and their relatively high productivity, salt marshes are important landscape features for many birds in the Northeast. Differences in avian species distributions in the Northeast result in differences in the avian com-munities of salt marshes within regions that may influence the response of the bird community to tidal restoration and other habitat restoration projects (Benoit and Askins 1999; Shriver et al. 2004). In this section the basic ecology and habitat re-quirements are reviewed for a suite of bird species that may be affected by tidal marsh restoration projects. Because different species require different types of habitat for breeding, foraging, or other activities (open water, mudflats, low marsh, and high marsh), their predicted response to tidal marsh restoration will depend on the objectives of the restoration action and, importantly, the time since implementation (table 7.2).

Breeding Species Specialized on Tidal Marsh

In the North Atlantic region, four avian species are entirely restricted to salt marshes—seaside sparrow, salt marsh sparrow, clapper rail, and willet, all of which are considered high conservation priorities (table 7.2). Given their dependence on tidal marsh habitats for all aspects of their annual cycle, these species should be given high priority when designing, implementing, and monitoring the ecological effects of tidal marsh restoration projects.

Seaside sparrows breed almost exclusively in tidal marshes along the Gulf and Atlantic coasts of the United States. The northern subspecies (*A. m. maritimus*) occurs from northern Virginia through New Hampshire and nests in both high and low marsh (Post and Greenlaw 1994). Post (1974) found that seaside sparrows nesting in unaltered *Spartina* marsh maintained clusters of small nesting territories of approximately 0.1 hectares and jointly used feeding areas, whereas in ditched marshes territories were all purpose and much larger at approximately 0.9 hectares. Nests are usually located in *Spartina alterniflora* but have also been observed in *Spartina patens* (salt meadow cordgrass) and *Distichlis spicata* (spikegrass) (Marshall and Reinert 1990; Post and Greenlaw 1994). Saltmarsh and seaside sparrows often nest within the same habitat and are closely related to one another (Zink and Avise 1990). However, saltmarsh sparrows occur only in the mid-Atlantic and New England coasts, but tend to be more common in the northern portion of their range, whereas seaside sparrows become numerically dominant in the mid-Atlantic estuaries (Greenlaw and Rising 1994). As specialists of tidal marsh systems, they are highly sensitive to habitat disturbance, loss, or fragmentation. The nest survival, and therefore fecundity, of saltmarsh and seaside sparrows are both influenced by tidal flooding, but each species uses a different strategy to reduce nest failure caused by tidal flooding (Gjerdrum et al. 2005). Seaside sparrows nest in taller vegetation, above the water level during the highest tides (Marshall and Reinert 1990; Gjerdrum et al. 2005), while saltmarsh sparrows nest only a few centimeters off the marsh surface and synchronize their nesting cycle with the tide cycle (Gjerdrum et al. 2005; Shriver et al. 2007). Restoration projects that significantly increase the height of the water level on the marsh have the potential to flood sparrow nests. This will only occur in potential restoration sites that support breeding populations of sparrows prior to restoration. For sites

TABLE 7.2

Bird species that could be influenced by salt marsh tidal restoration

Family	Species (common name)	Time Period			Seasons				Guild	BCR 30 priority[1]
		Early	Mid	Late	Breeding	Migrating	Wintering	Resident		
Anatidae	*Branta canadensis* (Canada goose)				X	X	X		Marsh generalist	Highest
	Cygnus olor (mute swan)				X	X	X		Marsh generalist	
	Anas rubripes (American Black Duck)	↓	↑	↑	X	X	X		Marsh specialist	Highest
	Anas platyrhynchos (mallard)	↑	↑	↑	X	X	X		Marsh generalist	High
	Anas discors (blue-winged teal)	↑	↑	↑	X	X	X		Marsh generalist	Moderate
	Anas strepera (gadwall)	↑	↑	↑		X	X		Marsh generalist	
	Aythya collaris (ring-necked duck)	↑	↑	↑		X	X		Marsh generalist	
	Anas discors (blue-winged teal)	↑	↑	↑		X	X		Marsh generalist	Moderate
	Anas crecca (green-winged teal)	↑	↑	↑		X	X		Marsh generalist	High
	Somateria mollissima (common eider)	↑	↑	↑	X		X		Marsh generalist	
	Lophodytes cucullatus (hooded merganser)	↑	↑	↑		X	X		Marsh generalist	Moderate
	Mergus merganser (common merganser)	↑	↑	↑		X	X		Marsh generalist	
	Mergus serrator (red-breasted merganser)	↑	↑	↑		X	X		Marsh generalist	
	Aix sponsa (wood duck)	↑	↑	↑		X	X		Marsh generalist	High
Podicipedidae	*Podilymbus podiceps* (pied-billed grebe)				X	X	X		Marsh specialist	
Phalacrocoracidae	*Phalacrocorax auritus* (double-crested cormorant)	↑	↓	↓	X	X	X		Marsh generalist	

Family	Species	Trend 1	Trend 2				Guild	Level
Ardeidae	*Botaurus lentiginosus* (American bittern)	↓	↑	X	X	X	Marsh specialist	Moderate
	Ixobrychus exilis (least bittern)	↓↑	↑	X X	X X	X X	Marsh specialist	Moderate
	Ardea herodias (great blue heron)	↑	↑	X X	X X	X X	Long-legged wader	
	Ardea alba (great egret)	↑	↑	X	X	X	Long-legged wader	Moderate
	Egretta thula (snowy egret)	↑	↑	X	X	X	Long-legged wader	Moderate
	Egretta tricolor (tricolored heron)	↑	↑	X	X	X	Long-legged wader	Moderate
	Egretta caerulea (little blue heron)	↑	↑	X	X	X	Long-legged wader	Moderate
	Bubulcus ibis (cattle egret)	↑	↑	X	X	X	Long-legged wader	
	Butorides virescens (green heron)	↑	↑	X	X	X	Long-legged wader	
	Nycticorax nycticorax (black-crowned night-heron)	↑	↑	X	X	X	Long-legged wader	Moderate
	Nyctanassa violacea (yellow-crowned night-heron)	↑	↑	X	X	X	Long-legged wader	Moderate
	Plegadis falcinellus (glossy ibis)	↑	↑	X	X	X	Long-legged wader	High
Pandionidae	*Pandion haliaetus* (osprey)	↑↓	↑	X X	X X	X X	Marsh generalist	
Accipitridae	*Haliaeetus leucocephalus* (bald eagle)	↑	↑	X X	X X	X X	Marsh generalist	
	Circus cyaneus (northern harrier)	↓	→	X	X	X	Marsh generalist	
	Falco sparverius (American kestrel)	↓	↑	X	X	X	Marsh generalist	

TABLE 7.2

Continued

Family	Species (common name)	Time Period			Seasons				Guild	BCR 30 priority[1]
		Early	Mid	Late	Breeding	Migrating	Wintering	Resident		
Rallidae	*Laterallus jamaicensis* (black rail)	↓	↓	↑	X	X	X		Marsh specialist	Highest
	Rallus limicola (Virginia rail)	↓	↓	↑	X	X	X		Marsh specialist	
	***Rallus longirostris* (clapper rail)**	↓	↓	↑	X	X	X		Marsh specialist	High
	Gallinula chloropus (common moorhen)	↓	↑	↑	X	X	X		Marsh specialist	
	Fulica americana (American coot)	↓	↑	↑	X	X	X		Marsh specialist	
Charadriidae	*Pluvialis squatarola* (black-bellied plover)	↑	↓	↓		X			Shorebird	High
	Charadrius semipalmatus (semipalmated plover)	↑	↓	↓		X			Shorebird	Moderate
	Charadrius vociferus (killdeer)	↑	↓	↓		X			Shorebird	Moderate
Haematopodidae	*Haematopus palliates* (American oystercatcher)	↓	↑	↑	X	X	X		Shorebird	Highest
Recurvirostridae	*Himantopus mexicanus* (black-necked stilt)	↑	↑	↑	X	X	X		Shorebird	
	Recurvirostra americana (American avocet)	↑	↑	→		X			Shorebird	Moderate
Scoloplacidae	*Tringa melanoleuca* (greater yellowlegs)	↑	↑	↑		X			Shorebird	High
	Tringa flavipes (lesser yellowlegs)	↑	↑	↑		X			Shorebird	Moderate
	***Tringa semipalmatus* (willet)**	↑	↑	↑	X	X	X		Marsh specialist	High
	Numenius phaeopus (whimbrel)	↑	→	→		X			Shorebird	Highest
	Calidris pusilla (semipalmated sandpiper)	↑	→	→		X			Shorebird	High

128

Family	Species								Category	Priority
	Calidris mauri (western sandpiper)	↑	↑				X		Shorebird	
	Calidris minutilla (least sandpiper)	↑	↑				X		Shorebird	Moderate
	Calidris fuscicollis (white-rumped sandpiper)	↑	↑				X		Shorebird	High
	Calidris bairdii (Baird's sandpiper)	↑	↑				X		Shorebird	
	Calidris alpine (dunlin)	↑	↑			X	X		Shorebird	
	Limnodromus griseus (short-billed dowitcher)	↑	↑			X	X		Shorebird	High
Laridae	Larus atricilla (laughing gull)	↓	↑	X					Gulls and terns	
	Larus philadelphia (Bonaparte's gull)	↓	↑			X			Gulls and terns	
	Larus delawarensis (ring-billed gull)	↓	↑	X		X			Gulls and terns	
	Larus argentatus (herring gull)	↓	↑	X		X	X		Gulls and terns	
	Larus marinus (great black-backed gull)	↓	↑		X				Gulls and terns	
	Sterna dougallii (roseate tern)	↑	↑	X					Gulls and terns	Highest
	Sterna fosteri (Forster's tern)	↑	↑	X					Gulls and terns	High
	Sterna antillarum (least tern)	↑	↑	X					Gulls and terns	High
	Sterna hirundo (common tern)	↑	↑	X					Gulls and terns	Moderate
	Sterna nilotica (gull-billed tern)	↑	↑	X					Gulls and terns	
	Sterna paradisaea (Arctic tern)	↑	↑	X					Gulls and terns	
	Rynchops niger (black skimmer)	↓	↑	X					Gulls and terns	Moderate
Tytonidae	Tyto alba (barn owl)	↓	↑	X			X		Marsh generalist	
Strigidae	Asio flammeus (short-eared owl)	↓	↑	X		X	X	X	Marsh generalist	Moderate
Alcedinidae	Ceryle alcyon (belted kingfisher)	↑	↑		X		X		Marsh generalist	

129

TABLE 7.2

Continued

Family	Species (common name)	Time Period			Seasons				Guild	BCR 30 priority[1]
		Early	Mid	Late	Breeding	Migrating	Wintering	Resident		
Tyrannidae	*Empidonax alnorum* (alder flycatcher)	↕	↕	↕	X				Marsh generalist	
	Empidonax traillii (willow flycatcher)	↕	↕	↕	X				Marsh generalist	High
	Tyrannus tyrannus (eastern kingbird)	↕	↕	↕	X				Marsh generalist	High
Corvidae	*Corvus brachyrhynchos* (American crow)	↕	↕	↕				X	Marsh generalist	
	Corvus ossifragus (fish crow)	↑	↑	↑				X	Marsh generalist	
Hirundinidae	*Tachycineta bicolor* (tree swallow)	↑	↑	↑	X	X			Marsh generalist	
	Stelgidopteryx serripennis (northern rough-winged swallow)	↑	↑	↑	X	X			Marsh generalist	
	Petrochelidon pyrrhonota (cliff swallow)	↑	↑	↑	X	X			Marsh generalist	
	Hirundo rustica (barn swallow)	↑	↑	↑	X	X			Marsh generalist	
Troglodytidae	*Cistothorus palustris* (marsh wren)	↓	↑	↑	X	X	X		Marsh generalist	
Parulidae	*Denidroica petechia* (yellow warbler)	↓	↑	↑	X	X			Marsh generalist	
	Geothlypis trichas (common yellowthroat)	↑	↑	↑	X	X			Marsh generalist	

Family	Species	Restoration response[1]		Seasons using tidal marshes		Guild membership	BCR priority[1]
Emberizidae	*Ammodramus nelson* (Nelson's sharp-tailed sparrow)	↓	↑	X	X	Marsh specialist	Moderate
	***Ammodramus caudacutus* (salt-marsh sparrow)**	↓	↑	X	X	Marsh specialist	Highest
	***Ammodramus maritimus* (seaside sparrow)**	↓	↑	X	X	Marsh specialist	Highest
	Melospiza georgiana[2] (swamp sparrow)	↓	↑	X	X	Marsh specialist	High[3]
	Melospiza melodia (song sparrow)						
	Passerculus sandwichensis (savannah sparrow)	↓	↑	X	X	Marsh generalist	
Icteridae	*Dolichonyx oryzivorus* (bobolink)	↓	↑	X	X	Marsh generalist	Moderate
	Agelaius phoeniceus (red-winged blackbird)	↔	→	X	X	Marsh generalist	Moderate
	Sturnella magna (eastern meadowlark)			X		Marsh generalist	Moderate
	Quiscalus quiscula (common grackle) and	↓		X	X	Marsh generalist	
	Quiscala major (boat-tailed grackle)			X			

Note: Table presents predicted response in three relative time periods of restoration (↑increase, ↓decrease, ↔ no change; blank indicates it is unknown how species will respond), seasons the species uses tidal marshes, guild membership, and bird conservation region (BCR) priority rankings. Bolded species are tidal marsh obligates.

[1] See BCR 30 report at http://www.acjv.org/bird_conservation_regions.htm.

[2] The subspecies *M. g. nigrescens* is a tidal marsh endemic restricted to mid-Atlantic tidal marshes for breeding.

[3] Coastal plain swamp sparrow *M.g. nigrescens*

131

that have breeding populations of sparrows prior to restoration, care should be taken in the design of the restoration implementation to reduce (if possible) during the breeding season (May–August).

Clapper rails are largely restricted to salt marshes in this region (although they are found in mangrove swamps in their subtropical range (Eddleman and Conway 1998). Unlike the Pacific coast population, clapper rail abundance seems relatively high along the East Coast, although recent quantitative monitoring data are absent. Little is known about annual productivity or adult and juvenile survival of clapper rails even though they are considered a game bird by most states where they occur (Eddleman and Conway 1998). Clapper rails would respond positively to tidal restoration projects that increase decapod biomass.

North America's smallest rail species, *Laterallus jamaicensis* (black rail), is found in freshwater and tidal marshes (Eddleman et al. 1994), but the greatest abundance of breeding birds is found in the latter. A preference for shallow water habitats makes it sensitive to marsh ditching and draining. However, it is unclear (given the species distribution) how much it depends on tidal flow. Without adequate data is it difficult to assess the status of black rail populations; however, most evidence suggests that the species has experienced drastic declines throughout its range in North America.

The willet, a medium-sized shorebird, is the only North American sandpiper with a breeding range that extends south of the north temperate region (Lowther et al. 2001). Willets breed most commonly on salt marshes, barrier islands, and barrier beaches (Bent 1929; Burger and Shisler 1978; Howe 1982). The eastern subspecies (*T. s. semipalmatus*) nests only in or near coastal salt marshes (Lowther et al. 2001). Mosquito ditching resulted in replacement of natural grass associations with shrubby growths and reduced invertebrate prey of shorebirds (Bourn and Cottam 1950). Draining and impoundment of salt marshes in the Maritime Provinces reduced habitat quality in some locales (Erskine 1992).

Sterna fosteri (Forster's tern) is the only tern restricted almost entirely to North America throughout the year. While feeding in a variety of open-water habitats, this species breeds primarily in marshes, including tidal marshes, where it is often associated with wrack.

Generalist Species Associated with Tidal Marshes

Anas rubripes (American black duck) is a species that is strongly associated with tidal marshes and present year-round in northeastern salt marshes from the Gulf of Maine south to coastal Virginia. This large dabbling duck breeds in a variety of wetland habitats including salt marshes and is a high conservation priority species due to declining populations (table 7.2). Presently, the continental American black duck population is half its historical size (Longcore et al. 2000). The American black duck is mainly a freshwater breeder but winters mostly in salt marshes along the Atlantic coast (Stotts and Davis 1960; Reed and Moisan 1971; Krementz et al. 1992). Black ducks have been shown to respond positively to freshwater wetland restoration on the breeding grounds (Stevens et al. 2003), but little is known about how this species responds to tidal restoration.

Anas discolor (blue-winged teal) nest in small numbers in tidal marshes from New Brunswick to North Carolina. A subspecies (*A. d. orphan*) restricted to Atlantic tidal marshes and characterized by darker coloration and heavier spotting has been described (AOU 1957) but may not be distinguishable from other subspecies. In addition to the American black duck and blue-winged teal, the generalized mallard and *Branta canadensis* (Canada goose) as well as the introduced *Cygnus olor* (mute swan) may breed in tidal marshes but tend to avoid more saline habitats. Tidal marshes in the Northeast are often used by molting waterfowl, making them important habitats for postbreeding. Tidal restoration projects that restore *Spartina*-dominated marshes may increase the extent and quality of molting habitat for waterfowl.

Ixobrychus exilis (least bittern) occurs in freshwater but is more strongly associated with brackish marshes throughout the eastern half of the United States (Gibbs and Melvin 1992). The least bittern is among the most inconspicuous of all North American marsh birds and is heard more frequently than seen. Least bitterns breed in freshwater and brackish marshes with dense, tall growths of aquatic or semiaquatic vegetation (particularly *Typha*, *Carex*, *Scirpus*, *Sagittaria*, or *Myriscus*) interspersed with clumps of woody vegetation and open water (Gibbs and Melvin 1992). Overwintering birds occur mainly in brackish and saline swamps and tidal marshes (Palmer 1962; Hancock and Kushlan 1984), but little is known about wintering habitats. *Botaurus lentiginosus* (American bittern) is found primarily in freshwater wetlands but occasionally breeds in tidal brackish marshes (Gibbs and Melvin 1992). Their coastal wintering distribution (Root 1988) also results in more individuals migrating through or wintering in brackish marshes (Hancock and Kushlan 1984). Given that least bitterns breed primarily in freshwater wetlands, tidal restoration projects will likely not alter, either to enhance or to degrade, least bittern habitat.

Circus cyaneus (northern harrier) is strongly associated with tidal marsh systems. This species occupies and nests in marshes, grasslands, and tundra throughout much of North America, but it is unclear how important tidal marshes are to the recovery of its populations.

While salt marshes support few species of perching birds (Passeriformes), such as the specialized sparrows discussed earlier, *Aegalius phoeniceus* (red-winged blackbird), *Cistothorus palustris* (marsh wren), and *Geolthlypis trichas* (common yellowthroat), a large variety of Passeriformes can be found in brackish tidal marshes, particularly those with emergent shrubs. These brackish upper marsh species include *Dendroica petechia* (yellow warbler), *Empidonax traili* (willow flycatcher), and *Melospiza melodia* (song sparrow). Marsh wrens that inhabit the tidal marshes of the North Atlantic region are undifferentiated from inland-breeding populations. But from New Jersey south, the species is restricted to tidal marshes, and several morphologically distinct subspecies of marsh wrens are recognized. Marsh wrens are particularly abundant in the presence of large grasses and reeds, where they often depredate the nests of other songbirds. *Quiscalus major* (boat-tailed grackle) is largely restricted to breeding in tidal wetlands through most of their range along the mid- to southern Atlantic and Gulf coasts. They occur within the Northeast Atlantic region along the shores of Long Island, but populations are spreading northward (Post et al. 1996).

Most of the twelve species of herons, egrets, and bitterns (Ardeidae) use tidal marshes extensively for both breeding and foraging (table 7.2). *Ardea herodias* (great blue heron) is one of the most widespread and adaptable wading birds in North America (Butler 1992). This species nests mostly in colonies often located on islands or in wooded swamps. In winter, great blue herons move to estuaries (Butler 1991), favoring coastal marine habitats along the US East Coast of, especially salt marshes.

In addition to the two aforementioned rail species, two other rails regularly occur in tidal marshes but primarily in the nonbreeding season. *Rallus limicolla* (Virginia rail) inhabits fresh and tidal wetlands (where they may breed in freshwater to brackish conditions) throughout North America and the northeast (Conway and Eddleman 1994). *Rallus elegans* (king rail), closely related to the clapper rail, is found in primarily freshwater marshes and rice fields but may breed in brackish wetlands. Although widely distributed, populations have declined over the last thirty years (Poole et al. 2005). King rails are also considered game birds in several states.

A wide variety of shorebirds, gulls and terns (Charadriiformes), and cormorants forage in the open waters and on the exposed mudflats within tidal marshes (Burger and Olla 1984). Many of these same species rely upon the cover of tidal marsh for roosting during periods of the night and day, particularly when high tides make nearby beaches and mudflats unavailable for foraging. While shorebirds are not restricted to tidal environments, tidal flux has a positive effect on prey abundance and surface activity, and the birds feed preferentially during certain portions of the tidal cycles. Therefore, changes in the tidal exchange that increase benthic invertebrate biomass will likely benefit shorebirds. However, tidal flow may have the negative effect of increasing turbulence and turbidity (particularly in shallow water), but this has not been studied within salt marsh systems.

Temporal Effects of Tidal Restoration on Birds

Studies to date have provided some evidence that tidal flow influences habitat by favoring dominant marsh plants preferred by the marsh specialist species, particularly *S. alterniflora* and *S. patens* (Roman et al. 1984; Burdick et al. 1997; Roman et al. 2002). Similar successes have occurred in the marshes of California (Foin et al. 1997; Zembal and Hoffman 2002). The degree to which these vegetation changes actually amount to the restoration of adequate habitat remains controversial (Zedler 2000) and has not been widely evaluated with demographic data, but some studies show that the restored tidal vegetation is recolonized by the target species.

In addition to restoration of tidal flow and associated habitat changes there are other important features that can influence bird populations during tidal restoration, including the recruitment of appropriate plant species (including human planting of key species), invasive species, edge effects due to restricted marsh size, adjacent land use, and the availability of sediments to contribute to marsh accretion. Therefore, it is rarely possible to assess the effects of tidal flow alone. However, it is useful to at least consider the possible short-term and long-term effects that tidal restoration might have on bird populations.

Short-term effects include the following:

- With enhanced tidal flow, water funneled into ponds will provide habitat for different waterfowl species depending upon pool depth. Increases in shallow water can increase the area of submerged aquatic vegetation for herbivorous waterfowl, and the formation of tidal flats can provide opportunities for foraging shorebirds, but these are also probably important to terrestrial feeding songbirds (such as sparrows), as well as rails. Tidal currents through deeper channels can increase the availability of fish for piscivorous diving species but may (particularly in shallow water systems) increase turbidity.
- Changes in the microspatial, seasonal, and diurnal patterns of flooding of marsh vegetation will determine what parts of the marsh can be used for nesting. Tidal flooding may increase the patchiness of available nesting sites, causing birds to clump nesting territories while using a broader area of the marsh to forage (Johnston 1956; Post 1974). Seasonal changes in tidal cycles can influence the timing of breeding. The presence of tidal flooding can filter out less specialized species.
- Increases in tidal flow can enhance the diversity and abundance of prey species (e.g., Roman et al. 2002).

Long-term effects include the following:

- Changes in floristics with tidal restoration, including increased heterogeneity in vegetation structure, favor avian species adapted to increased tidal flow (Zedler et al. 2001; Warren et al. 2002).
- The potential for gradual increase in marsh area is positive given that most tidal marsh endemic birds are associated strongly with the marsh vegetation and not the open water resources.

Monitoring and Research Needs

The relatively small scale of most salt marsh restoration projects and the variation inherent in avian community dynamics require development and implementation of standardized monitoring programs to allow better understanding of bird community responses to tidal marsh restoration activities.

Population Monitoring Techniques

Ecological restoration is defined as the process of assisting the recovery of an ecosystem that has been degraded, damaged, or destroyed (SER 2004) with the ultimate goal to create a self-supporting ecosystem that is resilient to perturbation without assistance. Measuring the success of these actions is a challenge, with some recommending specific metrics (Ruiz-Jaen and Aide 2005) and others a more integrated approach (Neckles et al. 2002). Given the inherent difficulty of detecting faunal signals, the relatively small size of tidal marsh restoration projects, and the changes in vegetation over time, future surveys should (1) attempt to adhere to standardized protocols (Konisky et al. 2006) including guild definitions, (2) be conducted for longer time intervals postrestoration (two, five, seven, ten, twelve, fifteen years), and (3) integrate recently developed analysis techniques, like occupancy modeling (MacKenzie et al. 2006). Neckles et al. (2002) recommended twenty-minute fixed radius point counts located within the marsh to detect differences in passerine and secretive marsh bird species. Adapting this approach to include four, five-minute time intervals where detected individuals are counted in each time interval would allow for the use of occupancy modeling techniques that would provide estimates of occupancy and abundance and would simultaneously estimate detection probabilities. Another simple adaptation that may improve the avian estimates and reduce sampling variation would be to count birds within and outside the 50-meter (164-foot) radius point. This would allow for a binary (near–far) distance sampling analysis (Buckland et al. 2001) that would estimate density, provide effective detection distance for each species, and account for variation in detection probabilities.

To increase detection rates for secretive marsh birds surveys should integrate a call-broadcast component (Conway 2008). This methodology uses audio broadcast calls of secretive marsh birds to elicit vocalizations and increase detection rates.

The restoration of tidal flow can impact the overall diversity and shift the dominant guild of birds. Because birds are directly linked to ecosystem condition, changes in avian community composition can be indicative of ecological integrity and can provide a robust monitoring tool (Bradford et al. 1998; O'Connell et al. 1998; Canterbury et al. 2000; Bryce et al. 2002; DeLuca et al. 2004). Sites with low ecological integrity are expected to support a low-integrity bird community and vice versa. Avian species can serve as good indicators because they are sensitive to changes in habitat structure (Riffell et al. 2001) and changes in trophic dynamics (Pettersson et al. 1995). The measurable avian metrics (occupancy, abundance by species) can be integrated into a community-level integrity index to monitor tidal marsh condition (DeLuca et al. 2004). This condition index approach, or others that consider the bird community assemblage (Bradford et al. 1998; O'Connell et al. 1998; Canterbury et al. 2000; Bryce et al. 2002; DeLuca et al. 2004), may provide more robust techniques to assess differences between tidal marsh restoration sites before and after restoration or between reference and restored sites.

Restoration Site Monitoring Network

Given the more than one hundred tidal restoration sites that now exist in the North Atlantic region, it would be very effective to sample the bird community at these sites and a series of reference sites to estimate the response of the bird community to tidal marsh restoration. Sites could be visited two to three times within a season (breeding, migration, winter) to count birds (Neckles et al. 2002), with site (restored or reference, time since restoration, size, dominant vegetation, etc.) and survey (date, time, wind speed, precipitation, etc.) covariates. These could then be integrated into an occupancy analysis that would first determine the effects of specific covariates on the detection probability for each species and then estimate the relative effects of different site covariates on occupancy.

Research Opportunities

Actual data on the trade-offs associated with nesting and feeding under varying tidal regimes are scant, and this is probably the area where the most research is needed. Although the consequences of improper management of water flow have not been documented for any tidal marsh specialist, the lessons learned from the impact of a large-scale imbalance of water flow and its negative effect on the endangered *Ammodramus maritimus mirabilis* (Cape Sable seaside sparrow; Curnutt et al. 1998) should direct our attention to more research on the direct impact of tidal management on marsh bird populations.

Research and management paradigms for tidal regimes should move beyond comparing free tidal flow to no flow to examine a gradient of tidal flow conditions (Ritter et al. 2008). Maximizing habitat for avian species that require different tidal conditions may require active management of varying levels of tidal influence. Presumably, intact estuarine systems had enough natural variation in the distribution of tidal influence to support a broad range of species and their specific requirements. But in fragmented and altered systems, tidal flow may need to be managed in different ways.

As argued in this chapter, tidal marsh species may be attempting to optimize reproductive outputs based on costs and benefits associated with average tidal regimes. They may be less able to cope with infrequent, but catastrophic, floods. Nest loss due to flooding tends to be episodic and may not even occur every year (Greenberg et al. 2006). But when it does it eliminates reproduction for a substantial portion of the breeding season. Storm surges are increasing, probably as a result of global warming, and models predict that these events will continue to increase. Therefore, it is worth considering tidal management regimes that reduce the impact of catastrophic flooding events while also considering the effect of such a management scenario on other biotic components and processes. More detailed demographic studies and modeling are necessary to assess the impact of increased flooding events on the survival probability of salt marsh breeding species, particularly the endemic species occurring in fragmented systems.

Few complete demographic studies of tidal marsh populations have been published, and even fewer (e.g., DiQuinzio et al. 2001) have focused on alternative marsh management approaches. Studies that rely upon estimating occupancy or abundance alone will not be able to determine the trajectory and hence the population survival probability. This is particularly crucial for tidal marsh endemic populations. Issues with reproductive failure and poor survivorship point to management issues that may take much longer to detect by relying upon the presence of a "standing crop" of adult birds. Additionally, partitioning demographic changes into reproductive failure (and its causes), fledging condition, and juvenile and adult survival provides researchers with the ability to search more specifically for causes of population declines. Tidal marsh restoration projects that target areas invaded by *Phragmites* with the primary objective to restore native vegetation should have the greatest benefit to endemic bird species. Continued and additional research into the effects of restoring tidal flow on the demographic parameters (fecundity, survival) will provide a better understanding of the variation associated with these restoration projects. Projects that incorporate sampling before and after implementation on multiple sites will certainly provide the most informative results but will also be the most expensive and logistically challenging. Given the number of restoration sites that now exist along the North Atlantic coast, a retrospective study could be implemented to estimate the key demographic parameters on a series of restored sites, which would provide meaningful information associated with the effects of time since restoration, extent of the implementation, and landscape context on the response by tidal marsh birds.

REFERENCES

American Ornithologists' Union (AOU). 1957. *Checklist of North American Birds.* 5th ed. Baltimore, MD: Lord Baltimore Press.

Benoit, L. K., and R. A. Askins. 1999. "Impact of the Spread of *Phragmites* on the Distribution of Birds in Connecticut Tidal Marshes." *Wetlands* 19:194–208.

Bent, A. C. 1929. *Life Histories of North American Shore Birds, Pt. 2: Order* Limicolae. United States National Museum Bulletin 146.

Bourn, W. S., and C. Cottam. 1950. *Some Biological Effects of Ditching Tide Water Marshes.* Research Report 19. Washington, DC: US Fish and Wildlife Service.

Bradford, D. F., S. E. Franson, G. R. Miller, A. C. Neale, G. E. Canterbury, and D. T. Heggem. 1998. "Bird Species Assemblages as Indicators of Biotic Integrity in Great Basin Rangeland." *Environmental Monitoring and Assessment* 49:1–22.

Brawley, A. H., R. S. Warren, and R. A. Askins. 1998. Bird Use of Restoration and Reference Marshes within the Barn Island Wildlife Management Area, Stonington, Connecticut, USA." *Environmental Management* 22:625–33.

Bryce, S. A., M. R. Hughes, and P. R. Kaufmann. 2002. "Development of a Bird Integrity Index Using Bird Assemblages as Indicators of Riparian Condition." *Environmental Management* 30:294–310.

Buckland, S. T., D. R. Anderson, K. P. Burnham, and J. L. Laake. 2001. *Introduction to Distance Sampling.* New York: Oxford University Press.

Burdick, D. M., M. Dionne, R. M. Boumans, and F. T. Short. 1997. "Ecological Response to Tidal Restorations of Two Northern New England Salt Marshes." *Wetlands Ecology and Management* 4:129–44.

Burger, J., and B. Olla. 1984. *Shorebirds: Migration and Foraging Behavior.* New York: Plenum.

Burger, J., and J. K. Shisler. 1978. "Nest-Site Selection of Willets in a New Jersey Salt Marsh." *Wilson Bulletin* 90:599–607.

Butler, R. 1991. "Habitat Selection and Time of Breeding in the Great Blue Heron (*Ardea herodias*)." Phd thesis, University of British Columbia, Vancouver.

Butler, R. W. 1992. *A Review of the Biology and Conservation of the Great Blue Heron* (Ardea herodias) *in British Columbia.* Canadian Wildlife Service Technical Report No. 154. Delta, BC: Canadian Wildlife Service.

Canterbury, G. E., T. E. Martin, D. R. Petit, L. J. Petit, and D. F. Bradford. 2000. "Bird Communities and Habitat as Ecological Indicators of Forest Condition in Regional Monitoring." *Conservation Biology* 14:544–58.

Conway, C. J. 2008. *Standardized North American Marsh Bird Monitoring Protocols.* Wildlife Research Report 2008–01. Tucson, AZ: US Geological Survey, Arizona Cooperative Fish and Wildlife Research Unit.

Conway, C. J., and W. R. Eddleman. 1994. "Virginia Rail." Pp. 460–501 in *Management of Migratory Shore and Upland Game Birds in North America*, edited by T. C. Tacha and C. E. Braun. Washington, DC: International Association of Fish and Wildlife Agencies in cooperation with the Fish and Wildlife Service, US Dept. of the Interior.

Cornelisen, C. D. 1998. *Restoration of Coastal Habitats and Species in the Gulf of Maine.* Gloucester, MA: Gulf of Maine Council on the Marine Environment and NOAA Coastal Services Center.

Curnutt, J. L., A. L. Meyer, T. M. Brooks, L. Manne, O. L. Bass, D. M. Fleming, M. P. Nott, and S. L. Pimm. 1998. "Population Dynamics of the Endangered Cape Sable Seaside Sparrow." *Animal Conservation* 1:11–21.

DeLuca, W. V., C. E. Studds, L. L. Rockwood, and P. P. Marra. 2004. "Influence of Land Use on the Integrity of Marsh Bird Communities of the Chesapeake Bay, USA." *Wetlands* 24:837–47.

DiQuinzio, D. A., P. W. C. Paton, and W. R. Eddleman. 2001. "Site Fidelity, Philopatry, and Survival of Promiscuous Saltmarsh Sharp-tailed Sparrows in Rhode Island." *Auk* 118:888–99.

DiQuinzio, D. A., P. W. C. Paton, and W. R. Eddleman. 2002. "Nesting Ecology of Saltmarsh Sharp-tailed Sparrows in a Tidally Restricted Salt Marsh." *Wetlands* 22:179–85.

Eddleman, W. R., and C. J. Conway. 1998. "Clapper Rail (*Rallus longirostris*)." In *The Birds of North America Online*, edited by A. Poole. Ithaca, NY: Cornell Lab of Ornithology.

Eddleman, W. R., R. E. Flores, and M. Legare. 1994. "Black Rails (*Laterallus jamaicensis*). In *The Birds of North America Online*, edited by A. Poole. Ithaca, NY: Cornell Lab of Ornithology.

Erskine, A. J. 1992. *Atlas of Breeding Birds of the Maritime Provinces.* Halifax, Nova Scotia: Nova Scotia Museum.

Foin, T. C., E. J. Garcia, R. E. Gill, S. D. Culberson, and J. N. Collins. 1997. "Recovery Strategies for the California Clapper Rail (*Rallus longirostris obsoletus*) in the Heavily Urbanized San Francisco Estuarine Ecosystem." *Landscape and Urban Planning* 38:229–43.

Gibbs, J. P., and S. M. Melvin. 1992. "American Bittern." Pp. 71–88 in *Migratory Nongame Birds of Management Concern in the Northeastern United States*, edited by K. Schneider and D. Pence. Newton Corner, MA: US Fish and Wildlife Service.

Gjerdrum, C., C. S. Elphick, and M. Rubega. 2005. "Nest Site Selection and Nesting Success in Salt Marsh Breeding Sparrows: The Importance of Nest Habitat, Timing, and Study Site Differences." *Condor* 107:849–62.

Greenberg, R., J. E. Maldonado, S. Droege, and M. V. McDonald. 2006. "Tidal Marshes: A Global Perspective on the Evolution and Conservation of Their Terrestrial Vertebrates." *BioScience* 56:675–85.

Greenlaw, J. S., and J. D. Rising. 1994. "Sharp-Tailed Sparrow (*Ammodramus caudacutus*). In *The Birds of North America, No. 112*, edited by A. Poole and F. Gill. Philadelphia, PA: Academy of National Science and American Ornithologists Union.

Hancock, J., and J. Kushlan. 1984. *The Herons Handbook*. Kent, England: Croom Helm.

Hansen, A. R., and W. G. Shriver. 2006. "Breeding Birds of Northeast Salt Marshes: Habitat Use and Conservation." *Studies in Avian Biology* 32:141–54.

Howe, M. A. 1982. "Social Organization in a Nesting Population of Eastern Willets (*Catoptrophorus semipalmatus*)." *Auk* 99:88–102.

Johnston, R. F. 1956. "Population Structure in Salt Marsh Song Sparrow, I: Environment and Annual Cycle." *Condor* 58:24–44.

Konisky, R. A., D. M. Burdick, M. Dionne, and H. A. Neckles. 2006. "A Regional Assessment of Salt Marsh Restoration and Monitoring in the Gulf of Maine." *Restoration Ecology* 14:516–25.

Krementz, D. G., D. B. Stotts, G. W. Pendleton, and J. E. Hines. 1992. "Comparative Productivity of American Black Ducks and Mallards Nesting on the Chesapeake Bay Islands." *Canadian Journal of Zoology* 70:225–28.

Longcore, J. R., D. G. McAuley, G. R. Hepp, and J. M. Rhymer. 2000. "American Black Duck (*Anas rubripes*)." In *The Birds of North America Online*, edited by A. Poole. Ithaca, NY: Cornell Lab of Ornithology.

Lowther, P. E., H. D. Douglas, I. Gratto-Trevor, and C. Gratto-Trevor. 2001. "Willet (*Tringa semipalmatus*)." In *The Birds of North America Online*, edited by A. Poole. Ithaca, NY: Cornell Lab of Ornithology.

MacKenzie, D. I., J. D. Nichols, J. A. Royle, K. H. Pollock, L. L. Bailey, and J. E. Hines. 2006. *Occupancy Estimation and Modeling: Inferring Patterns and Dynamics of Species Occurrence*. Amsterdam: Elsevier.

Marshall, R. M., and S. E. Reinert. 1990. "Breeding Ecology of Seaside Sparrows in a Massachusetts Salt Marsh." *Wilson Bulletin* 102:501–13.

Neckles, H. A., M. Dionne, D. M. Burdick, C. T. Roman, R. Buchsbaum, and E. Hutchins. 2002. "A Monitoring Protocol to Assess Tidal Restoration of Salt Marshes on Local and Regional Scales." *Restoration Ecology* 10:556–63.

O'Connell, T. J., L. E. Jackson, and R. P. Brooks. 1998. "A Bird Community Index of Biotic Integrity for the Mid-Atlantic Highlands." *Environmental Monitoring and Assessment* 51:145–56.

Palmer, R. S. 1962. *Handbook of North American Birds*. Vol. 1. New Haven, CT: Yale University Press.

Pettersson, R. P., J. P. Ball, K. Renhorn, P. Esseen, and K. Sjoberg. 1995. "Invertebrate Communities in Boreal Forest Canopies as Influenced by Forestry and Lichens with Implications for Passerine Birds." *Biological Conservation* 74:57–63.

Poole, A. F., L. R. Bevier, C. A. Marantz, and B. Meanley. 2005. "King Rail (*Rallus elegans*)." In *The Birds of North America Online*, edited by A. Poole. Ithaca, NY: Cornell Lab of Ornithology.

Post, W. 1974. "Functional Analysis of Space-Related Behavior in the Seaside Sparrow." *Ecology* 55:564–75.

Post, W., and J. S. Greenlaw. 1994. "Seaside Sparrow (*Ammodramus maritimus*)." In *The Birds of North America Online*, edited by A. Poole. Ithaca, NY: Cornell Lab of Ornithology.

Post, W., J. P. Poston, and G. T. Bancroft. 1996. "Boat-Tailed Grackle (*Quiscalus major*)," In *The Birds of North America Online*, edited by A. Poole. Ithaca, NY: Cornell Lab of Ornithology. http://bna.birds.cornell.edu/bna/species/207doi:10.2173/bna.207.

Raposa, K. B. 2008. "Early Ecological Responses to Hydrologic Restoration of a Tidal Pond and Salt Marsh Complex in Narragansett Bay, Rhode Island." *Journal of Coastal Research* 55:180–92.

Reed, A., and G. Moisan. 1971. "The *Spartina* Tidal Marshes of the St. Lawrence Estuary and Their Importance to Aquatic Birds." *Natural Canada* 98:905–22.

Riffell, S. K., B. E. Keas, and T. M. Burton. 2001. "Area and Habitat Relationships in Great Lakes Coastal Wet Meadows." *Wetlands* 21:492–507.

Ritter, A. F., K. Wasson, S. I. Lonhart, R. K. Preisler, A. Woolfolk, K. A. Griffith, S. Connors, and K. W. Heiman. 2008. "Ecological Signatures of Anthropogenically Altered Tidal Exchange in Estuarine Ecosystems." *Estuaries and Coasts* 31:554–71.

Roman, C. T., W. A. Niering, and R. S. Warren. 1984. "Salt Marsh Vegetation Change in Response to Tidal Restriction." *Environmental Management* 8:141–50.

Roman, C. T., K. B. Raposa, S. C. Adamowicz, M. J. James-Pirri, and J. G. Catena. 2002. "Quantifying Vegetation and Nekton Response to Tidal Restoration of a New England Salt Marsh." *Restoration Ecology* 10:450–60.

Root, T. 1988. *Atlas of Wintering North American Birds, an Analysis of Christmas Bird Count Data*. Chicago: University of Chicago Press.

Ruiz-Jaen, M. C., and T. M. Aide. 2005. "Restoration Success: How Is It Being Measured?" *Restoration Ecology* 13:569–77.

Schwarzbach, S. E., J. D. Albertson, and C. M. Thomas. 2006. "Effects of Predation, Flooding, and Contamination on the Reproductive Success of California Clapper Rails (*Rallus longirostris obsoletus*) in San Francisco Bay." *Auk* 123:45–60.

Seigel, A., C. Hatfield, and J. M. Hartman. 2005. "Avian Response to Restoration of Urban Tidal Marshes in the Hackensack Meadowlands, New Jersey." *Urban Habitats* 3:87–116.

Shriver, W. G., T. P. Hodgman, J. P. Gibbs, and P. D. Vickery. 2004. "Landscape Context Influences Salt Marsh Bird Diversity and Area Requirements in New England." *Biological Conservation* 119:545–53.

Shriver, W. G., P. D. Vickery, T. P. Hodgman, and J. P. Gibbs. 2007. "Flood Tides Affect Breeding Ecology of Two Sympatric Sharp-Tailed Sparrows." *Auk* 124:552–60.

Society for Ecological Restoration International (SER). 2004. *The Society for Ecological Restoration International Science and Policy Working Group International Primer on Ecological Restoration*. Tucson, AZ: Society for Ecological Restoration International.

Stevens, C. E., T. S. Gabor, and A. W. Diamond. 2003. "Use of Restored Small Wetlands by Breeding Waterfowl in Prince Edward Island, Canada." *Restoration Ecology* 11:3–12.

Stotts, V. D., and D. E. Davis. 1960. "The Black Duck in the Chesapeake Bay of Maryland: Breeding Behavior and Biology." *Chesapeake Science* 1:127–54.

Warren, R. S., P. E. Fell, R. Rozsa, A. H. Brawley, A. C. Orsted, E. T. Olson, V. Swamy, and W. A. Niering. 2002. "Salt Marsh Restoration in Connecticut: 20 Years of Science and Management." *Restoration Ecology* 10:497–513.

Zedler, J. B. 2000. "Progress in Wetland Restoration Ecology." *Trends in Ecology and Evolution* 15:402–7.

Zedler, J. B., J. C. Callaway, and G. Sullivan. 2001. "Declining Biodiversity: Why Species Matter and How Their Functions Might Be Restored in Californian Tidal Marshes." *BioScience* 51:1005–17.

Zembal, R., and S. M. Hoffman. 2002. *A Survey of the Belding's Savannah Sparrow (Passerculus sandwichensis beldingi) in California, 2001.* California Department of Fish and Game, Habitat Conservation Planning Branch, Species Conservation and Recovery Program Report 2002-01. Sacramento: California Dept. of Fish and Game.

Zink, R. M., and J. C. Avise. 1990. "Patterns of Mitochondrial DNA and Allozyme Evolution in the Avian Genus *Ammodramus*." *Systematic Zoology* 39:148–61.

PART III

The Practice of Restoring Tide-Restricted Marshes

Coastal managers from government agencies, conservation organizations, and environmental and engineering firms are at the forefront of facilitating tidal restoration projects. They identify and prioritize sites for restoration, engage in project design, establish performance criteria, implement environmental monitoring programs and adaptive management, seek productive partnerships, and secure appropriate funding. Given the fundamental and essential role that coastal managers play in tidal marsh restoration, it is important to seek their perspective on the challenges of developing and maintaining restoration programs and learn of achievements and program approaches. The New England and Atlantic Canada region, with a broad diversity of individual state and provincial approaches and experiences, offers a range of lessons learned with the intent of their being readily transferable to agencies or organizations that are developing programs aimed at restoration of tidal wetlands.

With a northward progression, this part of the book describes the genesis, challenges, organization, and achievements of programs aimed at restoring tidal flow to coastal wetlands in Connecticut (Rozsa, chap. 8), Rhode Island (Chaffee and others, chap. 9), Massachusetts (Durey and others, chap. 10), New Hampshire (Diers and Richardson, chap. 11), Maine (Kachmar and Hertz, chap. 12), and Canada's Atlantic Maritime Provinces (Bowron and others, chap. 13).

Chapter 8

Restoration of Tidal Flow to Degraded Tidal Wetlands in Connecticut

RON ROZSA

Connecticut's tidal wetlands, ranging from salt marsh to freshwater tidal wetlands, occur along the shores of Long Island Sound and border the tidal portions of the state's rivers (e.g., Connecticut, Quinnipiac, and Housatonic). Today, approximately 5900 hectares of tidal wetland occur in Connecticut, two thirds of which is *Spartina* (cordgrass)-dominated salt marsh (Warren et al. 2002). Tide gates, dikes, impoundments, road and railroad crossings, and undersized culverts contribute significantly to the hydromodification of tidal flow to Connecticut's tidal wetlands (Rozsa 1995), a leading cause of wetland degradation. The systematic restoration of these degraded tidal wetlands is a program that is over thirty years old and was a vision of the Coastal Area Management Program (now the Office of Long Island Sound Programs) of the Connecticut Department of Environmental Protection (DEP). This successful program is an example of the implementation of all aspects of restoration from planning through construction by a state agency. This chapter highlights program benchmarks, funding, methodology, monitoring, project examples, and future challenges regarding the restoration of salt and brackish marshes.

Key Restoration Program Benchmarks—Program Overview

The Connecticut Tidal Wetland Act of 1969 required that any permitted activities preserve tidal wetlands, but this act had no provision to reverse degradation caused by historical activities. This was remedied by a new state policy in the Coastal Management Act of 1980 "to encourage the rehabilitation and restoration of degraded tidal wetlands," marking the beginning of DEP's wetland restoration program. In 1981, the Coastal Area Management Program provided a grant to

Connecticut College to develop restoration recommendations for fifteen de-graded tidal wetlands, an important step toward implementation of a statewide tidal wetlands restoration program.

In 1985 the Department of Health Services invited DEP to assist with their Mosquito Control Program to develop an application for federal permits that would allow for the continuance of mosquito control activities in tidal wetlands. This networking led to several key restoration benchmarks, namely (1) the use of open marsh water management and the abandonment of maintenance ditching, (2) the requirement for the creation of a site plan review committee to evaluate all new mosquito control projects, and (3) the creation of a DEP partnership with the Mosquito Control Program for marsh restoration. Abandonment of ditching is leading to the restoration of pre-ditching habitat such as pannes in many coastal marshes. The site plan review committee brings together federal and state re-source experts and permit staff as well as wetland scientists to review Mosquito Control Program proposals for open marsh water management.

The DEP and Mosquito Control Program partnership for marsh restoration was initiated when DEP recommended to Mosquito Control an experimental ap-proach to the restoration of the 80 hectare Hammock River Marsh in Clinton, commencing with the opening of one of four tide gates to increase tidal flow. Veg-etation monitoring would be used to evaluate the need for additional gate open-ing. The resulting flows produced a flooding regime matching that of the low marsh habitat; thus mosquito production declined and the anticipated imple-mentation of open marsh water management was rendered unnecessary. Mos-quito Control realized that tidal flow restoration at subsided marshes could be an effective mosquito abatement technique.

In 1994, the Mosquito Control Program was eliminated from the Health De-partment, but staff and equipment (e.g., low ground pressure excavators) were transferred to DEP to form a dedicated restoration program for inland and tidal wetlands. In the same year, the Long Island Sound Study (an Environmental Pro-tection Agency national estuary program) provided funding to DEP to hire a full-time restoration coordinator. The coordinator assists in technical studies, project design, drafting permit applications, and securing funding. The coordinator inter-acts with a tidal marsh restoration project team to establish an annual workplan, discuss the status of ongoing projects, and provide staff support where needed.

Funding and Partnerships

From 1985 to 1993, tidal flow restoration projects were done by the Mosquito Control Program. In 1985 the Connecticut legislature established the Coves and Embayment Restoration Program, which would provide coastal municipalities a reimbursement of up to 50 percent for design and construction costs. A shortcoming of this approach was that municipalities had to provide the restoration funding and then be reimbursed. This program was modified in 1989 with the creation of the Long Island Sound Cleanup Account allowing the state to fund up to 100 percent of the project costs. Eligible expenses for both programs include preliminary engineering, design, and construction. An unanticipated outcome and benefit of these dedicated funds is their use as nonfederal match as the number of federal restoration programs grew in the 1990s.

The majority of restoration projects are designed and implemented by DEP, including the drafting of permit applications. In these instances, funds are only required to support construction. DEPs Office of Long Island Sound Programs takes the lead for the complex restoration projects that require consulting services to evaluate the flooding consequences of tidal flow restoration to low-lying properties and infrastructure. The implementation of these complex projects typically uses private contractors for the construction, as the required activities typically exceed the ability of DEP-owned equipment for implementation.

The chief role of partners, outside of the site plan review committee, is to assist in securing funding. DEP has sought and secured funding for most of the federal restoration fund programs. Two unconventional funding sources include the Environmental Protection Agency's Section 319 Nonpoint Source Management Program and the Federal Highway Administration's Transportation Enhancement Activities under the Intermodal Transportation Efficiency Act of 1991. Apparently Connecticut is the only state to have used these funds for restoration, and this program offers an excellent match formula (80 percent federal).

Restoration Approach

Roman et al. (1984) examined the biophysical characteristics of six diked and drained marshes and provided critical insight about the degradation process. Reducing or eliminating tidal flow was shown to cause the following changes: (1) the water table drops and aerates the peat, (2) peat decomposition occurs and causes submergence, (3) creek water and soil halinity decline to fresh or oligohaline levels, and (4) nonnative invasive *Phragmites australis* (common reed) displaces the native plants.

From these observations and other studies it was concluded that tides are the primary abiotic factor organizing tidal marsh communities, and it was hypothesized that restoration of tidal flow should arrest degradation and reset the wetland on a trajectory to becoming a self-maintaining ecosystem. Absent a detailed blueprint of the pre-degradation marsh condition (e.g., elevation, water table depth, and distribution of biotic communities) or knowledge of how the system would have changed in response to natural processes like sea level rise, restoration to a precise historic condition is not feasible. The restoration goal has never been to return a marsh to its pre-disturbance condition but rather to its former dominant ecosystem complex (e.g., salt, brackish, or fresh tidal marsh), thus avoiding setting unrealistic goals such as a specific percent high marsh and low marsh.

The typical methods for tidal flow restoration include manually opening or abandoning tide gates (when there is no threat of flooding low-lying lands), replacement of undersized culverts with larger culverts, and removal of fill from former marshes that have been buried. The risk of flooding low-lying properties is always evaluated. If flooding is potentially an issue, a consultant is hired to model tidal flow and determine if some type of control structure is needed (e.g., self-regulating tide gate, combination slide/flapper gate) to allow for tidal flow restoration while providing flood protection.

Most restoration projects are constructed by DEP, which has dedicated staff and currently owns low-ground pressure equipment (e.g., excavators, dozers) to install culverts, install or restore tidal creeks, and remove fill. These are autogenic restoration projects, which use a minimalistic restoration approach. Tidal flow is restored (being careful to match the hydrology to the existing elevation), but there is no planting, application of fertilizer, or soil augmentation. Spontaneous vegetation restoration has occurred on all substrate types, including peat, fine-textured dredged sediments, sandy sediments, and even landfill sediments. Experience has demonstrated that there is no need to create complex channel cross sections; thus simple straight-sided creeks are constructed. These will widen or narrow depending upon the size of the tidal prism and the associated current velocity in the creeks. DEP contacts the wetland property owners to obtain permission (permit requirement) and rarely needs any further community contact or support. However, there are reluctant property owners who are not interested in restoring ecological services. In those instances, the following list of restoration services have been shown to garner project support:

- Restoration of salt marshes will reduce or eliminate fire hazard by reducing the areal extent of *Phragmites*.
- Restoration of diked and drained marshes, regardless of halinity, creates low marsh habitat and thus reduces or eliminates the habitat of mosquitoes.
- Restoration eliminates water quality problems, specifically acid sulfate soil (e.g., low pH and release of aluminum from clay particles, which is toxic at low concentrations) (Dent 1986).
- Restoration re-creates scenic vistas through the replacement of tall *Phragmites* with native salt and brackish meadow plants.

Monitoring, Adaptive Management, and Research

In the absence of dedicated funds for restoration, long-term research was concentrated on a select number of sites with different degradation histories to evaluate the efficacy of tidal flow restoration. A particularly important research site is the Barn Island Wildlife Management Area, where the restoration of five impoundments occurred at different times, allowing for the identification of trajectories. The results of long-term monitoring and research at Barn Island and other benchmark sites are described in Warren et al. (2002) and will not be presented except to highlight some of the results:

- Hydroperiod and salinity are important factors associated with vegetation recovery following tidal restoration. *Phragmites* declines rapidly at the lowest elevations or areas of greater hydroperiods, but at the highest elevations, approximately high marsh, restoration to meadow vegetation can occur over one to two decades.
- The salt marsh snail, *Melampus bidentatus*, can take up to two decades to reach densities that are comparable to nearby reference marshes, but other fauna, such as amphipods and fish, may recover more rapidly.
- When restoring former salt marsh (polyhaline), *Phragmites* will be replaced by salt marsh vegetation following tidal flow restoration. However, restoration of tidal flow to brackish tidal marshes (mesohaline and oligohaline) dominated by *Phragmites* will often not restore native brackish vegetation. Following over twenty years of tidal restoration to a former brackish meadow at Barn Island, *Phragmites* remains abundant albeit over 2 meters shorter than prior to reintroduced flow. It appears that this restoring impoundment is still adapting to tidal flow restoration as *Phragmites* continues to decline and brackish meadow increases.

This latter point resulted in adoption of a standardized classification system for Connecticut marshes that could be used to better predict vegetation responses to tide restoration, including the post restoration abundance of *Phragmites*. The classification incorporates the vegetation concepts of Nichols (1920) and halinity conventions of Cowardin et al. (1979) as follows: (1) salt marsh complex—vegetation of the salt marsh series; polyhaline is dominant with inclusions of brackish meadows, especially at the upland border; (2) brackish marsh complex—vegetation of the brackish marsh series (mesohaline and oligohaline) is dominant; brackish meadows are abundant at the higher halinities, and brackish reed marshes occur at the intermediate to lowest halinities.

Connecticut has used 18 parts per thousand as a guide to estimating the nature of *Phragmites* abundance following tidal flow restoration. In retrospect, this number appears too high, but it does serve as a useful guide.

Restoration Achievements

DEP maintains a record of tidal wetland restoration projects, and the decadal trends for project number and acreage are presented in figure 8.1. Excluded are the few experimental projects to restore tidal hydrology through ditch plugging and the long-term restoration experiment for all tidal marshes when routine maintenance of mosquito ditches ceased in 1984. The 1970 data include five degraded marshes where the Mosquito Control Program opted to abandon the use of tidal gates for mosquito control and the restoration of Great Harbor marsh in Guilford where the tide gates were destroyed by the hurricane of 1954. The data for post-1980 are principally projects that the Coastal Area Management Program had an active role in planning, designing, and constructing. Most of these projects are ecological restoration but a few include fill removal as a compensation requirement for public agency projects. To date, tidal flows have been increased to over 730 hectares of tidal wetlands, restoring 12 percent of the total acres of brackish and salt marshes in Connecticut. There are a total of eighty-three projects.

Some project highlights follow:

- Great Harbor Marsh in Guilford is a 57 hectare salt marsh that was diked and drained. The hurricane of 1954 restored tidal flow. Initially much of the

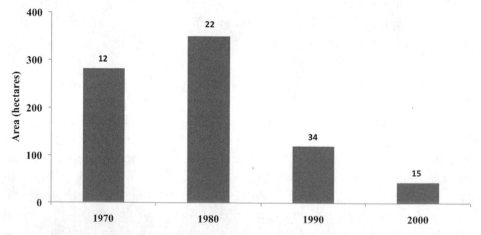

FIGURE 8.1. Area restored by tidal flow restoration per decade in Connecticut, and number of restoration projects conducted by decade (indicated above each decade bar). The 1970 statistic includes one restoration from 1954.

marsh was converted to peat flat. Then, for nearly sixty years peat flat would gradually convert to *Spartina alterniflora* (smooth cordgrass) low marsh. The question remains as to how long it will take for the original dominant vegetation, high marsh, to return.

- Barn Island demonstrates the successful restoration of five impounded marshes and that, even after thirty years, the marsh is still adapting to the tidal flow restoration. Hammock River in Clinton was initiated in 1985 through the opening of one of four tide gates. The conversion from *Phragmites* to salt marsh vegetation at this 81 hectare site is phenomenal, and it currently has very high densities of *Ammodramus maritimus* (seaside sparrow).

- Fairfield restoration sites are renowned for the development of the self-regulating tide gate by Thomas Steinke to address the need for tidal flow restoration while maintaining flood protection for low-lying development.

- Fletchers Creek at Silver Sands State Park in Milford is the first trash (landfill) to energy (vegetation) project (fig. 8.2). This included tidal flow restora-

FIGURE 8.2. Fletchers Creek (Milford, CT) restoration site. The narrow band of stunted *Phragmites* extending to the middle of the photograph marks the boundary of excavated landfill material (1997) on the right and wetland restored solely by tidal flow restoration in the upper left corner (1995). Prior to flow restoration the marsh was a dense monoculture of *Phragmites*, now returned to *Spartina* marsh. (Photo courtesy of Ron Rozsa)

tion and restoration through removal of former landfill material. Even though this appeared to be a *Phragmites* monoculture at the initiation of tidal flow restoration, the diversity of native plants is rather remarkable, especially given that no planting was done.

Lessons Learned and Future Challenges

Connecticut is recognized as a national leader in the restoration of degraded tidal marshes. Key to this success has been the leadership role played by the Coastal Area Management Program of the Connecticut DEP, which focused on all aspects of restoration, including active participation in restoration design by scientists, resource managers, and regulators; permit streamlining at the state and federal level; creation of a state funding program; and creation of a wetland classification system to help predict the outcome of tidal flow restoration. Approaches such as this are called a community of practice.

As in most states, funding of monitoring programs has proved to be a challenge. To address this, instead of investing funds in long-term monitoring at each restoration site, it was recognized that there were distinct classes of restoration (e.g., tidal flow restoration to diked marshes versus impounded marshes, fill removal, and others). Thus limited funds were invested to periodically assess the restoration of key marshes in each of these categories.

The fundamental lesson learned from monitoring and research is that tidal flow restoration was key to success — match the flows to the present-day marsh elevation and these tides would then organize the plant and animal communities. Planting, soil augmentation, and application of fertilizers were not necessary regardless of substrate type. Illustrating this is the restoration of Fletchers Creek, a marsh that had been drained, used as a landfill, and restored via tidal flow restoration to the subsided marsh and removal of landfill materials to match the elevation grades to the tides (fig. 8.2).

Projected accelerated sea level rise is the single greatest challenge for tidal marsh restoration. DEP is keenly aware that restoration projects of the past, especially the subsided marshes that are at lower elevations, are likely to be the most vulnerable to sea level rise. Considerations for new projects include life expectancy, costs, benefits, and opportunity for marine transgression. Practitioners need to realize that the use of self-regulating tide gates, while providing short-term benefits, may be an evolutionary dead end. As sea level rises, the frequency of flooding of low-lying lands will also increase, requiring managers to adjust the gates to reduce tidal inflow. Ultimately, even if the gates are closed to prevent tidal flooding, sea level rise will increase groundwater elevations, and the restoration marsh will become submerged. Connecticut is already incorporating new goals for marsh restoration projects including the recognition that, in the long term, the emergent restoration marsh is likely to become an embayment.

Even though there is uncertainty as to the sea level rise rate that will cause drowning of marshes, it is time to focus on creating the opportunities for future wetland restoration practitioners to restore tidal marsh, assuming humankind is successful at mitigating climate change and slowing sea level rise. Examples of possible adaptation strategies being discussed in Connecticut include identifying vulnerable plant species and creating a seed bank to assure the preservation of genetic diversity, identifying and protecting future marine transgression zones (wetland refugia) for wetlands to migrate to in the future, and cutting dead and dying trees in the marine transgression zones to eliminate shade, thereby promoting migration of wetland herbaceous vegetation.

References

Cowardin, L. M., V. Carter, F. C. Golet, and E. T. LaRoe. 1979. *Classification of Wetlands and Deepwater Habitats of the United States.* FWS/OBS 79/31. Washington, DC: US Fish and Wildlife Service, Biological Services Program.

Dent, D. 1986. *Acid Sulphate Soils: A Baseline for Research and Development.* Wageningen, Netherlands: International Institute for Land Reclamation and Improvement.

Gleason, H. A., and A. Cronquist. 1991. *Manual of Vascular Plants of Northeastern United States and Adjacent Canada.* New York: New York Botanical Gardens.

Nichols, G. E. 1920. "The Vegetation of Connecticut, VII: The Associations of Depositing Areas Along the Seacoast." *Bulletin of the Torrey Botanical Club* 47:511–48.

Roman, C. T., W. A. Niering, and R. S. Warren. 1984. "Salt Marsh Vegetation Change in Response to Tidal Restriction." *Environmental Management* 8:141–50.

Rozsa, R. 1995. "Human Impacts on Tidal Wetlands: History and Regulations." Pp. 42–50 in *Tidal Marshes of Long Island Sound—Ecology, History and Restoration,* edited by G. D. Dreyer and W. A. Niering. Bulletin No. 34. New London: Connecticut College Arboretum.

Warren, R. S., P. E. Fell, R. Rozsa, A. H. Hunter, A. C. Orsted, E. T. Olson, V. Swamy, and W. A. Niering. 2002. "Salt Marsh Restoration in Connecticut: 20 Years of Science and Management." *Restoration Ecology* 10:497–513.

Chapter 9

Salt Marsh Restoration in Rhode Island

CAITLIN CHAFFEE, WENLEY FERGUSON,
AND MARCI COLE EKBERG

Historically, Rhode Island experienced significant salt marsh loss as large areas were filled to create upland for development such as that in the urban center of Providence. It is estimated that 60 percent of the state's salt marshes have been lost as a result of filling. Loss and degradation of salt marshes resulting from human impacts continued to occur through the twentieth century. From the 1950s to the 1990s alone, Rhode Island experienced a net loss of over 120 hectares or 10 percent of its estuarine marshes. As of the mid-1990s, salt marshes in Rhode Island were estimated to represent over 1400 hectares of the state's estuarine habitat as embayment and fringing marshes throughout the Narragansett Bay estuary and associated with the coastal ponds located along the south shore barrier island region. Of that area, nearly half, or over 690 hectares, have been impacted by human activities such as ditching and impoundments (Tiner et al. 2003, 2004).

Throughout the past decade or so three studies were conducted to evaluate the status of salt marshes in Rhode Island. The collective purpose of these studies was to promote a comprehensive and coordinated approach to the restoration, protection, and monitoring of coastal wetlands, and to provide detailed information to resource managers that would allow them to prioritize projects throughout Rhode Island.

The first study, conducted in 1998 by the Narragansett Bay Estuary Program in partnership with the University of Massachusetts Natural Resources Assessment Group, the US Fish and Wildlife Service, the University of Rhode Island Environmental Data Center, and Save The Bay, identified and mapped degraded coastal wetlands and potential restoration sites within Narragansett Bay (Tiner et al. 2003). The study identified over eight hundred potential restoration sites comprising over 1600 hectares. The majority of the identified Narragansett Bay sites were classified as those degraded by human impacts such as ditching, filling (often with sediment from navigational dredging projects), and tidal restrictions. The remaining sites were no longer wetlands due to filling or draining, or presently existed as submerged or freshwater wetland habitat.

A complementary study conducted by the US Army Corps of Engineers identified 11 hectares of potential restoration sites along the south coastal region of Rhode Island. Save The Bay also conducted a salt marsh assessment in 1996 with over ninety volunteers trained to conduct salt marsh evaluations throughout Narragansett Bay. This effort produced important information about impacts such as buffer zone degradation, tidal restrictions, and stormwater discharges (fig. 9.1). It was found that about 70 percent of Rhode Island's remaining salt marshes—over 1100 hectares—are impacted by restrictions to tidal flow.

Restoration Program Overview: A Collaborative Effort

At the state government level in Rhode Island there is no centralized staff dedicated to salt marsh restoration planning and project management. Rather, restoration efforts occur as collaborations among various entities. The Rhode Island Habitat Restoration Team, established in 1998, is made up of representatives from various federal and state agencies, as well as regional conservation districts, local watershed groups, and other nonprofit organizations. The stated goals of this team are to promote habitat restoration, undertake restoration planning, coordinate restoration activities, implement restoration projects, provide technical assistance for

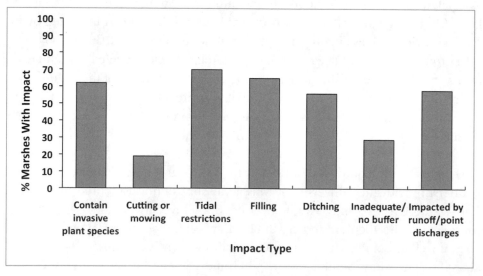

FIGURE 9.1. Impacts to Narragansett Bay salt marshes. (Data source: Save The Bay).

restoration, undertake restoration outreach and education, and promote habitat research and monitoring. Leadership of the team is shared among the Rhode Island Coastal Resources Management Council, Save The Bay, and the Narragansett Bay Estuary Program. The team has developed a general statewide strategy for habitat restoration that is periodically updated by the Coastal Resources Management Council and is currently developing detailed strategies for restoration of specific habitat types, including anadromous fish habitat, salt marshes, and seagrass beds.

Individual restoration projects in Rhode Island often begin as community-based efforts at the local level and gain planning and technical assistance from Habitat Restoration Team member organizations. Project management tends to be taken on by a local partner, such as a municipality or watershed organization. State and federal agencies such as the National Oceanic and Atmospheric Administration, Natural Resources Conservation Service, US Fish and Wildlife Service, Rhode Island Department of Environmental Management, and Coastal Resources Management Council provide funding and technical assistance, while organizations such as Save The Bay and the Narragansett Bay Estuary Program provide an array of additional support in the form of technical assistance, public outreach, grant-writing, pre- and postrestoration monitoring, and volunteer labor. The exceptions to this project management approach are capital projects implemented by the US Army Corps of Engineers. In these cases, while multiple stakeholders are represented on the project team and contribute to various aspects of the project, project management is assumed by the Corps, with the Coastal Resources Management Council acting as the local project sponsor.

Funding Approaches

The existence of a consistent source of state funding dedicated to habitat restoration with which additional federal funds can be leveraged has been vital to the success of salt marsh restoration projects in Rhode Island. The Coastal and Estuarine Habitat Restoration Program and Trust Fund, created and administered by the Coastal Resources Management Council since 2003, allocates $225,000 of the fees collected under the state's Oil Spill Prevention and Response statute to habitat restoration projects each year. Habitat Restoration Trust Fund monies are dispersed in accordance with Rhode Island General Law § 46-23.1-5(2) for design, planning, construction, or monitoring of habitat restoration projects. The legislation stresses the importance of collaboration among the sponsoring partners of habitat restoration projects, and the level and extent of collaboration among municipalities, nongovernmental organizations, watershed councils, and federal agencies are strongly considered before funding is granted. Project funding is determined by the Technical Advisory Committee consisting of members of the Rhode Island Habitat Restoration Team. Among the criteria considered during project evaluation are the long-term sustainability of the proposed project, the area of habitat to be restored, and the proposed project's consistency with state and regional habitat restoration planning initiatives such as the Rhode Island Coastal and Estuarine Habitat Restoration Strategy. To date, Trust Fund monies have been allocated to more than sixty projects and have leveraged more than $18 million in matching funds, which have supported over 77 hectares of salt marsh restoration.

Restoration Achievements and Challenges

Since the mid-1990s, twenty-one salt marsh restoration projects have been completed and three are in progress, comprising about 140 hectares, representing 10 percent of the state's 1400 hectares of salt marshes. The majority of these projects have involved restoring historical site hydrology through the removal or modification of hydrologic barriers, a relatively straightforward method of salt marsh restoration in Rhode Island from a construction and permitting perspective. There have been a smaller number of projects involving excavation of sediments to reestablish marsh elevations altered by filling, such as the Allin's Cove restoration project in Barrington led by the US Army Corps of Engineers. Projects involving excavation tend to present a greater number of permitting and implementation challenges related to dredged material disposal.

The collaborative approach to salt marsh restoration planning and implementation has been successful in Rhode Island, though it presents specific challenges. Diverse project teams can make efficient project management difficult. Local partners often have limited capacity for project management, making involvement of state, federal, and nonprofit partner organizations crucial to project success. However, decentralized project management can lead to gaps in project data, which hinders the accurate statewide assessment of salt marsh restoration status and trends. The Habitat Restoration Team developed a Rhode Island Restoration Portal website (www.edc.uri.edu/restoration/index.htm), where specific project information is stored. This portal was meant to provide a clearinghouse of restoration project information. However, much of the information on the site needs to be updated for the portal to be an effective restoration planning tool.

Other challenges to salt marsh restoration in Rhode Island include private ownership of marshes and adjacent lands. Activities on the marsh or in the upland buffer, such as cutting or mowing, dumping of yard waste or other debris, stormwater runoff from impervious surfaces, and filling on the marsh surface, can significantly degrade marsh health and detract from the potential success of tidal restoration projects. Outreach efforts, such as those conducted by Save The Bay, aim to educate landowners about the value of salt marshes and coastal buffers. However, these efforts often have a limited effect on changing landowner behavior, and they need to be complemented by additional measures such as conservation easements, tax incentives for restoration and preservation, and enforcement of regulatory protections.

The challenge of impacts from adjacent land use activities is compounded by the problem of global climate change and sea level rise. Climate change will likely impact completed tidal flow restoration projects if salt marsh accretion rates cannot keep pace with accelerated rates of sea level rise. This will result in a loss of salt marsh area as habitat types shift and low marsh converts to open water and intertidal mudflat habitat. This trend has already been observed in the Narragansett Bay Estuary Program study of wetland trends from the 1950s to 1990s (Tiner et al. 2004), with vegetated coastal wetland converting to open water, perhaps related to the impact of rising sea level associated with global warming. On the landward side, there are many places in Rhode Island where heavily developed shorelines, public infrastructure, and topography will prevent the landward migration of salt marshes. As a result, large areas of salt marsh will be lost as low-lying areas become increasingly inundated. For this reason, protection of upland areas adjacent to marshes will be necessary for their future survival. Sea level rise and its effect on project sustainability should also be taken into consideration for the planning and design of future tidal flow restoration projects.

Walker Farm Case Study: Partnerships and Adaptive Management

The Walker Farm salt marsh restoration project in Barrington, Rhode Island, is an example of both a successful restoration effort benefiting from many collaborations and an effective use of the adaptive management approach. Challenges to the project also highlight the difficulties that can arise in working with private landowners.

Walker Farm Marsh is a historic salt marsh that was altered by a number of roads and dam structures that restrict the amount of inflowing saltwater. Restricted tidal flow, decreased salinity in the marsh, and impoundment of open water resulted in the invasion of *Phragmites australis* (common reed) throughout the wetland, and flooding of the historic marsh. Saltwater flow into the marsh was inhibited by five tidal restrictions (the two major tidal restrictions are identified by arrows in the aerial photo, fig. 9.2).

The southern inlet restriction was a permanent dam with a culvert structure that restricts tidal flow. This culvert allowed tidal flow into a small portion of the marsh between the dam and the Walker Farm entrance road off Route 114. *Spartina alterniflora* (smooth cordgrass) is the dominant plant in this section of

FIGURE 9.2. Aerial view of Walker Farm salt marsh restoration project site, Barrington, Rhode Island.

the marsh. The next tidal restriction to the north was a small culvert under the access road allowing minimum tidal flow into the interior marsh. Two overgrown farm roads cut across the marsh and prevented further tidal inundation into the interior marsh. At the northern end of the marsh, a flap-gated historic earthen dam was built approximately sixty years ago to allow agricultural use of the marsh. Sometime between 1965 and 1970, the Town of Barrington made this original dam permanent to establish waterfowl habitat. This dam flooded the salt marsh and created a brackish pond that received tidal flow only during extreme high tides and storms. The dam also impounded freshwater from a tributary that discharges into the northeast corner of the pond, running under Route 114.

Since 1996, Save The Bay and the Barrington Conservation Commission's Salt Marsh Working Group have been involved in advocating for the restoration of Walker Farm marsh and engaging many partners. The Natural Resources Conservation Service provided funding for the construction of the restoration plan through the Wetlands Reserve Program. The National Oceanic and Atmospheric Administration provided funding for site assessment, prerestoration monitoring and design, engineering, and construction. The Rhode Island Coastal and Estuarine Habitat Restoration Trust Fund also provided funds for restoration construction. Ducks Unlimited provided design engineering, construction oversight, and construction funding. A local environmental consulting firm (ESS Group, Inc.) has conducted the permitting and the University of Rhode Island has been involved with site assessments and environmental monitoring.

Construction occurred in the summer of 2005, following almost a decade of planning, permitting, and prerestoration studies. The restoration project included modifying three existing tidal restrictions to improve flushing of the salt marsh. The southern structure includes water control structures to increase tidal flow. The structure under the access road was increased in size to allow for more tidal flow into the marsh interior. The former farm roads were removed from the marsh surface to improve tidal circulation within the salt marsh. The northern restriction also includes a water control structure in case large amounts of sediment were lost after initial restoration, as predicted by sediment studies.

Monitoring of porewater salinity, groundwater elevation, vegetation, and fish usage was conducted for several summers before and after restoration following common methods (Neckles et al. 2002; Roman et al. 2002). Results confirmed that plant and animal communities were significantly impacted by tidal restrictions and road fill. Prerestoration vegetation monitoring showed that *Phragmites* was the dominant plant species and had replaced native salt marsh vegetation. High water temperatures and reduction in the duration of tidal inundation of the marsh surface limited the use of this historic salt marsh by many fish species prior to restoration. Monitoring following restoration has shown promising results. *Phragmites* height decreased, likely in response to the introduction of saltwater. The area covered by *Phragmites* began to decrease. Native low and high marsh plants continue to colonize the marsh in greater density. The diversity of fish species using the marsh nearly doubled following tidal restoration.

Project partners have developed an adaptive management plan for the site and reviewed the plans with the permitting agency. Nekton monitoring results showed that following tidal restoration only the smallest fish remained in the marsh at low tide, since nearly all the water was drained. In response, the project partners agreed to make changes to the tide gate to allow some water to remain as a refuge for fish and other nekton. Monitoring the following summer showed that larger fish were indeed utilizing the marsh at low tide. A permit was received for cutting and mulching the dense stand of *Phragmites* in the southern cell. Based upon field inspections and monitoring data of the restoration area, it was determined that additional fill needed to be removed adjacent to the new creek to allow for sheet flow across the marsh. The contractor left a berm adjacent to the creek, which was not included in the design plans. The Town used the Rhode Island Department of Environmental Management mosquito abatement coordinator's low-pressure ground equipment to remove the berm, allowing sheet flow of tidal waters to facilitate restoration of native marsh grasses.

Lessons Learned from Rhode Island Restoration Experiences

After a decade of implementing salt marsh restoration projects, the Rhode Island habitat restoration community has gained considerable insight into what makes a project successful. Fundamental keys to success include development of a comprehensive statewide inventory of salt marshes, human impacts, and potential restoration projects that can be used for restoration planning and project prioritization; a consistent and reliable source of state habitat restoration funds that can be used to leverage federal funding; community-based partner organizations that can garner local support for projects; and an adaptive approach to projects that is based on monitoring results.

REFERENCES

Neckles, H. A., M. Dionne, D. M. Burdick, C. T. Roman, R. Buchsbaum, and E. Hutchins. 2002. "A Monitoring Protocol to Assess Tidal Restoration of Salt Marshes on Local and Regional Scales." *Restoration Ecology* 10:556–63.

Roman, C. T., K. B. Raposa, S. C. Adamowicz, M. J. James-Pirri, and J. G. Catena. 2002. "Quantifying Vegetation and Nekton Response to Tidal Restoration of a New England Salt Marsh." *Restoration Ecology* 10:450–60.

Tiner, R. T., I. J. Huber, T. Nuerminger, and A. L. Mandeville. 2003. *An Inventory of Coastal Wetlands, Potential Restoration Sites, Wetland Buffers, and Hardened Shorelines for the Narragansett Bay Estuary*. National Wetlands Inventory Cooperative Interagency Report. Hadley, MA: US Fish and Wildlife Service, Northeast Region.

Tiner, R. W., I. Huber, T. Nuerminger, and A. Mandeville. 2004. *Coastal Wetland Trends in Narragansett Bay Estuary During the 20th Century*. Narragansett Bay Estuary Program Report No. 04-124. Providence, RI: US Fish and Wildlife Service, Northeast Region.

Chapter 10

Restoration of Tidal Flow to Salt Marshes:
The Massachusetts Experience

HUNT DUREY, TIMOTHY SMITH, AND MARC CARULLO

Massachusetts has lost an estimated 41 percent of its precolonial salt marshes (Bromberg and Bertness 2005), and much of the remaining salt marsh is affected by a variety of stressors, including restricted tidal hydrology, anthropogenic eutrophication, and rising sea levels. Today, Massachusetts contains 21,200 hectares of salt marsh (MassGIS 2007; USFWS 2007).[1] Although statewide comprehensive numbers of confirmed tidal restrictions are not available, the state's Wetlands Restoration Program (WRP) has developed atlases for the majority of the state's coastal regions that identify over six hundred potential tidal restrictions. Anecdotal assessment of confirmed and suspected restrictions from these atlases and other sources suggests that over 4000 hectares of existing coastal wetlands are impacted by restricted or blocked tidal flow.

Massachusetts has a long history of stressors on coastal wetlands. For example, in nineteenth-century Boston, which, like most port cities of its time experienced great wealth and expansion, vast areas of salt marsh were destroyed to improve harbor access, expand upland development, and build transportation infrastructure. The most severe alterations occurred in the Boston Harbor region where hundreds of acres were filled to support commercial, residential, and industrial development. Marsh losses continued into the twentieth century at a rapid pace when the Boston Harbor region experienced a 62 percent loss of salt marsh—approximately 1346 hectares from 1893 to 1995—with the greatest losses occurring during the first half of the 1900s (Carlisle et al. 2005). In comparison, Cape Cod and the North Shore lost 23 percent (1860 hectares) and 14 percent (1218 hectares), respectively, during the 1900s (Carlisle et al. 2005). The total salt marsh loss in the Boston region is estimated at a very significant 81 percent (Bromberg and Bertness 2005).

Coastal development pressure remained high in the second half of the twentieth century following the passage of state and federal wetlands protection laws, and overall salt marsh area continued to decline during this period, albeit at a much slower rate. It is important to note, however, that Cape Cod salt marshes declined by nearly 800 hectares during this period, mostly resulting from conversions of salt marsh to upland and brackish/freshwater wetlands (Carlisle et al. 2005). Those conversions are likely the direct result of tidal restrictions, as new roads, cranberry bog dikes, and other infrastructure were built to accommodate Cape Cod's exponential population and land use expansion.

The Massachusetts Wetlands Restoration Program: Program Overview

In 1994, the Massachusetts Executive Office of Environmental Affairs established the Wetlands Restoration Program, one of the first state programs in the country dedicated to proactive wetland restoration. The founding mission of the program was to help people and communities voluntarily restore the state's degraded and former coastal wetlands and the important services they provide. In 2003, WRP was integrated into the state's Office of Coastal Zone Management and focused on the holistic restoration of degraded and former freshwater and tidal wetlands within the state's coastal watersheds. In July 2009, WRP merged with the state's Riverways Program to form a new Division of Ecological Restoration within the Department of Fish and Game. The Division's mission is to restore and protect the Commonwealth's rivers, wetlands, and watersheds for the benefit of people and the environment.

Presently, the state wetland restoration database contains records of over 110 potential, active, and completed salt marsh restoration projects. Projects range in size from less than 1 acre to more than a thousand acres; with costs ranging from less than $20,000 to more than $40,000,000. Ninety-one projects primarily involve restoration of tidal influence. The remaining projects involve other types of restoration actions such as removal or regrading of dredge spoil or other fill from former salt marshes, but these project types are less common, primarily due to higher construction costs relative to tidal restoration.

Salt marsh restoration projects that alleviate tidal restrictions are by far the most common practice in Massachusetts. The majority of these projects involve replacement of undersized or failing road and railroad culverts that historically were designed with little consideration of tidal hydraulics or fish and wildlife passage. Culvert replacement projects involve salt marshes ranging in size from less than 1 to more than a thousand acres. Costs vary widely and are influenced by factors such as the size and type of transportation crossing, traffic intensity, presence of underground or aboveground utilities, low-lying properties or infrastructure, flow volume and velocity, and geotechnical considerations. Nonetheless, culvert replacements are generally considered very cost-effective restoration treatments that provide benefits to many acres of upstream aquatic habitats with relatively straightforward hydraulic modifications in a small construction footprint.

Several Massachusetts towns and other restoration proponents have recently implemented extremely cost-effective measures to improve salt marsh and estuarine habitat conditions by simply changing the management regime of existing tide gates and other water-control devices. In some cases, the gate was no longer necessary or the reason for installing it was long forgotten, and the gate was able to be removed or permanently fixed in a fully open position. Several other tide gates have been structurally modified or replaced to achieve greater tidal influence while maintaining flood protection. More recently, new tide gate designs in the form of self-regulating tide gates have been implemented in several locations to provide automated control of tidal exchange and water levels.

Partnerships are the backbone of successful tidal restoration efforts in Massachusetts. Without the strong and diverse network of federal, state, local, corporate, and nonprofit partners, restoration projects simply would not get done. At its core, the Division of Ecological Restoration is a network-based program that collaborates with many others to help achieve restoration results.

To enhance the state's habitat restoration network, the Division chairs the Partnership to Restore Massachusetts Aquatic Habitats, a group of over thirty-five organizations involved in aquatic habitat restoration across Massachusetts. The Partnership meets biannually and communicates via an e-mail Listserv to discuss all aspects of habitat restoration, including science, policy, planning, permitting, funding, monitoring, and project implementation.

Federal partners bring extensive expertise, technical assistance, and the majority of funding to the table for Massachusetts projects. Close partnerships with federal restoration programs and staff are crucial to leveraging restoration dollars and advancing projects through construction. State and local government partners also play many key roles by assisting with all aspects of project development, permitting, and implementation.

Nonprofit partners contribute significantly to individual projects as well as to policy, political, and strategic planning needs at the state level. Finally, the corporate role in Massachusetts habitat restoration—through the first-in-the-nation Corporate Wetlands Restoration Partnership—has been exceptional and ranges from voluntary contributions of cash and technical services to advocacy for increasing restoration capacity at the state level.

Achievements and Challenges

Since its founding in 1994, the WRP (now part of the Division of Ecological Restoration) has helped partners complete seventy-five projects with 465 hectares of

wetlands under restoration. Over 450 hectares involved restoration of tidally restricted coastal marshes. WRP is currently working with partners to advance over
forty active tidal wetland projects that represent over 1200 hectares of future restoration potential. Working with many partners, the state has made great progress
restoring coastal wetlands, but there is much work that remains to be done. While
WRP has helped partners achieve significant restoration results, the program has
faced many challenges along the way. The most significant challenges—past and
present—include maintaining and enhancing support for wetland restoration and
increasing program capacity.

Coastal wetland restoration is still an emerging field. When WRP was formed
in the early 1990s, wetland restoration was a relatively novel activity. Creation of a
formal state restoration program at that time was a major accomplishment and a
significant advancement for the field as a whole. Since then, the program has
done much to increase public and political support and awareness for restoration, primarily by helping partners produce tangible, on-the-ground results and
benefits.

Massachusetts is in the vanguard of aquatic habitat restoration in New England. Major factors in the state's successful implementation of over seventy projects are the support and facilitation provided to local proponents by the WRP. Salt
marsh restoration projects in Massachusetts, regardless of size, are frequently complicated and typically involve multiple landowners, stakeholders, and jurisdic-
tions. They also often require contributions from multiple funders to cover high
construction costs and must navigate multilayered permitting requirements at the
local, state, and federal levels. Coordination of these numerous elements is often
beyond the capacity of any single entity. One of WRP's primary strengths is its
ability to bring together the partners, expertise, and resources needed to advance
restoration projects from concept to completion.

A few tidal restoration projects in Massachusetts have been problematic, and
the common thread between them has been the incorporation of water control
devices without adequate planning for postconstruction management. These devices include flash boards, stop logs, weirs, and adjustable tide gates designed to
restrict flood tide elevation, limit ebb tide drainage, and/or control sediment transport. In these problematic cases, water control devices were included in restoration project designs, but their management goals, operational protocols, and oversight requirements were never documented or made enforceable.

To avoid similar water control structure problems in the future, WRP is working with local project proponents, regulatory agencies, and other restoration partners to require detailed operational and maintenance plans for any water control
device proposed as part of a tidal restoration project. These plans specifically identify the purpose and objectives of the structure and the roles and responsibilities of

a multistakeholder oversight committee. Operational and maintenance plans for water control structures are now routinely incorporated into local, state, and federal permit conditions and grant contracts.

Site Selection Criteria, Monitoring, and Adaptive Management: Fundamental to Success

Two primary categories of criteria—ecological and practical—drive the evaluation and selection of restoration sites in Massachusetts. Both categories are equally important and must be carefully considered for each potential restoration project. Ecological considerations for tidal restoration projects are usually fairly straightforward and can be summed up as a qualitative cost-benefit analysis of the environmental effects of restoration treatments. Example criteria include the size and severity of wetland degradation, the potential to improve degraded ecological conditions, the trade-offs of converting one habitat type to another, and the effects of restoration treatments on endangered species.

Practical considerations cover an extremely broad and complex array of issues, including social, political, financial, engineering, and aesthetic. These are the issues that most often turn into project "show-stoppers." Examples of some potential show-stoppers include when restoration of tidal range would flood low-lying properties or infrastructure, when historical fill on salt marsh is contaminated, and when removal of a tidal restriction would alter impoundment water levels, causing a major change in the aesthetics and recreational uses of an open water area.

WRP relies on several different means to identify new projects and advance them through the typical phases of restoration. Many projects are brought to the program's attention via an annual call for Priority Project nominations. WRP uses a competitive process through the state's procurement system to solicit nominations of new restoration projects for state funding and technical support. WRP also conducts ongoing internal restoration planning activities using a combination of geographic information system (GIS), historical data sources, and information from local sources and site visits to identify, assess, and prioritize new restoration opportunities. All of these efforts incorporate the ecological and practical site selection criteria discussed earlier. Based on assessment of these criteria and current program capacity, each project is evaluated for acceptance. Once accepted, projects are eligible for technical assistance and grant funding.

Massachusetts restoration partners recognize that projects are not finished when the heavy equipment departs and the final bills are paid. Some practitioners (including WRP) hesitate to refer to projects as "complete," and instead use terms such as "under restoration" to describe postconstruction status. This term recognizes that completion of restoration treatments simply sets in motion a series of complex ecological processes with intended outcomes that have been carefully planned, yet remain uncertain. Long-term commitment to postconstruction monitoring and assessment is crucial to fully understand project trajectories and outcomes, and to inform corrective actions if needed.

Massachusetts salt marsh monitoring is dependent on rigorous and credible chemical, physical, and biological data collection tailored to measure and document systemic responses to restoration actions. WRP has found project monitoring and data management to be one of its biggest challenges. With more than seventy projects completed and many others in development, the effort required to adequately monitor all projects is far beyond the existing resources of both the program and local project sponsors.

To address this, WRP has developed a regional, volunteer-based salt marsh restoration monitoring network. The goal of this network is to enhance and utilize local and regional partner capabilities to help meet project monitoring needs. Since 2003, the program has provided grant funding and technical support to regional nongovernmental organizations that recruit, train, and manage volunteers for field data collection. Uniform data collection protocols have been developed, along with standardized data sheets and data management tools to promote statewide consistency and data transferability. These tools include a proprietary software program for data entry, management, and transfer, and a set of Microsoft Office Excel–based templates for analysis and reporting of salt marsh data.

Funding Alternatives

Massachusetts salt marsh restoration partners have spent well over $30 million since 1994 on over seventy constructed projects. The majority of this funding has come from federal agencies, followed, in descending order of relative contribution, by state, municipal, corporate, and other private investments.

As in other states with significant coastal wetland restoration activity, Massachusetts enjoys strong funding support from its federal partners, most notably the National Oceanographic and Atmospheric Administration Restoration Center, US Fish and Wildlife Service, Natural Resources Conservation Service, and US Army Corps of Engineers. However, the strength of these partnerships is bolstered significantly by sustained funding and project management capacity at the state level. The continued support of WRP staff and investment of state project funding are crucial factors in attracting large infusions of federal restoration dollars into Massachusetts. Since 2003, WRP has received a budget allocation for grants and technical services, and the program uses these resources to hire environmental consultants that perform project development tasks such as field survey, engineering design, and permitting. WRP funds are also dispersed to project sponsors and regional nongovernmental organizations through a competitive grants program for construction-related tasks and project monitoring.

WRP's direct funding for projects is particularly useful for early development tasks, such as initial project feasibility screening and conceptual design, that prepare projects for federal construction grants but that are more difficult to fund through competitive federal sources. In almost every case, WRP staff and funding contributions are also used to help fulfill the nonfederal matching requirements of federal grant programs, greatly leveraging the state's investment.

Several other state programs, such as the Massachusetts Department of Conservation and Recreation and the Department of Fish and Game, also contribute to proactive coastal wetland restoration projects. In addition, cities and towns make significant contributions of cash and in-kind services to projects, especially where elements of the restoration project are closely tied to town infrastructure. These additional state and local contributions provide another important piece of the match needed to secure federal grants.

The Massachusetts Corporate Wetland Restoration Partnership (MA-CWRP) plays a major role in the state's wetland restoration efforts, bringing in significant amounts of private funds and in-kind technical services and conducting important stewardship and public outreach activities. Established in 1999, the MA-CWRP was the first state CWRP in the nation and has been an important partner in advancing the mission of WRP. With over $1.7 million raised so far in cash and in-kind services for Massachusetts projects, the MA-CWRP also provides significant match for federal grants.

Other private funds have been raised from foundations, nongovernmental organizations, and private landowners. A major portion of this funding has come from Massachusetts' two largest private conservation landowners, the Massachusetts Audubon Society and The Trustees of Reservations, both implementing restoration projects on their own properties.

Building a Successful State Wetland Restoration Program: Some Guidance for Others

Developing and growing a stable, long-term state program in a nascent field such as habitat restoration can be challenging, especially with tight budgets and competing priorities. WRP has achieved some success in these important areas by staying focused on its core mission and pursuing the following key priorities:

• Establish and nurture strong partnerships and collaboration with public, private, and nonprofit entities, especially those who have (or could have) financial, political, and/or regulatory influence on restoration efforts.

- Build and maintain program and staff reputations for integrity, professionalism, reliability, and competence.
- Establish solid relationships and reputation with regulators.
- Maximize the leveraging of state investments to bring in local, federal, private, and other funding to support projects.
- Produce successful results and publicize those results with partners, senior agency officials, politicians, and the general public.

The Massachusetts experience has proven without question that successful restoration efforts require active partnerships at the individual project level and, more broadly, sustained collaboration at state and regional levels. Individual projects need to have buy-in and input from many key players, including landowners, neighbors, regulators, government officials, nonprofit organizations, and funders. At the state level, it is very helpful for scientists, regulators, practitioners, and others involved in restoration to get to know each other, communicate on important issues, discuss active and potential projects, and collaborate to advance the field of restoration.

References

Bromberg, K. D., and M. D. Bertness. 2005. "Reconstructing New England Salt Marsh Losses Using Historical Maps." *Estuaries and Coasts* 28:823–32.

Carlisle, B. K., R. W. Tiner, M. Carullo, I. K. Huber, T. Nuerminger, C. Polzen, and M. Shaffer. 2005. *100 Years of Estuarine Marsh Trends in Massachusetts (1893 to 1995): Boston Harbor, Cape Cod, Nantucket, Martha's Vineyard, and the Elizabeth Islands.* Cooperative Report, Boston, MA: Massachusetts Office of Coastal Zone Management; Hadley, MA: U.S. Fish and Wildlife Service; Amherst, MA: University of Massachusetts.

MassGIS DEP Wetlands [computer file]. 2007. Boston, MA: Commonwealth of Massachusetts, Executive Office of Energy and Environmental Affairs, Office of Geographic and Environmental Information (MassGIS). http://www.mass.gov/mgis/.

USFWS National Wetlands Inventory [computer file]. 2007. Washington, D.C.: U.S. Department of the Interior, Fish and Wildlife Service. http://www.fws.gov/wetlands.

Note

1. MassGIS DEP Wetlands and USFWS National Wetland Inventory data were processed by the authors using the methods described in Carlisle et al. 2005. Source dates for these data vary by locale; the most current data available were used.

Chapter 11

Restoration of Tidal Flow to Salt Marshes: The New Hampshire Experience

TED DIERS AND FRANK D. RICHARDSON

Although it has a short coastline, New Hampshire contains two regionally signifi-cant estuaries that support tourism, commerce, and industry. Most of New Hamp-shire's salt marsh area is located in a back barrier estuary (Hampton-Seabrook) that has experienced extensive development for tourism over the past century. Our other significant estuary is a drowned river valley called the Great Bay Estu-ary. Great Bay is less developed and has tidal marshes fringing an extensive series of linked bays and rivers. Along the coastline between our two major estuaries are small marsh-dominated systems landward of barrier beaches.

It has been difficult to determine how much salt marsh exists in New Hamp-shire given the number of small estuarine and fringe marshes, but it appears to be approximately 2400 hectares if all tidal marshlands are included (PREP 2009). This represents a loss of approximately 20 percent of the estimated acreage at the time of settlement (Odell et al. 2006).

Program Development

The New Hampshire Coastal Program encourages and supports wetlands restora-tion projects, and salt marsh restoration efforts in particular. Further review and support are given by the Department of Environmental Services, which requires a Dredge and Fill Permit for all wetlands restoration projects. Proactive wetlands restoration projects, those not associated with compensatory mitigation or en-forcement requirements, are only charged a small fee. The US Army Corps of En-gineers and the Department of Environmental Services have an agreement that assures a thorough review of salt marsh restoration projects by the federal resource agencies.

Tidal restoration began in New Hampshire in the early 1990s with small projects at Awcomin Marsh in Rye and Stuart Farm in Stratham. These small projects not only set the stage for future projects but also, in the case of Stuart Farm, began the creation of monitoring protocols that were used in scientific research to better understand environmental responses in a restoring system. Prior to that, restoration projects were few and uncoordinated. The trend in the state has been for salt marsh projects to include more partners, greater interrelationship with other efforts, and multiresource objectives. Almost all the projects have been proactive and led by a partnership of state, federal, and local government, often coordinated through the coastal zone management agency, the New Hampshire Coastal Program.

New Hampshire has a robust partnership for restoration that includes the Natural Resources Conservation Service, New Hampshire Coastal Program, New Hampshire Department of Environmental Services, US Fish and Wildlife Service, National Oceanographic and Atmospheric Administration, Jackson Estuarine Laboratory, University of New Hampshire, Rockingham County Conservation District, and local communities. Specific projects typically needed local partners. For example, the Little River Project included all the major partners just listed, plus two town governments, New Hampshire Department of Transportation, New Hampshire Fish and Game, Ducks Unlimited, Audubon Society of New Hampshire, and residents of North Hampton. A corporate partner, the Seabrook Station nuclear power plant, was also actively involved in a restoration project.

The hallmark of New Hampshire restoration projects is collaboration. The partners listed earlier have consistently worked to eliminate "turf" issues and to achieve project goals. Meetings are held often with all interested parties and partners to review the design and comment early on recommendations. Oftentimes this will be an onsite visit to familiarize everyone with the proposed location(s) to be followed with a meeting to make whatever revisions may be needed on the site plans. This approach continues through fundraising and on to the final ribbon cutting ceremony, and often follow-up monitoring and site assessment.

Salt marsh restoration has generally been popular with the residents of the coast with few exceptions. This is not to say that the effort has been without controversy. However, the controversies have come down to differences in specific project goals and design rather than whether or not restoration is desirable.

Achievements and Challenges

In 1994, the Natural Resources Conservation Service (NRCS 1994) identified thirty-one sets of tidal restrictions on the coast. Of these, seventeen have been eliminated, restoring approximately 260 hectares of marsh. This calculation includes all the areas upstream of a tidal restriction that was removed or improved. Also included are about 30 hectares of fill removal and mosquito control projects. The Piscataqua Regional Estuaries Partnership (PREP 2009) tracks restoration projects as an environmental indicator and reports that 113 hectares of salt marsh were restored by tidal restriction removal between 2000 and 2008. Prior to 2000 about 150 hectares were restored.

Of the remaining fourteen restrictions, three are in the planning stages and six more are possible but present severe challenges. The remaining five restrictions are probably permanent due to unavoidable flooding of property or infrastructure, political, or historical issues. That said, most people concluded that the restriction at the Taylor River in Hampton was permanent because it is crossed by Interstate 95 and has a dam associated with it. But the tide-restricting culvert was recently identified as problematic by the New Hampshire Department of Transportation, so dam removal and marsh restoration are being seriously examined. This points to the value in having a target statewide restoration plan and longevity in staffing to overcome challenges to restoration.

Limited staff for project management is a key issue. On average, salt marsh restoration projects take three to five years from inception to construction and have many steps (e.g., fund raising, permitting, design, and local decision making). Thus sustained project management is critical. Given the nature of small communities in New Hampshire (primarily governed by volunteer boards who advocate for and manage projects), the need for assistance from state or federal partners to provide long-term institutional support is essential.

Although funding for salt marsh restoration projects may come from a number of sources, the project applicant is most often a local conservation commission. Therefore, effective partnering becomes very important in bringing restoration projects from an idea to implementation. Local conservation commissions in the New Hampshire seacoast region have been proactive in identifying areas in need of restoration as well as working closely with the state and federal resource agencies involved in project development and management.

Contractor selection and oversight constitute another restoration challenge. Usually once a project concept is developed and funding secured, a request for proposals is advertised. Contractors submit proposals and make bids on the estimated cost based on the scope of work. Careful diplomacy is required in the selection process. Selection of the best contractor at the best price can be a difficult process. During economic downturns, contractors might apply for work projects for which they have little or no experience, but "have all the equipment needed to get the job done." Few contractors actually specialize in wetland restoration, and fewer still, in our experience, have expertise with tidal wetlands. The best contractor is the one who listens and is willing to take direction from the wetland scientists, engineers, and regulators who developed the project. At least one project suffered the effects of an overzealous contractor who made incorrect assumptions and decisions, leading to the need for further remediation.

Tidal Restoration Projects in New Hampshire

The first tidal restoration in New Hampshire occurred in 1994 at Stuart Farm in Stratham. This small project not only set the stage for future projects but also began the creation of monitoring protocols and was used in scientific research to better understand environmental responses in a restoring system (Burdick et al. 1997). Since that time, dozens of projects throughout the seacoast and around Great Bay have been implemented. The Little River Salt Marsh Restoration in North Hampton provides a typical example.

The Little River Salt Marsh is a back barrier marsh lying between two rocky headlands. US Department of Agriculture soil maps indicate that originally the marsh was approximately 78 hectares in size. Until recently the original marsh had been reduced to 17 hectares of healthy salt marsh because saltwater exchange had been inadequate for over a hundred years. In 1890, residents installed a large ditch, known locally as "the trunk," through the beach. By 1948, the 6 meter open ditch had been reduced to a 120 centimeter culvert. This culvert proved too small to allow adequate tidal flow into the marsh, leading to two problems: local flooding and loss of salt marsh habitat.

Over the decades, most of the original salt marsh turned into freshwater marsh and shrub swamp dominated by invasive species such as *Phragmites australis* (common reed) due to ponding of freshwater runoff and lack of tidal flushing. In addition, flooding of nearby houses occurred regularly as the small culvert restricted outflow during and after storms. Local residents expressed a keen interest in restoring the marsh to reduce their flooding problems. Preliminary studies indicated that it was possible to restore the marsh and solve flooding problems simultaneously. Hydrologic modeling showed that, for a large storm, a properly sized culvert would only allow water to rise to 2.4 meters as opposed to the 3.5 meter level in the marsh that caused extensive flooding.

In 2001, based on recommendations by the Natural Resources Conservation Service and the US Army Corps of Engineers, the 120 centimeter culvert was replaced with two adjacent 2 by 4 meter box culverts running 76 meters under a state highway and beach cottages (fig. 11.1). Dredging of tidal creeks was needed due to years of siltation and to reduce the expected erosion, and two undersized culverts under other roads crossing the marsh were replaced, to complete the restoration at a total cost of approximately $1.2 million.

FIGURE 11.1. Little River Salt Marsh Restoration Project, 2000. (Photo courtesy of US Army Corps of Engineers).

Currently, tidal flow has been restored to approximately 69 hectares of salt marsh. Approximately 40 hectares of marsh have begun to revert back to salt marsh from shrub swamp. Salinity levels have increased significantly throughout the marsh (fig. 11.2), and systematic monitoring is being conducted to evaluate environmental changes. The primary unintended consequence of the restoration was to dramatically increase the mosquito population in some areas due to a transformation from breeding freshwater to saltwater mosquitoes. There were few open bodies of water big enough to support larvivorous fish, but many small puddles exist, supporting vast numbers of mosquitoes. Mosquito monitoring over the past four years has shown that this situation is extremely dynamic, with mosquito breeding areas moving around the marsh as the vegetation and surface hydrology change. An adaptive management project is under way to improve fish access to some of the breeding areas by improving tidal circulation (restoring and expanding creeks and removing standing dead wood).

New Hampshire projects have had their share of controversy, largely related to the amount of surface water left on the marsh after tidal restoration. The Landing

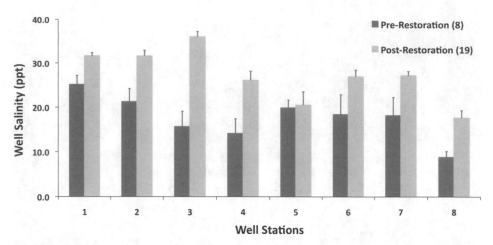

FIGURE 11.2. Pore water salinity measured from wells (5 to 25 centimeter depth) showing means (+ standard deviation) of sampling dates for each station. (Data source: New Hampshire Coastal Program)

Road project in Hampton is a good example. Completed in the mid-1990s, the project removed fill and a roadbed blocking flow to a portion of a salt marsh. A perimeter ditch was dug to better drain road runoff and reduce *Phragmites* growth. Over time a creek formed and expanded, completely draining a prominent salt pool. This pool was one of the original features that the project hoped to improve since it supported a high number of wading birds. As yet, there is no consensus among resource professionals for appropriate actions to restore the pool to a flooded condition.

Similarly, the issues surrounding open marsh water management have been loudly debated. Ten restoration projects have been completed for waterfowl habitat improvement and mosquito control. One of these, Pickering Creek in Greenland, is considered a success in project design and implementation. Originally designed with ditch plugs installed at the main tidal creek (the typical technique), the design was altered by moving the plugs closer to the head of the ditches. In this way the minimum amount of water needed to control mosquitoes was maintained behind the plug rather than the maximum amount of water possible. This was more palatable to most agencies and accomplished the mosquito control goal of the project, as demonstrated by monitoring. The Fairhill Marsh in Rye, on the other hand, was largely a pool creation project. It was designed to create pools on a vegetated high marsh surface in areas of high mosquito breeding and to provide open water habitat for marsh fauna (i.e., insectivorous fish). Anecdotal evidence suggests that vegetated high marsh loss has continued years after project completion due to supersaturation and subsidence of the sediment. While the mosquito control objectives of this project have been met, the impacts to the salt marsh are considered by many to be less desirable.

The controversy around open water marsh management projects led directly to the creation of the New Hampshire Salt Marsh Advisory Team. Composed of regulators, scientists, and consultants, this review team meets periodically to review projects as they move through the design phase. The New Hampshire Coastal Program organizes the meetings around specific projects being considered. This team serves an essential role, making certain that all design elements are considered, including the potential impacts of climate change on restoration projects.

Site Selection Criteria, Monitoring, and Adaptive Management

Initially, little thought was given to site selection priorities; instead restoration project selection was based on opportunity. This began to change around 2006 with the Great Bay Restoration Compendium, which identified and mapped restoration opportunities across four different habitat/species types (Odell et al. 2006). The idea was to look for projects that would have multiobjective restoration potential. The compendium led directly to the 2009 creation of The Partnership to Restore New Hampshire's Estuaries, a formal memorandum of understanding–based organization of the agencies and nongovernmental organizations dedicated to habitat restoration.

The Department of Environmental Services Wetlands Permit for a salt marsh restoration project requires environmental monitoring and photodocumented reports following established protocols for specific environmental parameters. In addition, a reference marsh with similar characteristics to the restoration site is monitored so comparisons can be drawn and the restoration performance evaluated. The salt marsh restoration monitoring protocol used for most projects is the Gulf of Maine Salt Marsh Monitoring Protocol (Neckles et al. 2002). Establishing a reference marsh site in New Hampshire has been problematic since humans have altered almost every significant marsh. Two reference sites are used for the monitoring program: one in Great Bay Estuary and the second in a relatively unimpacted part of a back-barrier marsh. The New Hampshire Coastal Program maintains an inventory of all salt marsh restoration projects, which includes aerial photographs of each site. An attempt is being made to create a long-term photo file for each site that will be perpetuated, thus giving future researchers a view of the various projects over time. Presently, these data are archived at the New Hampshire Coastal Program.

The New Hampshire Volunteer Salt Marsh Monitoring program gives citizens an opportunity to monitor the physical and biological condition of salt marshes.

Volunteers receive classroom and field training in preparation for each monitoring season. Volunteers then work side by side with scientists and managers to collect information on fish, vegetation, salinity, water quality, and groundwater at local salt marsh restoration sites. Scientists are contracted for support and highly technical monitoring. The program began in 2003 and is led by the New Hampshire Coastal Program with help from several partners. A volunteer training packet is provided to each potential volunteer that includes a volunteer monitoring handbook (Drociak and Bottitta 2005) and a field guide (Drociak 2005).

Although we strive to support efforts that restore natural processes to tidal marshes so further human interference is minimized, restoration efforts are not a one-time fix. Even a project deemed successful needs to be observed for potential changes due to erosion or invasive species. Local assistance from volunteer groups and conservation commissions can provide invaluable observations. Also desirable are long-term monitoring plans with contingency funds to address unanticipated and undesirable outcomes. This is a more complex issue than it appears since most project funding timelines are less than five years. A portion of the restoration funding put in escrow might be helpful. Otherwise, projects will have to qualify for follow-up funding required for adaptive responses, which are not as likely to be approved as new proposals. Therefore, the need for follow-up activities must be emphasized in original project proposals.

Funding Alternatives

New Hampshire has been quite fortunate to have a wide variety of funding sources for salt marsh restoration, including the following:

- Coastal Zone Management Act (both Sections 306A and 309), administered by the National Oceanographic and Atmospheric Administration Office of Ocean and Coastal Resource Management. New Hampshire Coastal Program also used Section 309 funds for project management (one staff person for the past nine years).
- Direct congressional appropriation within the National Oceanographic and Atmospheric Administration appropriations committee.
- Environmental Protection Agency Section 319: Nonpoint Source Program.
- Department of Agriculture: Wetland Reserve Program and Wildlife Habitat Incentive Program of the Natural Resource Conservation Service. Alan Ammann, a wildlife biologist with Natural Resources Conservation Service, was the first person in the country to convince the Service that, because salt marsh haying was an agricultural activity, salt marsh restoration was eligible for funding. Equally important, that designation allowed the Service engi-

neering staff to survey and design projects, representing a significant and essential contribution.

- US Fish and Wildlife Service through the North American Wetlands Conservation Act and help from local Fish and Wildlife Service staff through Partners for Wildlife.
- Local funding: Seacoast communities have contributed to many restoration projects through conservation funds and specific appropriations. For example, local residents contributed over $70,000 as nonfederal match for the Little River restoration.

Lessons Learned, Guidance for Building a Restoration Program

The history of tidal restoration in New Hampshire is rich and extends over fifteen years. Over 10 percent of the marshes in the state have been restored or improved. The state is nearing completion of restoring tidal flow to all salt marshes where tidal restoration is practical. The following are some of the important lessons we have learned over our years of salt marsh restoration:

- Seek out and cultivate diverse partnerships.
- Local support is critical.
- Build a diverse project team that meets throughout the project, from early planning to monitoring.
- Support from the public and funding agencies is easiest with multipurpose restoration objectives.
- Nontraditional partnerships, such as with the state Department of Transportation, are often quite fruitful.
- Design of projects must be well conceived but with an eye toward unfolding field conditions and the ability to make adjustments (adaptive management).
- Celebrate success!

REFERENCES

Burdick, D. M., M. Dionne, R. M. Boumans, and F. T. Short. 1997. "Ecological Responses to Tidal Restorations of Two Northern New England Salt Marshes." *Wetland Ecology and Management* 4:129–44.

Drociak, J. 2005. *Life in New Hampshire Salt Marshes: A Quick Reference Guide*. NH Department of Environmental Services publication no. WD-04-19.

Drociak, J., and G. Bottitta. 2005. *A Volunteer's Handbook for Monitoring New Hampshire's Salt Marshes*. NH Department of Environmental Services publication no. WD-04-21.

Natural Resources Conservation Service (NRCS). 1994. *Evaluation of Restorable Salt Marshes in New Hampshire.* Durham, NH: US Department of Agriculture.

Neckles, H. A., M. Dionne, D. M. Burdick, C. T. Roman, R. Buchsbaum, and E. Hutchins. 2002. "A Monitoring Protocol to Assess Tidal Restoration of Salt Marshes on Local and Regional Scales." *Restoration Ecology* 10:556–63.

Odell, J., A. Eberhardt, P. Ingraham, and D. Burdick. 2006. *Great Bay Estuary Restoration Compendium.* Report to the NH Coastal Program and the NH Estuaries Project. Arlington, VA: The Nature Conservancy.

Piscataqua Region Estuaries Project (PREP). 2009. *Piscataqua Region Estuaries Project.* State of the Estuary Report. www.prep.unh.edu/resources/pdf/2009_state_of_the-prep-09.pdf.

Chapter 12

Restoration of Tidal Flow to Salt Marshes:
The Maine Experience

JON KACHMAR AND ELIZABETH HERTZ

Maine is associated with rocky coastal features, but there are several large salt marsh complexes along its expansive coastline, including the Webhannet/Little River (1600 hectares) and Scarborough Marsh (1000 hectares) complexes in southern Maine (Taylor 2008). More than two thirds of the total area of salt marsh in Maine (7900 hectares) occurs in the southern region, with marshes developing behind barrier systems and in sheltered estuaries (Jacobson et al. 1987). Fringing marshes are perhaps a dominant salt marsh type in Maine, bordering the rock and cliff shores within protected areas (Jacobson et al. 1987; Morgan et al. 2009).

The types of human impacts to Maine's marshes are similar to impacts seen throughout New England, such as ditching, diking, filling, tidal flow restrictions, invasive species, hardening of the upland boundary from development, and increased freshwater runoff from impervious surface (Taylor 2008). Undersized culverts, tide gates, and other structures restricting tidal flow and diminishing the overall ecological health of salt marshes occur throughout Maine's coast (Bonebakker et al. 2000; Crain et al. 2009). In southern Maine, a region experiencing increased development pressure along the coast, 28 percent (902 hectares) of the salt marshes are threatened by tide restrictions (Crain et al. 2009). This chapter reviews the efforts in Maine aimed at restoring tide-restricted salt marsh ecosystems and discusses the many achievements and challenges.

Program Development and Coordination

Maine has no formal habitat restoration program within state government for developing, implementing, or monitoring salt marsh restoration projects; however, the Maine State Planning Office hosts a senior planner to act as a restoration coordinator for the State of Maine as well as other projects within the Gulf of Maine region. This position is also partially supported by the Gulf of Maine Council on the Marine Environment and the National Oceanic and Atmospheric Administration's (NOAA's) Restoration Center.

A partnership between the Gulf of Maine Council and the NOAA Community-Based Restoration Program supports much of the salt marsh restoration undertaken in Maine. The Gulf of Maine Council/NOAA Habitat Restoration Partnership has provided over $3.5 million in funding to ninety-four projects in the Gulf of Maine watershed (Maine, Massachusetts, New Hampshire, New Brunswick, Nova Scotia) since 2002, leveraging nonfederal funds and in-kind match. Over 35 percent of these Partnership funds have been committed to salt marsh restoration and monitoring projects in Maine. Grant recipients are primarily municipalities and nonprofit organizations, with restoration methods including replacement of undersized culverts, removal of fill, and modification or removal of dams.

Project selection for community-based restoration projects through the Partnership is based on review of four criteria: (1) potential for the project to restore degraded habitat to a self-sustaining, pre-degraded condition; (2) technical merit and project feasibility; (3) partnership/cooperation from community groups and other organizations; and (4) cost effectiveness and budget detail and consideration.

To determine the types of restoration projects that will contribute to meaningful habitat restoration projects at a regional scale throughout the Gulf of Maine, the Gulf of Maine Council/NOAA Habitat Restoration Partnership developed the *Gulf of Maine Habitat Restoration Strategy* (Gulf of Maine Council Habitat Restoration Subcommittee 2004). The strategy offers a regional approach to habitat restoration, providing the opportunity to address common restoration goals and objectives that encompass several US states and Atlantic Canada. The strategy identifies habitat targets for Gulf of Maine salt marshes and identifies regionally significant restoration projects, including a 200 hectare restoration of tide-restricted marsh associated with the West Branch Pleasant River, Addison, Maine.

Achievements and Challenges

Without a centralized restoration program, much of the restoration that has taken place in Maine has resulted from an opportunistic rather than a strategic process. Even so, important steps have been taken in the identification and restoration of Maine's salt marshes.

Inventories and Site Selection Criteria

To begin the essential process of developing a statewide inventory of restoration projects, the Maine State Planning Office contracted restoration inventories for four coastal watersheds, including three watersheds in southern Maine and one in the midcoast region. The intent was twofold: to develop a system of ranking restoration projects with existing capacity for on-the-ground implementation, and to develop a methodology for inventorying all of Maine's seventeen coastal watersheds as designated by the Maine Department of Environmental Protection.

The Habitat Restoration Inventories identify restoration potential in all water bodies and adjacent riparian areas, screen and prioritize restoration opportunities, and develop a database for storing information. An online, interactive database that allows users to search for restoration projects is available at http://restoration.gulfofmaine.org/. The inventory identifies areas that could potentially be enhanced through habitat restoration. Sites are categorized by town, water body, habitat, source of degradation, area of habitat affected, type of restoration needed, cost rank, and project status. This database is fundamental to selecting and prioritizing sites for restoration.

Starting in 2007, the US Fish and Wildlife Service (USFWS) in cooperation with other state and federal agencies began to inventory road–stream crossings in an effort to identify barriers to stream connectivity (Abbott 2008). More than 4000 sites were visited, with severe barriers to aquatic organism passage/connectivity noted for about 40 percent of the sites. A simple extrapolation based on multiplying that percentage (40 percent) by the number of road–stream crossings within the historical range of diadromous fish in Maine (23,664) yields an estimate of over 9229 severe barriers.

Environmental Monitoring

Maine salt marsh restoration projects are encouraged to include a monitoring component to assess project effectiveness. The Gulf of Maine Council/NOAA partnership recommends using the salt marsh monitoring protocols developed for the Gulf of Maine region (Neckles and Dionne 2000; Neckles et al. 2002). Core variables to be monitored include hydrology (tidal signal and surface elevations), soils and sediments (pore water salinity), vegetation (abundance, composition, height of species of concern, stem density of species of concern), nekton (density, length, biomass, species richness), and birds (abundance, species richness, feeding and breeding habits). In conjunction with the core variables, there are several additional variables that may be implemented if deemed necessary.

To evaluate compliance with this monitoring approach and effectiveness, a study was conducted to assess thirty-six salt marsh restoration projects in the Gulf of Maine region, including eight projects in Maine (Konisky et al. 2006). This study revealed that adherence to the standard methods was sometimes marginal, thereby hampering regional comparisons, but practitioners were clearly supportive of using standard monitoring protocols, and it was expected that consistent application of the protocols will be evident in future monitoring.

Successful Salt Marsh Restorations

There have been a wide variety of salt marsh restoration projects in Maine, including ditch plugging, culvert replacement, tide gate installation, and fill removal, to restore form and function to these important coastal habitats. Each of the tidal restoration projects highlighted here showcases certain elements that helped ensure their success. The Pemaquid Culvert Replacement is an example of how important a "local champion" is to successful completion of a project that is many years in the making. Sherman Marsh is a unique example, where capitalizing on an unexpected event resulted in an important restoration project.

PEMAQUID CULVERT REPLACEMENT

The Pemaquid Marsh restoration in the town of Bristol, located in midcoast Maine, took place in summer 2005 consisting of replacement of corrugated steel culverts that were undersized and had been crushed; considerably restricting tidal flow. Prior to restoration the 2.4 hectare tide-restricted marsh was ecologically degraded, and further, the integrity of the road infrastructure was threatened by the crushed culverts. The culverts were replaced with an 8 foot by 10 foot concrete box culvert with precast sections, sized based on hydrologic modeling to entirely remove the tidal restriction. This project successfully met dual objectives: transportation infrastructure maintenance and environmental restoration.

The Pemaquid project was primarily funded with monies and labor match from the town, a Gulf of Maine Council/NOAA Habitat Restoration Partnership grant, and additional support from the Maine Corporate Wetland Restoration Partnership. In-kind state technical assistance resources were provided by the Maine State Planning Office. This partnership of local, state, federal, and private resources successfully restored adequate tidal flow to the marsh to sustain the functions and values of the habitat.

The project's origins began many years earlier with a local resident overlooking the marsh. Her persistence as a local "champion" for the health of the marsh—

having seen the resource degrade over time—was instrumental in building local support from town government as well as residents surrounding the marsh. The typical perception that culverts cannot be replaced with a larger pipe in estuarine areas due to unintended flooding was abated through modeling and early education with landowners abutting the marsh.

Sherman Marsh

Formerly known as Sherman Lake, Sherman Marsh was originally a tidal marsh that was impounded in the 1930s in Newcastle, Maine. The impounded lake encompassed about 80 hectares until October 2005, when the existing dam failed during a storm event. The lake drained of freshwater within hours, and tidal flow was reintroduced. The Maine Department of Transportation owned the dam structure under busy Route 1, and the failure of the infrastructure was an immediate threat to the stability of the roadway. After several public forums, the Department of Transportation decided not to rebuild the dam and to rebuild the bridge infrastructure in a manner to minimize the impact on tidal flow, thereby allowing a full tidal restoration of the marsh. The state agencies with purview over freshwater resources (Department of Inland Fisheries and Wildlife) and marine resources (Department of Marine Resources) both agreed that the dam breach offered a unique opportunity to restore native salt marsh habitat. This project was precedent setting in that it created an unplanned tidal flow restoration.

Challenges

Challenges to salt marsh restoration in Maine fall into two main categories: coordination and capacity. Most salt marsh restoration in Maine has been opportunistic rather than strategic. Lack of a salt marsh restoration program at the state level to identify restoration needs and prioritize those needs in a coastwide assessment has created a mission-driven approach to salt marsh restoration rather than an ecosystem approach.

Salt marsh restoration in Maine has been driven primarily by the US Fish and Wildlife Service Gulf of Maine Program and the NOAA National Marine Fisheries Service. These federal agencies have committed substantial resources to project identification, development, and implementation. However, their efforts would be greatly enhanced if a statewide strategic salt marsh restoration plan were available that outlined restoration priorities and techniques.

As previously noted, the Gulf of Maine Council/NOAA partnership program provides partial funding for a habitat restoration coordinator in the Maine Coastal

Program at the State Planning Office, but securing the remaining funds for this position is an ongoing challenge. The lack of state funds carries over to the needed nonfederal project match, which puts the burden of finding matching funds fully on the project proponents—frequently nonprofit citizen or watershed groups. Nonfederal funding is particularly important for leveraging federal grant monies that typically require a 1:1 match of federal to nonfederal dollars. Without adequate match, federal grants are difficult to secure. In particular, nonfederal cash match, rather than in-kind services, results in a stronger application when applying for competitive grant sources.

Lack of funding at all levels has made it difficult to carry out a coastwide assessment and prioritization of salt marshes and restoration needs. This gap has reinforced a program-driven, opportunistic approach to salt marsh restoration.

Role of Partners

Partnering on projects is an essential aspect to successful restoration projects. There is no single entity with the capacity in human resources or in funds to implement projects. Partnering creates stronger projects by developing buy-in and support for project implementation.

Federal partners bring funds, technical expertise, and the skills necessary to implement complex restoration projects. The major federal partners on salt marsh restoration projects in Maine include the USFWS Gulf of Maine Program, USFWS Rachel Carson National Wildlife Refuge, NOAA's Restoration Center (Gloucester, MA), often in partnership with the nonprofit Gulf of Maine Council, the Natural Resources Conservation Service, and the US Army Corps of Engineers (New England District).

State partners bring local, site-specific expertise and on-the-ground knowledge. Agency personnel have the historical site knowledge necessary when developing a restorative solution. State funds brought by state agency partners in Maine have been and will continue to be limited. The support from state agencies is typically in the form of in-kind staff time to plan, design, or implement restoration projects. State agencies in Maine provide additional support for salt marsh restoration projects as opportunities arise, often working in conjunction with the Maine State Planning Office restoration coordinator for project planning and funding. The Maine Department of Environmental Protection, Maine Department of Marine Resources, Maine Department of Inland Fisheries and Wildlife, and Maine Department of Transportation's Environmental Office have all supported salt marsh restoration projects.

Local governments in Maine play an important role in restoration projects. They have served as project proponents, securing grants and applying for appropriate permits, and have provided in-kind support in the manner of equipment, time, and supplies.

Local project champions and volunteers passionate about environmental stewardship actively participate in project planning, acquiring matching funds, and collecting essential monitoring data. Local or regional partners in Maine have included watershed and friends groups, and nonprofit groups like Ducks Unlimited, Trout Unlimited, and The Nature Conservancy.

The Maine Corporate Wetland Restoration Partnership, part of the national Coastal America program, was created in 2000. It is a public–private partnership focused on restoring wetlands, river systems, and other aquatic habitats made up of state and federal officials, private businesses, and other environmental organizations. This program provides critically important nonfederal project matching funds from member businesses—in addition to pro bono services such as engineering design, legal services, and media relations. While Corporate Wetland Restoration Partnership funds have been a relatively small percentage of total wetland restoration funding in Maine, these cash and in-kind contributions have leveraged much larger projects into the implementation phase.

Lessons Learned for Others Building Restoration Programs

Maine has almost 6000 kilometers of tidally influenced coast and about 79 square kilometers of salt marsh (Jacobson et al. 1987). Restoration opportunities are prevalent along the Maine coast, especially in more-developed southern Maine, where nearly one third of the salt marshes are degraded by restricted tidal flow (Crain et al. 2009). Salt marsh restoration is moving forward in Maine with substantial support from NOAA, USFWS, Natural Resources Conservation Service, the Army Corps of Engineers, and the Gulf of Maine Council/NOAA Habitat Restoration Partnership, along with dedicated commitments from the Maine State Planning Office, other state agencies, local government, nonprofit organizations, and volunteer citizens. However, salt marsh restoration efforts in Maine would greatly benefit from a state program that is dedicated to all aspects, including statewide inventory of restoration sites and priority setting, coordinating grant writing, securing funds, promoting partnerships, designing and implementing projects, and monitoring and assessing projects pre- and postrestoration.

REFERENCES

Abbott, A. 2008. *Maine Road-Stream Crossing Survey Manual*. Falmouth, ME: Gulf of Maine Coastal Program, US Fish and Wildlife Service. http://www.state.me.us /doc/mfs/fpm/water/docs/stream_crossing_2008/MaineRoad-StreamCrossingSurvey Manual2008.pdf.

Bonebakker, E. R., P. Shelley, and K. Spectre. 2000. *Return the Tides Resource Handbook*. Rockland, ME: Conservation Law Foundation. http://www.transitterminal.com /wetlands/RTTRESBK%20FRONT.pdf.

Crain, C. M., K. B. Gedham, and M. Dionne. 2009. "Tidal Restrictions and Mosquito Ditching in New England Marshes; Case Studies of the Biotic Evidence, Physical Extent, and Potential for Restoration of Altered Hydrology." Pp. 149–70 in *Human Impacts on Salt Marshes: A Global Perspective*, edited by B. R. Silliman, E. D. Grosholz, and M. D. Bertness. Berkeley: University of California Press.

Gulf of Maine Council Habitat Restoration Subcommittee. 2004. *The Gulf of Maine Habitat Restoration Strategy*. Gulf of Maine Council on the Marine Environment. http://www2.gulfofmaine.org/habitatrestoration/documents/HabitatRestoration StrategyFinal.pdf.

Jacobson, H. A., G. L. Jacobson Jr., and J. T. Kelley. 1987. "Distribution and Abundance of Tidal Marshes along the Coast of Maine." *Estuaries* 10:126–31.

Konisky, R. A., D. M. Burdick, M. Dionne, and H. A. Neckles. 2006. "A Regional Assessment of Salt Marsh Restoration and Monitoring in the Gulf of Maine." *Restoration Ecology* 14:516–25.

Morgan, P. A., D. M. Burdick, and F. T. Short. 2009. "The Functions and Values of Fringing Salt Marshes in Northern New England, USA." *Estuaries and Coasts* 32:483–95.

Neckles, H. A., and M. Dionne, eds. 2000. *Regional Standards to Identify and Evaluate Tidal Wetland Restoration in the Gulf of Maine*. Wells National Estuarine Research Reserve Technical Report. Laurel, MD: US Department of the Interior, US Geological Survey, Patuxent Wildlife Research Center. http://www.pwrc.usgs.gov/resshow /neckles/gpac.htm.

Neckles, H. A., M. Dionne, D. M. Burdick, C. T. Roman, R. Buchsbaum, and E. Hutchins. 2002. "A Monitoring Protocol to Assess Tidal Restoration of Salt Marshes on Local and Regional Scales." *Restoration Ecology* 10:556–63.

Taylor, P. H. 2008. *Salt Marshes in the Gulf of Maine: Human Impacts, Habitat Restoration, and Long-Term Change Analysis*. Gulf of Maine Council on the Marine Environment. http://www2.gulfofmaine.org/saltmarsh/.

Chapter 13

Salt Marsh Tidal Restoration in Canada's Maritime Provinces

TONY M. BOWRON, NANCY NEATT, DANIKA VAN PROOSDIJ, AND
JEREMY LUNDHOLM

Salt marshes form an important component of the coastal landscape of the Canadian Maritimes. The characteristics of salt marshes are determined by a wide range of physical and biological controls. The coastal zone of the Maritime Provinces (Nova Scotia, New Brunswick, and Prince Edward Island) exhibits a diverse geologic and sea level history, sediment supply, tidal amplitude (micro- to macrotidal), and varying exposure to wave energy. These factors contribute to the development of three distinct biophysical regions of salt marsh: Bay of Fundy, Atlantic Coastal, and Gulf of St. Lawrence/Northumberland Strait (fig. 13.1; Hatcher and Patriquin 1981; Roberts and Robertson 1986; Wells and Hirvonen 1988). Most recent estimates indicate that there are approximately 287 square kilometers of salt marsh in the Maritimes (table 13.1; Hanson and Calkins 1996; Mendelsohn and McKee 2000). The majority of this (54 percent) occurs along the coast of Nova Scotia.

Biophysical and Geographical Setting

The Maritime Provinces encompass three distinct coastal regions, each characterized by different hydrology, geomorphology, and land use history.

Bay of Fundy

The Bay of Fundy includes the shore zones of both New Brunswick and Nova Scotia. Salt marshes located in this region cover an area of 153 square kilometers, accounting for 53 percent of the total marsh area in the Maritime region (Hanson and Calkins 1996; Neily et al. 2003). It is a high macrotidal estuary with

Salt marsh area in the Canadian Maritime Provinces

Province	Length of coastline (km)	Area of salt marsh (km^2)	% of total Maritime Province salt marsh area	Source
Nova Scotia (NS)	7578	154	54	Wells and Hirvonen 1988; Hanson and Calkins 1996; Neily et al. 2003
New Brunswick (NB)	2269	85	30	Hanson and Calkins 1996
Prince Edward Island (PE)	1107	48	17	Hanson and Calkins 1996

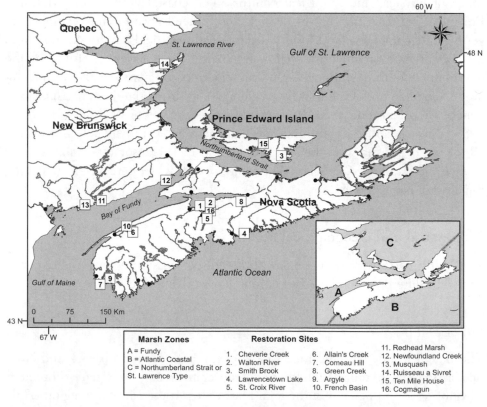

FIGURE 13.1. Location of salt marsh restoration projects within the Maritime Provinces.

semidiurnal tides that may exceed 16 meters at the head of the bay (Desplanque and Mossman 2004). Suspended sediment concentrations are high, ranging from 50 to 300 mg·l⁻¹ over the marsh surface (Gordon and Cranford 1994; van Proosdij et al. 2006a). This inorganic material contributes to the high sedimentation rates and elevation change recorded on marshes within the region (van Proosdij et al. 2006a,b) and their minerogenic nature.

Since the 1630s, salt marshes in the Fundy region have been diked primarily for agricultural use. It was estimated that prior to European settlement there were approximately 395 square kilometers of salt marsh in the region (Gordon and Cranford 1994). Compared to current wetland inventory estimates, 69 percent of salt marsh habitat was lost due to diking (Hanson and Calkins 1996). Currently, in Nova Scotia, 174 square kilometers of former marshland are protected by 241 kilometers of dikes (with 260 aboiteaux) (Milligan 1987). Today, Fundy marshes exposed to tidal flow occur in small pocket estuaries at river mouths or extend laterally along the edge of the dikes or the shoreline. In some areas new marsh has developed on accreting intertidal flats as a result of causeway construction (van Proosdij et al. 2009). In others, dikelands have reverted back to salt marsh where dikes have failed and have not been repaired or replaced (Milligan 1987; Hanson 2004). Restoration potential is high since large areas can be restored by removal of a single dike, causeway, or road structure.

Atlantic Coast

Salt marshes of the Atlantic Coastal Region primarily occupy the eastern shore of Nova Scotia and contain approximately 13 percent of total salt marsh area in the Maritimes (Hanson and Calkins 1996). This coast is exposed to high wave energy and has a tidal range of 2 meters (Wells and Hirvonen 1988). Marshes along this coastline are restricted to small pocket wetlands in protected areas or are part of a few large complexes associated with estuaries (Chagueé-Goff et al. 2001) and sandy or gravel barriers.

Historically salt marshes in this region were hayed and grazed without the use of dikes. The drowned coastline of the Atlantic Coast results in the land rising sharply from the shoreline in most areas, resulting in little infilling of salt marsh for construction of human infrastructure (Hanson 2004). Most restoration opportunities arise due to inadequate culvert size or placement.

Gulf of St. Lawrence/Northumberland Strait

This is a low-energy system. Tides range from 2 to 4 meters with mixed components of semidiurnal and diurnal influences, which can exert a profound influence on the coastal geomorphology of the region. For example, in the western section, tides are mainly diurnal with a period of twenty-five hours; on some days the tide can remain high for twelve hours, increasing the inundation period (Davis and Browne 1996). The Gulf of St. Lawrence consists of low-elevation plain and receives sandy sediments from the numerous barrier islands, dunes, and lagoons. Approximately 34 percent of the total salt marsh area falls within this region. Developmental pressures are high, with frequent infilling of salt marshes and alteration of adjacent habitat. Coastal marshes were hayed and grazed without the use of dikes, although some were ditched to drain salt pannes and ponds to create drier soils for livestock and equipment (Hanson 2004). The combination of relatively low land elevations, intensive coastal zone development, and erosive soils makes this area highly susceptible to sea level rise damage (Shaw et al. 1998).

Climate and Vegetation

The climate of the Maritimes falls within the humid continental type according to the Koppen Climate Classification system. As a result, salt marshes in this area are subject to ice and snow for at least four months of the year. Ice is important for local sediment transport, particularly into the high marsh region (van Proosdij et al. 2006b), and transport of rhizome material.

Salt marshes in this region are usually classified into high and low marsh, with low marsh being lower in species diversity and dominated by *Spartina alterniflora* (smooth cordgrass) (Hanson and Calkins 1996). Environmental conditions and vegetation zonation are generally similar to those in New England salt marshes (McKee and Patrick 1988). In the low marsh, as *S. alterniflora* tufts are removed by erosion or ice scour, *Salicornia* spp. (glasswort), *Sueda maritima* (sea-blite), *Atriplex* spp. (orach), and others colonize the bare patches. Pannes and ponds can contain stunted *S. alterniflora* and other short halophytic species such as *Salicornia* spp. and *Sueda maritima*, with *Ruppia maritima* (widgeongrass) and/or *Zostera marina* (eelgrass) in deep areas (Hatcher and Patriquin 1981; Chmura et al. 1997).

High marsh areas are dominated by *Spartina patens* (salt meadow cordgrass), with more potential subordinates than the low marsh, including *Pucinellia maritima* (seashore alkali grass), *Plantago maritima* (seaside plantain), *Triglochin maritima* (seaside arrow grass), *Solidago sempervirens* (seaside goldenrod), *Limonium carolinianum* (sea lavender), *Juncus gerardii* (black rush), *Hierochloe odorata* (sweet grass), *Glaux maritima* (sea milkwort), *Sueda maritima*, *Distichlis spicata* (spikegrass), and *Atriplex patula* (marsh orach) (Ganong 1903; Roberts and Robertson 1986; Hanson and Calkins 1996; van Proosdij et al. 1999). In Nova

Scotia Atlantic coast high marshes, the dominant is more often *Carex paleacea* (salt marsh sedge) (Hatcher and Patriquin 1981).

The presence of *Juncus balticus* (baltic rush) and *Puccinellia maritima* in the region is representative of a northern element not present south of the border (Chmura et al. 1997); otherwise, salt marsh vegetation is similar to those species found to the south in New England (Jacobson and Jacobson 1989; Boumans et al. 2002; Morgan and Short 2002; Roman et al. 2002; Crain et al. 2004). Although Adam (1990) considers Maritime salt marshes to belong to the boreal type, low marshes in the region are dominated by *S. alterniflora* and are thus more similar to the northern subtype of West Atlantic marshes that occur from Maine to Massachusetts.

Marshes in the Maritimes differ from those further south in the United States in that, even with tidal restrictions, the northern marshes have very little *Phragmites australis* (common reed). Nonnative, invasive forms of this grass are present in New Brunswick and Nova Scotia but not Prince Edward Island, are not abundant, and only occur at a few locations (Catling et al. 2004). It is not considered a direct threat to most salt marshes in the region, but existing populations are spreading with increased road construction and human disturbance (Catling et al. 2004) and should be monitored. While salt marshes in New England are characterized as having less low marsh (Argow and FitzGerald 2006) than those further south, macrotidal areas around the Bay of Fundy have large areas of low marsh. Some authors suggest that ice scour in northern New England results in smaller low marsh area (Ewanchuck and Bertness 2004), but ice is also present in the Bay of Fundy, which can have large areas of low marsh. Vertical range of both high and low marsh dominants increases with tidal range in the Bay of Fundy (Byers and Chmura 2007); thus the macrotidal situation may promote much greater low marsh coverage than in micro- and mesotidal regions, despite the disturbance presented by ice in these systems. Gaps in our knowledge of Maritimes salt marsh ecosystems include the role of ice as a disturbance and as a dispersal agent for plants, population differentiation in the dominant *Spartina* species, and the regional distribution of potential invaders, especially *Phragmites* and *Lythrum salicaria* (purple loosestrife).

Salt Marsh Restoration Activities in the Maritimes

Tidal wetlands are valued ecosystems, and their restoration has become a common practice in the Gulf of Maine over the past two decades (Short et al. 2000; Neckles et al. 2002). Conversely, salt marsh restoration is still in its infancy in Atlantic Canada. In Newfoundland and Labrador, which have not experienced the high rates of degradation or loss common to the other three Atlantic Provinces, salt marsh restoration does not exist. Therefore, this chapter focuses only on the Maritime Provinces. The majority of the intentional salt marsh restoration projects that have occurred in the region have been opportunistic and compensatory (required by legislation) in nature (reactive, rather than proactive).

Much of the salt marsh restoration activity to date in the Maritimes has focused on the marshes in the Bay of Fundy, the area of greatest historical loss. The unique hydrological and sediment conditions experienced in the Bay, combined with the historical significance of these marshes (dikelands), the severity of habitat loss (69 percent), the importance for migratory and endangered species, and importance as a significant component of the Gulf of Maine ecosystem have served to focus restoration efforts.

Prior to 2005, activities that resulted in the restoration of salt marsh habitat were either infrastructure development projects that resulted in the unanticipated establishment or redistribution of salt marsh (i.e., Windsor Causeway, Windsor, Nova Scotia; van Proosdij et al. 2009), or decommissioning projects conducted by nongovernmental organizations such as Ducks Unlimited Canada (i.e., removal of water control structures on freshwater impoundments built on tidal wetlands) and government departments such as the Nova Scotia Department of Agriculture (table 13.2). Projects with the primary goal of restoring salt marsh have only been undertaken since 2005, and of those projects, few have involved long-term comprehensive monitoring programs. To date there is no central agency or guiding program at either the provincial or the federal level directing salt marsh restoration efforts.

Restoration Methods

Salt marsh ecosystem function may be restored passively when a dike is breached during a storm with little to no human interference (Crooks et al. 2002), when a management decision is made to suspend maintenance of a dike and allow an area to "go out to sea" (Milligan 1987), or through active means by planned removal or modification of a barrier to restore hydrology (Klötzli and Grootjans 2001; van Proosdij et al. 2010). Active restoration of tidal wetlands in the Maritimes has involved installation of properly sized and placed culverts or bridges; the removal of aboiteaux, tide gates, or water control structures; and the partial or complete removal of causeways or dikes (table 13.2). For most of these projects, restoration efforts have focused on the reduction or elimination of the primary restriction to hydrology and have relied on natural processes to restore the native flora and fauna.

Given the history of agricultural use of many of the salt marshes in the region, particularly in the Bay of Fundy, a majority of marshes were heavily ditched,

TABLE 13.2

Tidal marsh restoration projects in Canadian Maritime Provinces

Marsh name	Marsh area (ha)[1]	Anticipated ecosystem	Date completed[2]	Dike breach	Culvert/aboiteau replaced	Creation	Other	Creek excavation	Panne creation	Pre	Post	Vegetation	Soils and sediments	Fish	Invertebrates	Birds	Hydrology	Geomorphology	Use of reference site
						Restoration type				Monitoring years					Monitoring program				
Nova Scotia																			
Cheverie	43	Salt marsh	2005		X					4	5	X	X	X	X	X	X	X	Y
Walton	9.3	Salt marsh	2005	X				X		1	5	X	X	X	X	X	X	X	Y
Smith Gut	2.3	Salt marsh	2006		X					1	5	X	X	X	X		X	X	Y
Lawrencetown	1.5	Salt marsh	2007		X					1	5	X	X	X	X		X	X	Y
St. Croix	19	Tidal brackish	2009	X		X		X	X	1	5	X	X	X	X		X	X	Y
Cognagun	4.8	Salt marsh	2009	X		X			X	1	5	X	X	X	X		X	X	Y
Allain's Creek	14.2	Salt marsh	2000	X						1	5	X	X	X	X		X	X	N
Comeau Hill	10.1	Salt marsh	2006/1989	X								X						X	
Green Creek	26	Salt marsh	Early 1980s	X															N
Arglye	20	Salt marsh	Early 1980s	X															N
French Basin	1.6	Tidal brackish	N/A[3]	X						1		X	X	X	X	X	X		Y

197

TABLE 13.2

Continued

Marsh name	Marsh area (ha)[1]	Anticipated ecosystem	Date completed	Restoration type						Monitoring years		Monitoring program							Use of reference site
				Dike breach	Culvert/aboiteau replaced	Creation	Other	Creek excavation	Panne creation	Pre	Post	Vegetation	Soils and sediments	Fish	Invertebrates	Birds	Hydrology	Geomorphology	
New Brunswick																			
Redhead marsh	71.2	Salt marsh/tidal brackish	2005	X						1	4	X		X		X			N
Newfoundland Creek	78.5	Salt marsh	2001	X						1	5	X							
Musquash	15.4	Salt marsh	2005	X						1	3	X	X		X	X	X	X	Y
Ruisseau à Sivret	16.2	Salt marsh	2005				X	X											
Prince Edward Island																			
Ten Mile House	42.5	Salt marsh	2002	X						1	5	X				X			N

[1]Postrestoration.

[2]Date completed refers to the year that the restoration (earthworks) occurred.

[3]Feasibility study completed March 2008, construction work not yet undertaken.

Sources: Funding for these restoration projects, the lead proponents, and those responsible for monitoring are represented by numerous agencies and organizations, including Wildlife Habitat Canada, Gulf of Maine Council, Ecology Action Centre, CB Wetlands and Environmental Specialists, Ducks Unlimited Canada, Nova Scotia Department of Agriculture, Clean Annapolis River Project, Nova Scotia Transportation and Infrastructure Renewal, Atlantic Coastal Action Program Saint John, Mount Allison University, Acadia University, New Brunswick Wildlife Trust Fund, and the Habitat Stewardship Fund.

drained, and land formed. Land forming is a method of draining dikeland soils by surface shaping, with surface sloped at 1 to 2 percent over a distance of 42 to 60 meters (Milligan 1987). To restore these types of sites (St. Croix River), more complex restoration activities are necessary, including the re-creation of tidal channel networks and ponds.

Achievements, Challenges, and Opportunities

An estimated 370 hectares of salt marsh habitat has been or is in the process of being restored in Nova Scotia (151.8 hectares, eleven projects), New Brunswick (181.3 hectares, five projects), and Prince Edward Island (42.5 hectares, five projects). Most restoration projects (76 percent) were primarily associated with dike breaching and 22 percent with culvert replacement (table 13.2). This does not include four completed or two planned causeway removal projects in the region where salt marsh recovery was not a primary goal.

The Ten Mile House restoration project in Prince Edward Island was formerly a tidal wetland and tributary to the Hillsborough River (fig. 13.2a). It was converted to a freshwater system in 1981 by Ducks Unlimited Canada for recreational purposes at the request of the community. The site was returned to full tidal influence in 2002.

The first intentional salt marsh restoration project to be undertaken in Nova Scotia was the Cheverie Creek Tidal River and Salt Marsh Restoration Project (Bowron et al. 2011). Cheverie Creek is a small tidal river that had a causeway across the mouth of the river with an undersized culvert that restricted tidal flow to 4 to 5 hectares of the marsh surface (fig. 13.2b). Collaborative efforts to restore tidal flow to the system began in 2002 and included prerestoration monitoring. Restoration involved the replacement of the old wooden box culvert in 2005 with a significantly larger elliptical aluminum culvert. This restored tidal flow to the full 43 hectares of former marsh surface. Postrestoration monitoring is planned through to 2012.

Restoration of a salt marsh along the Musquash River on the New Brunswick side of the Bay of Fundy began in 2005 (fig. 13.2c). This project is one of many ongoing initiatives to protect the Musquash estuary, one of the last ecologically intact estuaries in the Bay of Fundy and designated a Marine Protected Area in 2007. The site had been artificially impounded for decades by a railroad (built in 1882), a dike (1960s), and, more recently, the Trans Canada Highway. Recovery of this site involved the removal of 1.1 kilometers of rail bed, reconnection of the main creek to the main tidal channel, and construction of a protective dike and aboiteau around an adjacent property. Limited pre- and postrestoration monitoring was conducted as part of this project.

(a)

(b)

(c)

FIGURE 13.2. Examples of restoration project sites within the Maritime Provinces. (a) Ten Mile House Salt Marsh Restoration Project, Prince Edward Island. Converted to a freshwater habitat in 1981 by Ducks Unlimited Canada, the 42.5 hectare site was returned to full tidal influence when the water control structure and fishway were removed in 2002 (Photo courtesy of Ducks Unlimited Canada). (b) Cheverie Creek was the first intentional salt marsh restoration project with a comprehensive long-term monitoring program to be undertaken in Nova Scotia. The installation of the larger culvert in 2005 restored tidal flow to more than 40 hectares of tidal wetland (Photo courtesy of Tony Bowron, CBWES Inc.). (c) Musquash Salt Marsh Restoration Project, New Brunswick. The restoration of tidal flow to the 18.8 hectare salt marsh located at the headwaters of the Musquash River in 2005 is part of ongoing efforts to protect the Musquash estuary, which was designated a Marine Protected Area in 2007 (Photo courtesy of D. Meadus, Ducks Unlimited Canada).

As the practice of salt marsh restoration expands across the Maritimes, a number of challenges are arising related to a lack of federal or provincial mechanisms to support strategic or proactive salt marsh restoration, including the following:

- Inconsistent implementation and enforcement of legislation and policy within and between provinces
- Funding (particularly for activities beyond the earthworks such as feasibility, education, research, and monitoring)
- Significant gaps in our knowledge of the status of salt marsh habitats in the region and the opportunities for restoration
- A lack of a provincial or regional approach to identifying, prioritizing, and implementing restoration

Regional Approaches to Marsh Restoration

Several initiatives have been undertaken in the Maritimes on a provincial and a regional scale to identify salt marsh and tidal river systems as having "potential" for restoration. The most comprehensive of which was a series of tidal barrier audits conducted by the Ecology Action Centre (Nova Scotia) and the Conservation Council of New Brunswick between 2000 and 2004 to identify adverse effects of tidal crossings such as bridges, culverts, and aboiteaux on salt marshes and tidal rivers in the Bay of Fundy. The audits have been integrated as a comprehensive and readily accessible digital spatial database (van Proosdij and Dobek 2005; http://husky1.smu.ca/~dvanproo/Research_main.html). This exciting new tool contains information on geographical location of barriers, type and restriction (i.e., complete or partial), assessment of restoration potential, habitat observations, photographs and maps, as well as surrounding property information (van Proosdij and Dobek 2005). To date, inventories/audits have been completed for only the Bay of Fundy, making the identification of restoration sites outside the bay and the establishment of regional priorities for restoration difficult.

It has been recognized that sites for restoration need to be prioritized, bringing an ecosystem-based approach to the planning and implementation of salt marsh restoration initiatives (Musselman and Graham 2007). The expansion of the tidal barrier audit process to encompass the entire region would be an important step to initiating a more regional approach to restoration planning. The challenge is the focus of this prioritization and whether areas or systems that are severely degraded or smaller marshes within intact systems should be the priority (Musselman and Graham 2007). Either choice has funding implications that could dictate whether larger or smaller sites are restored, as well as the goals of the organization conducting the restoration.

Another regional or community-based initiative examined the issue of salt marsh restoration in New Brunswick as an adaptive response to sea level due to climate change (Marlin et al. 2007). The study sought to develop a systematic method that communities could use to assess a dike for possible removal. Assessment criteria and processes were developed based on climate change and adaptive capacity literature and community consultation. The project determined that assessing dikelands is a complex undertaking and that the specific process and criteria utilized by individual communities will vary, and the outcome for each community and dike will reflect this. However, the study stressed that communities must begin to reinforce existing protective structures or remove them and restore the original natural buffers in order to mitigate the impacts of climate change.

At the federal level, Fisheries and Oceans Canada applies the principle of "no net loss" and requires compensation for the harmful alteration, disruption, or destruction of fish habitat. Compensation involves offsetting impacts by undertaking additional activities, such as restoring or enhancing fish habitat in another area. In an effort to take a proactive approach to compensatory restoration, Fisheries and Oceans Canada has developed a restoration support tool, a database that will contain detailed information on potential restoration sites, to ensure regulatory requirements are being met, ensure greater collaboration with partners, and respond promptly to opportunities for habitat restoration (Hamilton, pers. comm., 2009).

Outside the Bay of Fundy, however, little is known about the full extent of loss and degradation of salt marsh in the Maritimes, and there has been no comprehensive initiative undertaken to locate and assess lost or degraded marshes and to identify potential restoration sites. Ongoing initiatives, like the Fisheries and Oceans Canada restoration support tool, and new initiatives, such as conducting additional tidal barrier audits and assessments of salt marsh habitat throughout the region, are necessary to identify and evaluate degraded habitats and to establish a foundation for future prioritization initiatives.

Funding

In the Maritimes, there is no level of government mandated to provide resources specifically to salt marsh restoration. Typically, salt marsh restoration is being advanced by enforcement of legislation that requires habitat compensation (such as the aforementioned requirement of Fisheries and Oceans Canada for harmful alteration, disruption, or destruction of fish habitat). The onus is then on the proponent to develop the compensation proposal and secure funding. Proactive restoration can be initiated by acquiring funds from nongovernmental organizations and

TABLE 13.3

Key federal and provincial mechanisms protecting and promoting the restoration of salt marshes in the Canadian Maritime Provinces

Legislation and policies	Government agency
Federal government	
Fisheries Act	Fisheries and Oceans
Federal Policy on Wetland Conservation	All federal departments
Migratory Birds Convention Act	Environment Canada
Nova Scotia	
Environment Act (Wetland Designation Policy)	Department of Environment
Nova Scotia Wetland Conservation Policy	Department of Environment
Agricultural Marshland Conservation Act	Department of Agriculture
New Brunswick	
Wetlands Conservation Policy	Department of Natural Resources
Coastal Areas Protection Policy	Department of Environment
Prince Edward Island	
Wetland Conservation Policy	Environment, Energy, and Forestry
Environmental Protection Act	Environment, Energy, and Forestry

programs (e.g., Gulf of Maine Council on the Marine Environment, www.gulfof maine.org: Habitat Stewardship Program; Adopt A Stream Program, www.nova scotiasalmon.ns.ca/) or through internal means. Table 13.3 provides an overview of the various provincial and federal policies and acts that directly or indirectly promote the protection and restoration of salt marshes in the Maritimes.

The steps involved in salt marsh restoration projects can include project identification, feasibility study, planning and implementation, and a pre-/postmonitoring program. There is a cost associated with each step in the restoration process; therefore, funding can dictate which steps are completed. For example, if funding is not available for monitoring, then no data will be collected on how the site responds. Similarly, some projects only secure funding for a feasibility study and baseline data collection, leaving the earthworks to proceed opportunistically. The steps completed in tidal restoration are largely dictated by the project type (proactive or compensatory) and the resources available.

For a compensation project, proponents are required to restore a specific ratio (typically 3:1) of restored habitat to lost habitat. Therefore, monitoring is required by permit to ensure that the required amount and quality of habitat has been restored. Monitoring is expensive, however, adding significantly to the cost of a restoration project and thus making restoration a less desirable option than avoiding damaging a pristine wetland in the first place. Despite the cost, the number of restoration projects in the region continues to increase. As restoration becomes more common practice, so does the need for research into restoration science and optimal monitoring practices, as well as to gain a better understanding of salt marshes in the region (Musselman and Graham 2007; Pett 2007).

The Gulf of Maine Council on the Marine Environment has provided funding to projects in the Bay of Fundy region to assist environmental nongovernmental organizations in the initial planning, feasibility, and baseline work of potential restoration sites, but then funding is required from other sources to bring the project to completion (e.g., Cheverie Creek; French Basin; Musquash). In addition, these funds are for the front work of proactive projects, not those slated for compensation. What makes this challenging is that regulatory agencies will not typically allow compensatory money to be used to identify potential restoration sites, proponents do not want to spend money on feasibility studies on sites, and funding agencies (i.e., GOMC) are reluctant to provide funding to community groups or environmental nongovernmental organizations for a project that may become compensatory in nature.

In addition, some proponents aim to capitalize on economies of scale, looking to larger (mitigation banking) projects that can be completed with greater ecological value per unit area (Pett 2007). The Nova Scotia Department of Transportation and Infrastructure Renewal has stated that the monitoring and construction costs associated with its compensation banking projects are $4.20 per square meter for larger projects and $17–$42 per square meter for smaller projects (Pett 2007). Restoration project costs can range from $8300 to upward of $988,000 (Cheverie Creek). This range can be due to the type of restoration required, such as dike breach (Allains Creek and Newfoundland Creek: $8300– $17,000) or culvert replacement (Cheverie Creek: $988,000), and the steps completed (e.g., monitoring at Cheverie Creek: $250,000).

Partnerships

Partnerships are essential to the success of salt marsh restoration projects and arise through funding, research, community involvement, and socioeconomic development. Some projects are completed by an individual organization (e.g., Ducks Unlimited Canada); whereas others are completed with many individuals and organizations involved (e.g., federal, provincial, and municipal governments, conservation groups, universities, private industry, and local communities). The importance of partnerships in these projects is due to the resources needed, including money, time, and knowledge. Certain knowledge is required for monitoring various indicators, for example, and identifying/analyzing the samples collected. This expertise can be found at local universities and in the community and may be provided at a cost, in-kind, or voluntarily.

Due to costs of a particular step in the restoration process or a longer time-frame, there may be several sources of funds required to complete a project. For example, the Cheverie Creek salt marsh restoration project began as a grassroots proactive project led by a nongovernmental organization with research supported between 2002 and 2005 by eight different sources. Restoration implementation and monitoring thereafter were funded by two sources.

Salt marsh restoration projects can provide the opportunity for remarkable partnerships. Proactive restoration projects can involve the community and have volunteers assist in monitoring, community outreach, and local school programs. These partnerships can extend the message about salt marshes and their restoration as well as stimulate the local economy (Pett 2007). In Cheverie, a community group called the Cheverie Crossway Salt Marsh Society partnered with local government, the business community, Dalhousie University, and the local school to develop a trail system and interpretive center that will highlight the restoration project. University involvement in both proactive and compensatory projects provides the opportunity to conduct monitoring and research activities beyond the scope of the original project or regulatory requirements.

Monitoring

Examination of passive restoration sites can provide long-term records about vegetation and geomorphic recovery (Crooks et al. 2002; French 2006; Byers and Chmura 2007); however, this provides limited information about driving or limiting variables within the first few years after the breach event that may influence the ultimate recovery of the system (e.g., Klötzli and Grootjans 2001; Able et al. 2008). In addition, both European and North American experiences have shown that salt marsh reestablishment is not inevitable at all sites following a breach (e.g., Haltiner et al. 1997; French et al. 2000; Klötzli and Grootjans 2001; Williams and Orr 2002; French 2006). Properly monitored restoration projects can provide additional information on constraints that may have caused a restoration project to proceed along pathways not initially anticipated (Klötzli and Grootjans 2001).

The six compensatory salt marsh restoration projects under way in Nova Scotia (Cheverie Creek, Walton River, Lawrencetown Lake, Smith Gut, St. Croix River, and Cogmagun River) represent the first intentional salt marsh projects to be undertaken in the region with associated pre- and postrestoration monitoring programs. The monitoring programs were based on the Global Programme of Action Coalition for the Gulf of Maine Regional Monitoring Protocol (Neckles et al. 2002; Bowron et al. 2011) and included complete data collection at a paired reference site. They involved a minimum of one year pre- and five years post-restoration data collection for a series of physical and biological parameters within five ecological indicator categories of hydrology, soils and sediments, vegetation, fish, and invertebrates. Other projects in the region have experienced different levels of monitoring as determined by project requirements, funding, and opportunity. The full-scale monitoring requirements of compensatory restoration projects is important in ensuring that these legislatively required projects are progressing along acceptable ecological trajectories or whether additional restoration activities (adaptive management) are required to ensure project success. Monitoring efforts also collect essential information on salt marsh ecology and response to restoration activities that can help improve the success and cost-effectiveness of future restoration efforts. Knowledge gained from larger projects can help in the development of smaller or proactive projects.

Lessons Learned for Building a Restoration Program

Legislatively required compensation projects are leading the way in the Canadian Maritimes, and there is every indication that this trend will continue into the near future. There is a need to progress from this case by case approach to a more strategic regional approach that encourages both compensation and net gain restoration projects. In order to achieve this, inventories of intact, degraded, and lost coastal wetland ecosystems, such as the tidal barrier audit for the Bay of Fundy, need to be conducted for each of the provinces. Such inventories can promote regional planning and implementation of restoration. Additionally, the restoration support tool being developed by Fisheries and Oceans Canada, which will be populated with information from community groups that have a potential salt marsh restoration project in mind, will assist in this progression to a regional proactive approach. However, the region continues to fall short on the provision of clear leadership and vision. The establishment of a lead agency, consistent and adequate funding source(s), and a regional restoration strategy is critical. In the absence of leadership and a strategic approach, restoration projects, however successful on an individual basis, will continue to occur in isolation and in ignorance of broader landscape needs.

Leadership and a regional strategy will support the interest and expertise in salt marsh restoration. It has been shown that strong partnerships have developed in the region, and there is much opportunity and growing support for primary research into the ecological form and function of salt marshes and the restoration of degraded and lost habitats. The high degree of historical loss of salt marshes due to reversible activities means that there are many opportunities in each province to engage in cost-effective and ecologically successful salt marsh restoration projects. With a growing awareness of the significance of salt marshes as a critical component of the coastal ecosystem and the severity of the legacy of loss that has plagued wetlands throughout the region, so grows our awareness of restoration as a viable habitat management option. Restoration offers the opportunity not only to compensate for unavoidable current and future losses of salt marsh, but to go further and begin to proactively restore marshes that have previously been lost.

REFERENCES

Able, K. W., T. M. Grothues, S. M. Hagan, M. E. Kimball, D. M. Nemerson, and G. L. Taghon. 2008. "Long Term Response of Fishes and Other Fauna to Restoration of Former Salt Hay Farms: Multiple Measures of Restoration Success." *Reviews in Fish Biology and Fisheries* 18:65–97.

Adam, P. 1990. *Saltmarsh Ecology*. Cambridge: Cambridge University Press.

Argow, B. A., and D. M. FitzGerald. 2006. "Winter Processes on Northern Salt Marshes: Evaluating the Impact of In situ Peat Compaction due to Ice Loading, Wells, ME." *Estuarine, Coastal and Shelf Science* 69: 360–69.

Boumans, R. M. J., D. M. Burdick, and M. Dionne. 2002. "Modeling Habitat Change in Salt Marshes after Tidal Restoration." *Restoration Ecology* 10: 543–55.

Bowron, T., N. Neatt, D. van Proosdij, J. Lundholm, and J. Grahams. 2011. "Macrotidal Salt Marsh Ecosystem Response to Culvert Expansion." *Restoration Ecology* 19: 307–22.

Byers, S. E., and G. L. Chmura. 2007. "Salt Marsh Vegetation Recovery on the Bay of Fundy." *Estuaries and Coasts* 30:869–77.

Catling, P. M., G. Mitrow, L. Black, and S. Carbyn. 2004. "Status of the Alien Race of Common Reed (*Phragmites australis*) in the Canadian Maritime Provinces." *Botanical Electronic News* no. 324. http://www.ou.edu/cas/botany-micro/ben/ben324.html.

Chagueé-Goff, C., T. S. Hamilton, and D. B. Scott. 2001. "Geochemical Evidence for the Recent Changes in a Salt Marsh, Chezzetcook Inlet, Nova Scotia, Canada." *Proceedings of the Nova Scotia Institute of Science* 41:149–59.

Chmura, G. L., P. Chase, and J. Bercovitch. 1997. "Climatic Controls on the Middle Marsh Zone in the Bay of Fundy." *Estuaries* 20:689–99.

Crain, C. M., B. R. Silliman, S. Bertness, and M. D. Bertness. 2004. "Physical and Biotic Drivers of Plant Distribution across Estuarine Salinity Gradients." *Ecology* 85:2539–49.

Crooks, S., J. Schutten, G. D. Sheern, K. Pye, and A. J. Davy. 2002. "Drainage and Elevation as Factors in the Restoration of Salt Marsh in Britain." *Restoration Ecology* 10:591–602.

Davis, D. S., and S. Browne. 1996. *The Natural History of Nova Scotia—Theme Regions*. Halifax, NS: Government of Nova Scotia and Nimbus Publishing.

Desplanque, C., and D. J. Mossman. 2004. "Tides and Their Seminal Impact on the Geology, Geography, History and Socio-economics of the Bay of Fundy, Eastern Canada." *Atlantic Geology* 40:1–130.

Ewanchuk, P. J., and M. D. Bertness. 2004. "Structure and Organization of a Northern New England Salt Marsh Plant Community." *Journal of Ecology* 92:72–85.

French, C. E., J. R. French, N. J. Clifford, and C. J. Watson. 2000. "Sedimentation-Erosion Dynamics of Abandoned Reclamations: The Role of Waves and Tides." *Continental Shelf Research* 20:1711–33.

French, P. W. 2006. "Managed Realignment—The Developing Story of a Comparatively New Approach to Soft Engineering." *Estuarine Coastal and Shelf Science* 67:406–23.

Ganong, W. F. 1903. "The Vegetation of the Bay of Fundy Salt and Diked Marshes: An Ecological Study." *Botanical Gazette* 36:161–86, 280–302, 349–69, 429–55.

Gordo, D. C., Jr., and P. J. Cranford. 1994. "Export of Organic Matter from Macrotidal Salt Marshes in the Upper Bay of Fundy, Canada." Pp. 257–64 in *Global Wetlands: Old World and New*, edited by W. J. Mitsch. New York: Elsevier.

Haltiner, J., J. B. Zedler, K. E. Boyer, G. D. Williams, and J. C. Callaway. 1997. "Influence of Physical Processes on Design, Functioning and Evolution of Tidal Wetlands in California (USA)." *Wetlands Ecology and Management* 4:73–91.

Hanson, A. 2004. *Breeding Bird Use of Salt Marsh Habitat in the Maritime Provinces.* Canadian Wildlife Service Technical Report Series no. 414, Atlantic Region.

Hanson, A., and L. Calkins. 1996. *Wetlands of the Maritime Provinces: Revised Documentation for the Wetlands Inventory.* Canadian Wildlife Service Technical Report Series no. 267, Atlantic Region.

Hatcher, A., and D. G. Patriquin. 1981. *Salt Marshes in Nova Scotia.* Halifax NS: Institute for Resource and Environmental Studies, Dalhousie University.

Jacobson, H., and G. L. Jacobson Jr. 1989. "Variability of Vegetation in Tidal Marshes of Maine, USA." *Canadian Journal of Botany* 67:230–38.

Klötzli, F., and A. P. Grootjans. 2001. "Restoration of Natural and Semi-natural Wetland Systems in Central Europe: Progress and Predictability of Developments." *Restoration Ecology* 9:209–19.

Marlin, A., J. Ollerhead, and D. Bruce. 2007. *A New Brunswick Dyke Assessment Framework: Taking the First Steps.* New Brunswick, Canada: New Brunswick Environmental Trust Fund, Mount Allison University.

Mckee, K. L., and W. H. Patrick. 1988. "The Relationship of Smooth Cordgrass (*Spartina alterniflora*) to Tidal Datums: A Review." *Estuaries and Coasts* 11:143–51.

Mendelsohn, I. A., and K. L. McKee. 2000. "Salt Marshes and Mangroves." Pp. 502–36 in *North American Terrestrial Vegetation*, 2nd ed., edited by M. G. Barbour and W. D. Billings. Cambridge: Cambridge University Press.

Milligan, D. C. 1987. *Maritime Dykelands.* Halifax: Nova Scotia Department of Government Services—Publishing Division.

Morgan, P. A., and F. T. Short. 2002. "Using Functional Trajectories to Track Constructed Salt Marsh Development in the Great Bay Estuary, Maine/New Hampshire, U.S.A." *Restoration Ecology* 10:461–73.

Musselman, R., and J. Graham. 2007. *Main Workshop Messages. Proceedings of the Workshop: Six Years in the Mud—Restoring Maritime Salt Marshes: Lessons Learned and Moving Forward.* Dartmouth, NS: Bedford Institute of Oceanography.

Neckles, H. A., M. Dionne, D. M. Burdick, C. T. Roman, R. Buchsbaum, and

E. Hutchins. 2002. "A Monitoring Protocol to Assess Tidal Restoration of Salt Marshes on Local and Regional Scales." *Restoration Ecology* 10:556–63.

Neily, P., E. Quigley, L. Benjamin, B. Stewart, and T. Duke. 2003. *Ecological Land Classification for Nova Scotia*. Vol. 1: *Mapping Nova Scotia's Terrestrial Ecosystems*. Truro: Nova Scotia Department of Natural Resources, Forestry Division.

Pett, R. J. 2007. *Habitat Banking for HADD and Wetland Compensation—New Partnership Opportunities and Significant Environmental, Economic and Community Benefits*. Proceedings of the 2007 Annual Conference of the Transportation Association of Canada (TAC), Saskatoon, Saskatchewan.

Roberts, B. A., and A. Robertson. 1986. "Salt Marshes of Atlantic Canada: Their Ecology and Distribution." *Canadian Journal of Botany* 64:455–67.

Roman, C. T., K. B. Raposa, S. C. Adamowicz, M-J. James-Pirri, and J. G. Catena. 2002. "Quantifying Vegetation and Nekton Response to Tidal Restoration of a New England Salt Marsh." *Restoration Ecology* 10:450–60.

Shaw, J., R. B. Taylor, S. Solomon, H. Christian, and D. L Forbes. 1998. "Potential Impacts of Global Sea-Level Rise on Canadian Coasts." *Canadian Geographer* 42:365–79.

Short, F. T, D. M. Burdick, C. A. Short, R. C. Davis, and P. A. Morgan. 2000. "Developing Success Criteria for Restored Eelgrass, Salt Marsh and Mudflat Habitats." *Ecological Engineering* 15:239–52.

van Proosdij, D., and P. Dobek. 2005. *Bay of Fundy Tidal Barriers Database Development*. Final report and interactive CD prepared for the Gulf of Maine Council and Bay of Fundy Ecosystem Partnership.

van Proosdij, D., J. Lundholm, N. C. Neatt, T. M. Bowron, and J. M. Graham. 2010. "Ecological Re-engineering of a Freshwater Impoundment for Salt Marsh Restoration in a Hyper-tidal System." *Ecological Engineering* 36:1314–1332.

van Proosdij, D., J. Ollerhead, R. G. D. Davidson-Arnott, and L. E. Schostak. 1999. "Allen Creek Marsh, Bay of Fundy: A Macro-tidal Coastal Salt Marsh." *Canadian Geographer* 43:316–22.

van Proosdij, D., R. G. D. Davidson-Arnott, and J. Ollerhead. 2006a. "Controls on the Spatial Patterns of Sediment Deposition across a Macro Tidal Salt Marsh over Single Tidal Cycles." *Estuarine, Coastal and Shelf Science* 69:64–86.

van Proosdij, D., J. Ollerhead, and R. G. D. Davidson-Arnott. 2006b. "Seasonal and Annual Variations in the Sediment Mass Balance of a Macro-tidal Salt Marsh." *Marine Geology* 225:103–27.

van Proosdij, D., T. Milligan, G. Bugden, and C. Butler. 2009. "A Tale of Two Macro Tidal Estuaries: Differential Morphodynamic Response of the Intertidal Zone to Causeway Construction." *Journal of Coastal Research* SI 56:772–76.

Wells, E. D., and H. E. Hirvonen. 1988. "Wetlands of Atlantic Canada." Pp. 249–303 in *Wetlands of Canada*, edited by National-Wetlands-Working-Group, Sustainable Development Branch. Ottawa, ON: Environment Canada and Polyscience Publications.

Williams, P. B., and M. K. Orr. 2002. "Physical Evolution of Restored Breached Levee Marshes in San Francisco Bay Estuary." *Restoration Ecology* 10:527–42.

PART IV

Integrating Science and Practice

Successful salt marsh restoration projects are dependent on an integration of science and practice, with coastal managers, scientists, and engineers, from diverse disciplines, collaborating toward a common goal of restoring degraded ecosystems. These teams must work together to document and quantify ecological responses to restoration, continually learning from experiences that will lead to development of more effective approaches to restoration, monitoring, and applied research. The chapters in this part offer science-based tools to facilitate tidal flow restoration projects. Buchsbaum and Wigand (chap. 14) describe the role of adaptive management and monitoring as essential practices. The concept of ecosystem services is discussed in chapter 15 (Chmura and others) and presented as a multiple-parameter method to assess the trajectory of restoration efforts. Ecosystem simulation models that integrate physical and biological processes, as described by Konisky (chap. 16), are used to guide the design of projects, evaluate expected responses to tidal restoration, facilitate the development of monitoring programs, and communicate restoration objectives and expected outcomes. Glamore (chap. 17) closes this part on integrating science and practice with an engineering perspective on modifications to tide-restricting structures (e.g., tide gates) that can be used to achieve desired hydrologic responses.

Chapter 14

Adaptive Management and Monitoring as Fundamental Tools to Effective Salt Marsh Restoration

ROBERT N. BUCHSBAUM AND CATHLEEN WIGAND

Adaptive management as applied to ecological restoration is a systematic decision-making process in which the results of restoration activities are repeatedly monitored and evaluated to provide guidance that can be used in determining any necessary future management actions (Salafsky et al. 2001; SER International 2004). In the setup phase, stakeholders agree upon the overall goals of the restoration project, plan restoration activities, and develop conceptual models that describe expected responses to the management actions (fig. 14.1). After implementation of management measures, the project enters an iterative phase with a focus on monitoring and assessment of ecological responses. These determine whether adjustments to management measures or the conceptual models are warranted. As applied to restoration of salt marshes and other estuarine habitats, an adaptive management framework should include targets for specific structural and functional components (e.g., vegetation and hydrology), a schedule for management and restoration activities, a monitoring plan to assess the progress of the project, interim criteria with triggers, and a commitment to alter the conceptual models, monitoring plan, and even project goals if interim criteria are not met.

This chapter discusses the role of adaptive management and monitoring as essential practices to guide projects that are designed to restore tidal flow to coastal wetlands. An underlying assumption of adaptive management is that ecosystems are complex and inherently variable, making it difficult to precisely forecast the outcome of any management action. Thus adaptive management and monitoring go hand in hand.

Gregory et al. (2006) described two types of adaptive management: active and passive. Active adaptive management is an experimental approach wherein different management actions are compared in a statistically sound experimental

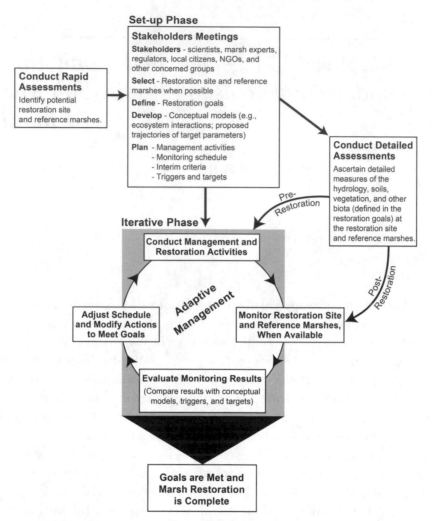

FIGURE 14.1. Adaptive management framework and monitoring approaches for effective salt marsh restoration (based in part on Teal and Weishar 2005 and Williams et al. 2007).

design to inform future management activities. Active adaptive management requires that the size of the project, the time frame, the existence of different potential management scenarios and outcomes, and the sensitivity of the habitat are such that an experimental approach is feasible. Similar to active adaptive management, Zedler (2006) uses the term "adaptive restoration" for restoration projects that incorporate experimentation into the original design. At the onset of a project, the optimal restoration targets and even the most appropriate parameters to be monitored are not necessarily known, but this eventually becomes clear through experimentation.

In passive adaptive management there is no experimental testing of different management options, though restoration activities may still be modified in response to monitoring results (Gregory et al. 2006). Passive adaptive management is more typical in New England salt marshes because there is usually only one management option, such as enlarging a culvert or removing a dike to increase tidal flow. Also, many tidal projects in New England are small, which may limit the ability to develop valid experimental tests of different management actions. Because active adaptive management provides a more direct way of enhancing future restoration practices, practitioners should take advantage of such opportunities when they arise.

Monitoring in Support of Effective Salt Marsh Restoration

Monitoring is essential to an adaptive management framework. Before any management action is taken, monitoring can provide information for identifying and prioritizing potential restoration sites, selecting appropriate restoration approaches, and setting goals. After the initiation of a restoration project, monitoring provides the basis for determining the accuracy of the conceptual models developed for the project and how well a project is meeting its planned targets.

Determining the Amount of Monitoring Necessary in a Project

Almost all monitoring programs recognize the need for different levels of monitoring intensity depending on the size, resources, goals, and complexity of the project. The US Environmental Protection Agency has recommended a three-level approach for assessment of wetlands in its Environmental Monitoring and Assessment Program (US EPA 2002a; Kentula 2007), an approach that can be applied toward restoration monitoring. The first level is an office-based assessment to provide an overview of the condition of the adjacent watershed and salt marsh of interest. Mapping is a major component of a first-level assessment and generally relies on readily available digital data. The second level is an on-site rapid assessment using a few core parameters (e.g., vegetation and hydrology indicators) that are relatively easy to measure, are responsive to anthropogenic stressors, and give an accurate indication of wetland condition that could help in prioritizing sites for restoration or evaluating some restoration activities (table 14.1). The third level consists of specialized, detailed studies to address distinctive site conditions and track certain restoration targets (table 14.2). A third-level assessment is the most costly and time intensive. All three assessment levels are needed in an active adaptive management framework, but only levels one and two may be necessary for passive adaptive management.

TABLE 14.1

Rapid assessment methods to assess coastal wetland condition and prioritize restoration sites

Procedures	Method for assessment	Reference
California Rapid Assessment Method (CRAM)	Buffers, hydrology, physical structure, biotic structure, stressors	Collins et al. 2008
Delaware Rapid Assessment Protocol (DERAP)	Stressors relative to their potential to effect hydrology, biogeochemical cycling, and biota	Jacobs et al. 2008
Global Programme of Action Coalition (GPAC) Baseline Habitat Mapping	Habitat mapping, vegetation, historic records and manipulations, stressors	Neckles and Dionne 2000
MA Coastal Zone Management Rapid Habitat Assessment Method	Buffers, hydrology, soils, vegetation, stressors	Carlisle et al. 2002
New England Rapid Assessment Method (NERAM)	Watershed buffers, habitat map, vegetation, soils, on-site and watershed disturbances and stressors	Carullo et al. 2007

Like the US Environmental Protection Agency, protocols developed by the Global Programme of Action Coalition for the Gulf of Maine (GPAC) consider mapping and site description as the basic first steps in salt marsh assessment (Neckles et al. 2002). These protocols distinguish certain core parameters that should be measured at all sites and others that are optional depending upon the goals of the project (table 14.2).

From a statistical perspective, the sampling intensity needed to assess a specific response depends on the size of the marsh being monitored, the variability of the parameter being measured, and the desired precision of the estimate. If preliminary sampling is possible, then the number of observations needed to achieve a certain statistical power can be based on the estimated population variance, a standard statistical procedure in field studies (e.g., Raposa et al. 2003; James-Pirri et al. 2007). The GPAC protocols propose a sampling regime for a certain size range of salt marshes in a specific region based on prior sampling and best professional judgment (Neckles and Dionne 2000).

Experimental Design in Monitoring

After some preliminary site information has been gathered, but before management activities are undertaken, the next step in restoration is to decide upon a monitoring design. This must provide the ability to distinguish changes caused by restoration activities from natural year-to-year variation and from any long-term regional trends. Comparison of a restored site to one or more reference marshes is a

TABLE 14.2

Detailed assessment methods for monitoring coastal wetlands in restoration and reference sites

Procedures	Method for assessment	Reference
GPAC Marsh Hydrology	Hydroperiod, tidal signal, surface elevations, and, when resources are available, tidal creek cross sections, water table depth, and extent of tidal flooding	Neckles and Dionne 2000
GPAC Soils and Sediments	Pore water salinity, and, when resources are available, organic matter, sediment accretion, sediment elevation, redox potential, and pore water sulfides	Neckles and Dionne 2000
GPAC Marsh Vegetation	Plant abundance, composition, height of species of concern, stem density of species of concern, and, when resources are available, plant aboveground biomass and stems that are flowering	Neckles and Dionne 2000
US EPA Using Vegetation to Assess Environmental Conditions	Plant species/taxa richness, stem density, percent cover, observations of hydrology and soils	US EPA 2002c
NPS (National Park Service) Monitoring Salt Marsh Vegetation	Plant cover, species composition and abundance, height of key species, and, when resources are available, average water table level, soil salinity, and soil sulfide	Roman et al. 2001
GPAC Nekton	Species composition, density, length, biomass and species richness, and, when resources are available, fish growth and diet, and larval mosquitos	Neckles and Dionne 2000
US EPA Developing an Invertebrate Index of Biological Integrity for Wetlands	Invertebrate abundance, species/taxa richness, tolerant species, trophic function, water quality, watershed land use/cover, and, when resources are available, condition of the invertebrates	US EPA 2002d
NPS Monitoring Nekton in Shallow Estuarine Habitats	Species composition, species richness, size structure, and abundance	Raposa et al. 2003
GPAC Birds	Bird abundance, species richness, feeding and breeding behavior, and, when resources are available, small passerines and cryptic birds	Neckles and Dionne 2000
US EPA Biological Assessment Methods for Birds	Extent and condition of structural attributes and bird habitat; bird species, presence, abundance, and frequency of occurrence; duration of wetland use, feeding and breeding behavior	US EPA 2002e

key element of the monitoring program (SER International 2004). Both restored and reference sites may change over the course of the project, so the reference is essential for determining regional changes that cannot be attributed to the management activities, such as responses to climate change, nutrient loading, and invasive species. Brinson and Rheinhardt (1996) describe the use of a series of reference wetlands to establish standards that typify wetland conditions in a region. Wigand et al. (2010) reported detailed measures of plants, invertebrates, and soils among salt marshes with similar hydrology and geomorphology, but varying watershed nitrogen loads to develop a reference set of salt marshes in southern New England. A salt marsh reference set provides templates for the development of conceptual models describing ecosystem structure and processes and establishes a framework for estimating changes in structure and function following restoration.

A study design that facilitates adaptive management in a restoration project is a before, after, control, impact (BACI) design (Underwood 1992; Stewart-Oaten and Bence 2001; US EPA 2002b). Studies in New England using this design compared the tide-restricted marsh, the same marsh after reintroduction of tidal flow, and a nearby, unrestricted reference marsh sampled at the same times as the restricted marsh (Raposa 2002; Roman et al. 2002; Buchsbaum et al. 2006). The BACI design allowed Buchsbaum et al. (2006) to distinguish changes in growth of the invasive *Phragmites australis* (common reed) due to the hydrologic restoration from year-to-year differences in rainfall. When possible, it is best to collect several years of prerestoration data on both restored and reference marshes in order to understand year-to-year variations. Monitoring several reference marshes along with the restored site allows for a better estimate of natural variability in targeted parameters and functions and will allow for a better assessment of the response of the restored site (Underwood 1992; Short et al. 2000).

Trajectories for Restoration Targets

A conceptual framework for evaluating restoration progress that flows naturally from a BACI design is that of trajectories. Trajectories are modeled pathways of changes over time in selected ecological parameters (e.g., increase in percent cover of desired vegetation) after restoration (Simenstad and Thom 1996; Craft et al. 1999; Zedler and Calloway 1999; Morgan and Short 2002). One of the challenges in using trajectories to assess restoration success is that different parameters move toward equivalency with reference marshes at different rates (Burdick et al. 1997; Zedler and Calloway 1999; Fell et al. 2000; Morgan and Short 2002; Warren et al. 2002). Fish are likely to return almost immediately to a hydrologically restored salt marsh (Teo and Able 2003), but they may not achieve the same level of growth as in a reference marsh for several seasons due to differences in the types of food available (Wozniak et al. 2006). Vegetation changes are likely to take five years or longer, and soil organic content could take decades before reaching functional equivalence with reference marsh sites (Zedler and Calloway 1999; Morgan and Short 2002). Since long-term (five to twenty-five years) monitoring is not often feasible for many projects, trajectories can be modeled using a space-for-time substitution. In this approach previous marsh restoration projects differing in the length of time since they were restored are used as a basis for modeling anticipated changes in various parameters (Morgan and Short 2002; Warren et al. 2002; Wozniak et al. 2006).

Over time the trajectories of the restored marsh are predicted to move toward the reference (fig 14.2). The slopes of the curves for different parameters allow scientists and regulators to evaluate whether a project is supporting the expected trajectory of ecosystem development. Triggers can be established based on these slopes. If the triggers are exceeded, further management actions may be implemented as consistent with the adaptive management framework.

Quantitative versus Qualitative Assessments and the Value of a Statistical Approach

Quantitative measurements are almost always needed to determine whether a marsh is reaching its restoration targets. Such data coupled with statistical tests provide an unbiased, standardized assessment of whether the predicted changes have actually occurred. Qualitative observations, such as repeated photographs from permanent photo stations, can be useful as supplemental information, but these are limited to only a few parameters that can be easily observed. Many key

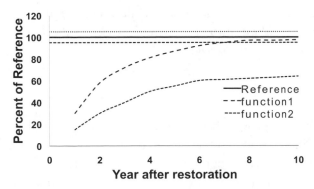

FIGURE 14.2. Hypothetical trajectories for two marsh functions in relation to an idealized reference marsh. Function 1 achieves equivalency with the reference defined as being within the 95 percent confidence limits in about seven years. Function 2 is about 60 percent of the reference marsh after ten years. Note that the reference marsh may also be changing over time.

marsh organisms and processes not captured in photographs are required for quantitative assessment. Examples include changes in flooding, pore water salinity and nutrients, and colonization of a restored marsh by smaller halophytes, invertebrates, and fish. Statistical analyses of quantitative data can also be used to understand the similarities and differences of marshes across a region and to determine relationships between key variables such as that between salinity and the types of plants in a marsh.

The use of statistical procedures allows marsh scientists to go beyond relying on "best professional judgment" in assessing whether a restoration project is meeting its targets and whether additional management is necessary. Analysis of variance has been used to examine changes before and after restoration in parameters such as salinity or the distribution of individual species (Burdick et al. 1997; Warren et al. 2001; Raposa 2002; Raposa and Roman 2003; Buchsbaum et al. 2006). Changes in the suite of marsh species have been analyzed using ordination procedures (Gauch 1982; Kent and Coker 1992) and analysis of similarities (Clarke and Warwick 2001; Raposa 2002; Roman et al. 2002; Raposa and Roman 2003; Buchsbaum et al. 2006).

Parameters to Monitor

In an adaptive management framework, the selection of parameters to monitor should flow logically from the site-specific goals of the restoration. In a tidal restoration project with a goal of enhancing biodiversity by managing the spread of an invasive plant, such as *Phragmites* or *Lythrum salicaria* (purple loosestrife), vegetation and hydrologic monitoring are obvious priorities. If the goal is enhancing the habitat for fish or birds, detailed biotic studies or the development of indirect indicators of biota are required. Parameters often monitored in salt marsh restoration projects are described in the following sections, and the sources of methods for their assessment are summarized in table 14.2.

Mapping

Baseline habitat maps provide basic information on the location, features, general ecological condition, and potential stresses on a site (Neckles et al. 2002; US EPA 2002a). The current use of geographic information systems allows access to landscape databases (e.g., land use/land cover, wetlands, elevations, soils) that are often available from state agencies. Baseline maps should include areas of wetland vegetation, marsh pannes, pools, creeks, mosquito ditches, roads, railroads, dikes, culverts, and any other features on the marsh and the surrounding upland that may have an impact on the marsh (Neckles and Dionne 2000). A contour map indicating marsh elevations is another valuable planning tool because it shows the potential for different marsh habitats to form after a hydrologic restoration. Light Detection and Ranging (LIDAR) data are becoming widely available for coastal systems and have been used to produce contour maps.

Time series of orthophotos or historical maps have been used to document changes in the extent of wetlands and the amount of open water versus vegetated marsh (Hartig et al. 2001; Bromberg and Bertness 2006). Depending upon the quality and scale of the photo, it may be possible to use maps to quantify changes in the vegetation after a tidal restriction is removed, such as revegetation by native plants or the decline in the amount of invasive species.

Hydrology

Hydrology is a fundamental control on the structure and function of a salt marsh and a key parameter subjected to management actions in tidal restoration projects. A number of hydrologic parameters can be monitored. Tidal range and hydroperiod (the frequency, duration, and depth that a marsh is flooded) can be monitored on both sides of a culvert to document the increase in tidal exchange after restoration (Neckles and Dionne 2000). The aerial extent of inundation reveals whether seawater is now reaching areas that may have previously been dry but were targeted for restoration (Warren et al. 2001; Smith and Warren 2007). Water table depth (along with pore water salinity), determines to a large extent the distribution and abundance of plant species in the marsh (Neckles et al. 2002). Hydrologic parameters often respond quickly to restoration, thus providing an early indicator of the likelihood of project success.

Pore Water Chemistry

Pore water salinity helps determine whether conditions are likely to support native salt marsh vegetation such as *Spartina* spp. (cordgrass), or plants more tolerant of brackish water, such as *Typha angustifolia* (narrowleaf cattail) or the exotic invasives *Phragmites* or *L. salicaria*. Pore water salinity is typically sampled by sinking wells into the marsh surface or by using a syringe to directly withdraw a pore water sample from the sediment through a stainless steel tube with a slotted point (after Portnoy and Giblin 1997). Pore water extracted with a syringe can also be examined for sulfides, pH, redox potential, and nutrients when these measures may influence restoration goals. Sulfides are of particular interest because of their effect on *Phragmites* growth (Chambers et al. 2003). In addition, sulfide concentrations can often be a more reliable measure of the oxygen status of the soils than redox (Neckles and Dionne 2000).

Marsh Development Processes

The vulnerability of salt marshes to increasing rates of sea level rise makes marsh accretion rate relative to sea level rise an important target in restoration projects (Morris et al. 2002; Cahoon and Guntenspergen 2010). In addition, vertical accretion and marsh elevation are specific targets in projects where prior diking or tidal restrictions lowered marsh elevations below that optimal for desired vegetation (e.g., Cornu and Sadro 2002), often transforming these marshes and associated tidal flats into shallow ponds. Marsh surface elevation, rates of sediment accretion, and soil organic matter (percent total and macro-organic matter) are useful parameters for assessing the responses of a formerly degraded marsh to hydrologic restoration, particularly in the context of sea level rise.

Vegetation

Vegetation is an obvious monitoring need in restoration projects since vegetation is often a restoration target as well as an indicator of marsh health. In their meta-analysis based on the GPAC protocols, Konisky et al. (2006) noted that 89 percent of all the monitoring projects in the Gulf of Maine region had collected plant species richness and abundance data. Vegetation is relatively simple to monitor, at least compared to mobile organisms. The most valuable vegetation parameter is the change in the cover or abundance of different plant species as a result of the restoration. A common metric derived from these in-marsh assessments is the ratio of the percent cover of invasive to native plants. More intensive studies might include stem density (shoots per square meter), productivity (end-of-season biomass per square meter), canopy height, and an estimate of the proportion of flowering shoots.

There are a number of standard methods for vegetation analysis, including point intercept, line intercept, and quadrat sampling (Barbour et al. 1999). Studies of New England marsh restoration projects have used quadrat sampling along randomized transects (Roman et al. 2002), line intercept (Buchsbaum et al. 2006), and belt transects (Warren et al. 2001). Transects are typically oriented to span the marsh elevation gradient from marsh creeks to the upland.

Mobile Fauna—Nekton and Birds

Nekton (fish and invertebrates that swim in the water column) and birds are often key restoration targets. Unlike vegetation, however, they are challenging to sample due to their mobility, which leads to high variation and low statistical power. Thus a monitoring program that includes mobile fauna needs to have adequate resources for repeated sampling.

NEKTON

Fish and macroinvertebrates in marshes serve as important ecological links with fisheries in coastal waters and integrate cumulative effects of multiple environmental stressors (e.g., changes in salinity, water quality, and flow). The expected response of nekton to tidal restoration depends to a large extent on the severity of the restriction (Raposa and Roman 2003). Ideally, measurements should yield both nekton density (numbers per area) as well as species diversity, but the former has been particularly challenging with the currently available sampling gear. Further, different gear is needed to sample different components of the nekton. For example, bottomless lift nets, fyke nets, flumes, or pit traps are used to sample species that inhabit the marsh surface at high tide (Rozas 1992; Burdick et al. 1997; Able and Hagan 2000; Fell et al. 2003). If the primary interest is in tidal creeks, pools, or ditches, then quantitative enclosure methods, like throw traps (Raposa 2002; Raposa et al. 2003) and ditch nets (James-Pirri et al. 2010) are an appropriate choice. By estimating the area of marsh features flooded at low tide, densities in creeks and pools can be extrapolated to calculate resident nekton populations (McKinney et al. 2010). Important variables for nekton are density and biomass by species and species richness (Neckles and Dionne 2000). With adequate resources, monitoring can be expanded to examine food webs, the growth rates of selected species, and the responses of different age classes.

BIRDS

Birds are highly visible organisms whose popularity with the general public makes them a frequent target for marsh restoration. They are relatively easy to identify compared to other taxa and can be monitored by knowledgeable volunteers using standardized point counts or transects (Ralph et al. 1995; Conway 2008). Birds in salt marshes respond to changes in the ratio of open water to vegetated marsh; the availability of nesting, feeding, and roosting sites; the abundance of their prey; and disturbances (US EPA 2002e). Another compelling reason to monitor birds is that certain species of marsh birds are of conservation concern and could be affected by hydrologic restoration (e.g., *Ammodramus caudacutus* [saltmarsh sparrow]). Many states have a list of species of conservation priority, which can be found on their wildlife agency's website.

Like nekton, the mobility of birds leads to high day-to-day and even hour-to-hour variability, requiring repeated sampling to determine whether a restoration project has had a measureable impact. An additional issue is that the size of most marsh restoration projects in New England is smaller than the home ranges of many bird species (Neckles et al. 2002). If the funding and expertise are not available for detailed bird monitoring, it may be necessary to derive indirect measurements of habitat quality (Zedler 2006).

Epibenthic Invertebrates

Some salt marsh monitoring programs include epibenthic invertebrates that inhabit the marsh surface (Calloway et al. 2001; Carlisle et al. 2002). Amphipods, isopods, snails, crabs, insects, and other marsh invertebrates are major transformers of marsh primary production to higher trophic levels. They are responsive to changes in salinities, the presence of pollutants, and increases in nutrients. The effects of invasive plants on invertebrates can be substantial, particularly if the habitat was previously unvegetated (Levin et al. 2006), or relatively minor (Warren et al. 2001; Weis and Weis 2003). Marsh surface invertebrates can be measured by direct counts within quadrats (Buchsbaum et al. 2009) through the use of litter bags (Scatolini and Zedler 1996), pitfall traps (Able and Hagan 2000), and D nets (Carlisle et al. 2002).

Example of Adaptive Management Applied to Tidal Restoration

One of the best examples of how adaptive management has been used in salt marsh restoration projects is a 4000 hectare salt marsh restoration on Delaware Bay carried out by Public Service Enterprise Group as mitigation for a discharge permit (Teal and Weishar 2005). The degraded marshes that were targeted for restoration included salt hay farms enclosed by dikes and brackish areas dominated by the invasive reed *Phragmites australis*. Initial stakeholder meetings to set restoration goals and monitoring priorities included regulators, scientists, staff from Public Service Enterprise Group, and local residents. Formal targets with timelines were set for the restoration of ecosystem structure and function (e.g., interim target of 45 percent cover of "desirable" salt marsh vegetation after seven years, final target of 76 percent cover after twelve years, less than 5 percent cover of *Phragmites* after twelve years, and evidence of similarity of fish utilization between the restored and nearby reference marshes).

In keeping with the adaptive management framework, criteria and thresholds were set that, if exceeded, could indicate acceptable progress or could trigger intervention and action. Some examples of these thresholds were excessive ponding of water on the marsh platform, excessive upland flooding, occlusion of the channels through which normal tidal exchange would occur, reestablishment of less *Spartina* cover than specified, and reduction of less invasive *Phragmites* cover than specified (Teal and Weishar 2005). Leaders of the monitoring program typically met twice a year with stakeholders and a committee of regulators and external scientists to review the monitoring data, to discuss triggers that might have been exceeded, and to suggest further management actions as needed. They also identified deviations of the ecological responses from the conceptual models that could be addressed before any formal regulatory triggers were exceeded.

The Public Service Enterprise Group restoration project proceeded well in meeting its targets after five years (Teal and Weishar 2005). As with many salt marsh restoration projects, the principle of "self-design" was incorporated into the restoration plan. This approach relies on the self-organizing ability of the ecosystem to develop an assemblage of plants, microbes, and animals that are best adapted for existing conditions (Mitsch and Wilson 1996; Mitsch 2000; Weishar et al. 2005). The substrate, hydrology, and plant propagules initially present or provided to a site at the outset of the project set the marsh on a course that restores ecological attributes and functions with a minimum of human intervention. In the Public Service Enterprise Group project, much of the drainage system evolved naturally into a system of creeks and channels after the tidal flow was restored by breaching the dikes and constructing primary and secondary channels. Vegetation goals were met far sooner than anticipated in the restoration plan even though the project relied almost entirely on natural reseeding and subsequent vegetative propagation by *Spartina alterniflora* (smooth cordgrass) rather than transplanting. Not surprisingly in a project of this magnitude, some adjustments were needed several years into the project. These included additional channelization to improve drainage in a few areas and the targeted use of herbicides and further flooding to eradicate *Phragmites* so that a more diverse assemblage of plants and animals could reestablish (Philipp and Field 2005; Teal and Peterson 2005; Teal and Weishar 2005). Such actions were based on monitoring results and therefore consistent with the adaptive management framework. Gregory et al. (2006) would likely call this an example of passive adaptive management, since the original conceptual model was not tested using a replicated experimental design.

Practical Guidance to Adaptive Management

Adaptive management and monitoring are essential components of an effective salt marsh restoration program. Stakeholders need to come together at the outset to agree on the overall project goals, conceptual models that describe the system and the management actions, and specific targets within general functional areas (e.g., hydrologic, soil, vegetative). There should be an understanding and a commitment from stakeholders that goals and targets may need to be adjusted during the course of the restoration if monitoring results are inconsistent with results anticipated from the conceptual model(s). This is a key component of adaptive management (Thom 2000); however, it is perhaps the most difficult step to implement.

There must be agreement on the conceptual models and schedules for monitoring specific parameters in the restored and reference marshes and for the triggers that will lead to a reexamination of management activities. Effective restoration requires a commitment to monitor the recovery of a restored site for a number of years, gauged to the anticipated trajectories of key parameters.

Use of standardized monitoring protocols among a number of restoration projects with managed and reference marshes allows for comparisons between restoration projects within a region. Regional analysis allows scientists to predict outcomes of restoration initiatives under varying conditions, thus informing future restoration efforts.

Including an experimental aspect (active adaptive management) in a subset of salt marsh restoration projects will afford restoration specialists and the scientific community opportunities to gain information on how to improve restoration practices and more efficiently meet targets and goals in restoration efforts. Sometimes field experiments are essential to determine optimum management activities in a current restoration project, but at other times the knowledge gained will benefit future restoration projects. Given past abuses (e.g., diking, filling, and ditching) and present problems of rising sea level, global warming, and cultural eutrophication, it is in the best interest of society to improve salt marsh restoration practices through an adaptive management framework with a strong monitoring component.

Acknowledgments

The authors thank the editors, Charles Roman and David Burdick, for their encouragement and guidance in producing this work. Kerstin Watson provided a very helpful review of the manuscript.

REFERENCES

Able, K. W., and S. M. Hagan. 2000. "Effects of Common Reed (*Phragmites australis*) Invasion on Marsh Surface Macrofauna: Responses of Fishes and Decapod Crustaceans." *Estuaries* 23:633–46.

Barbour, M. G., J. H. Burk, W. D. Pitts, F. S. Gilliam, and M. W. Schwartz. 1999. *Terrestrial Plant Ecology*. 3rd ed. Menlo Park, CA: Addison Wesley Longman.

Brinson, M. M., and R. Rheinhardt. 1996. "The Role of Reference Wetlands in Functional Assessment and Mitigation." *Ecological Applications* 6:69–76.

Bromberg, K. D., and M. D. Bertness. 2006. "Reconstructing New England Salt Marsh Losses Using Historical Maps." *Estuaries* 28:823–32.

Buchsbaum R. N., J. Catena, E. Hutchins, and M. J. James-Pirri. 2006. "Changes in Salt Marsh Vegetation, *Phragmites australis*, and Nekton in Response to Increased Tidal Flushing in a New England Salt Marsh." *Wetlands* 26:544–57.

Buchsbaum, R. N., L. A. Deegan, J. Horowitz, R. H. Garritt, A. E. Giblin, J. P. Ludlam, and D. H. Shull. 2009. "Effects of Regular Salt Marsh Haying on Marsh Plants, Algae, Invertebrates and Birds at Plum Island Sound, Massachusetts." *Wetlands Ecology and Management* 17:469–87.

Burdick, D. M., M. Dionne, R. M. Boumans, and F. T. Short. 1997. "Ecological Responses to Tidal Restorations of Two Northern New England Salt Marshes." *Wetlands Ecology and Management* 4:129–44.

Cahoon, D. R., and G. R. Guntenspergen. 2010. "Climate Change, Sea-Level Rise, and Coastal Wetlands." *National Wetlands Newsletter* 32:8–12.

Calloway, J. C., G. Sullivan, J. S. Desmond, G. D. Williams, and J. B. Zedler. 2001. "Assessment and Monitoring. Pp. 271–335 in *Handbook for Restoring Tidal Wetlands*, edited by J. B. Zedler. Boca Raton, FL: CRC Press.

Carlisle, B. K., A. M. Donovan, A. L. Hicks, V. S. Kooken, J. P. Smith, and A. R. Wilbur. 2002. *A Volunteer's Handbook for Monitoring New England Salt Marshes.* Boston: Massachusetts Office of Coastal Zone Management.

Carullo, M., B. K. Carlisle, and J. P. Smith. 2007. *A New England Rapid Assessment Method for Assessing Condition of Estuarine Marshes: A Boston Harbor, Cape Cod and Islands Pilot Study.* Boston: Massachusetts Office of Coastal Zone Management.

Chambers, R. M., D. T. Osgood, D. J. Bart, and F. Montalto. 2003. "*Phragmites australis* Invasion and Expansion in Tidal Wetlands. Interactions among Salinity, Sulfides, and Hydrology." *Estuaries* 26:398–406.

Clarke, K. R., and R. M. Warwick. 2001. *Change in Marine Communities: An Approach to Statistical Analysis and Interpretation.* 2nd ed. PRIMER-E: Plymouth. Plymouth, UK: Plymouth Marine Laboratory.

Collins, J. N., E. D. Stein, M. Sutula, R. Clark, A. E. Fetscher, L. Grenier, C. Grosso, and A. Wiskind. 2008. *California Rapid Assessment Method (CRAM) for Wetlands.* Version 5.0.2. http://www.cramwetlands.org/documents/2008-09-30_CRAM%205.0.2.pdf.

Conway, C. J. 2008. *Standardized North American Marsh Bird Monitoring Protocols.* Wildlife Research report no. 2008-01. Tucson, AZ: US Geological Survey, Arizona Cooperative Fish and Wildlife Research Unit.

Cornu, C. E., and S. Sadro. 2002. "Physical and Functional Responses to Experimental Marsh Surface Elevation Manipulation in Coos Bay's South Slough." *Restoration Ecology* 10:474–86.

Craft, C., J. Reader, J. N. Sacco, and S. W. Broome. 1999. "Twenty-five Years of Ecosystem Development of Constructed *Spartina alterniflora* L. Marshes." *Ecological Applications* 9:1405–19.

Fell, P. E., R. S. Warren, and W. A. Niering. 2000. "Restoration of Salt and Brackish Tidelands in Southern New England: Angiosperms, Macroinvertebrates, Fish, and Birds." Pp. 845–58 in *Concepts and Controversies in Tidal Marsh Ecology*, edited by M. P. Weinstein and D. A. Kreeger. Dordrecht: Kluwer Academic.

Fell, P. E., R. S. Warren, J. K. Light, R. L. Rawson Jr., and S. M. Fairley. 2003. "Comparison of Fish and Macroinvertebrate Use of *Typha angustifolia, Phragmites australis,* and Treated *Phragmites* Marshes along the Lower Connecticut River." *Estuaries* 28:534–51.

Gauch, H. G. 1982. *Multivariate Analysis in Community Ecology*. New York: Cambridge University Press.

Gregory, R., D. Ohlson, and J. Arvai. 2006. "Deconstructing Adaptive Management: Criteria for Applications to Environmental Management." *Ecological Applications* 16:2411–25.

Hartig, E. K., V. Gornitz, A. Kolker, F. Mushacke, and D. Fallon. 2001. "Anthropogenic and Climate Change Impacts on Salt Marshes of Jamaica Bay, New York City." *Wetlands* 22:71–89.

Jacobs, A., E. McLaughlin, and D. L. O'Brien. 2008. *Mid-Atlantic Tidal Wetland Rapid Assessment Method Version 1.0*. Dover: Delaware Department of Natural Resources and Environmental Control, Division of Water Resources.

James-Pirri, M. J., C. Roman, and J. Heltshe. 2007. "Power Analysis to Determine Sample Size for Monitoring Vegetation Change in Salt Marsh Habitats." *Wetlands Ecology and Management* 15:335–45.

James-Pirri, M. J., C. T. Roman, and J. L Swanson. 2010. "A Method to Quantitatively Sample Nekton in Salt-Marsh Ditches and Small Tidal Creeks." *Transactions of the American Fisheries Society* 139:413–19.

Kent, M., and P. Coker. 1992. *Vegetation Description and Analysis: A Practical Approach*. London: Bellhaven Press.

Kentula, M. E. 2007. "Foreword: Monitoring Wetlands at the Watershed Scale." *Wetlands* 27:412–15.

Konisky, R. A., D. M. Burdick, M. Dionne, and H. A. Neckles. 2006. "A Regional Assessment of Salt Marsh Restoration and Monitoring in the Gulf of Maine." *Restoration Ecology* 14:516–25.

Levin, L. A., C. Neira, and E. D. Grosholz. 2006. "Invasive Cordgrass Modifies Wetland Trophic Function." *Ecology* 87:419–32.

McKinney, R. A., K. B. Raposa, and T. E. Kutcher. 2010. "Use of Urban Marine Habitats by Foraging Wading Birds." *Urban Ecosystems* 13:191–208.

Mitsch, W. J. 2000. "Self-Design Applied to Coastal Restoration: An Application of Ecological Engineering." Pp. 554–64 in *Concepts and Controversies in Tidal Marsh Ecology*, edited by M. P. Weinstein and D. A. Kreeger. Dordrecht: Kluwer Academic.

Mitsch, W. J., and R. F. Wilson. 1996. "Improving Success of Wetland Creation and Restoration with Know-How, Time, and Self-Design." *Ecological Applications* 6:77–83.

Morgan, P. A., and F. T. Short. 2002. "Using Functional Trajectories to Track Constructed Marsh Development in the Great Bay Estuary, New Hampshire USA." *Restoration Ecology* 10:461–73.

Morris, J. T., P. V. Sundareshwar, C. T. Nietch, B. Kjerfve, and D. R. Cahoon. 2002. "Responses of Tidal Wetlands to Rising Sea Levels." *Ecology* 83:2869–77.

Neckles, H. A., and M. Dionne, eds. 2000. *Regional Standards to Identify and Evaluate Tidal Wetland Restoration in the Gulf of Maine: A GPAC Workshop Report*. Wells,

ME: Wells National Estuarine Research Reserve. http://www.pwrc.usgs.gov/resshow/neckles/gpac.htm.

Neckles, H. A., M. Dionne, D. M. Burdick, C. T. Roman, R. Buchsbaum, and E. Hutchins. 2002. "A Monitoring Protocol to Assess Tidal Restoration of Salt Marshes on Local and Regional Scales." *Restoration Ecology* 10:556–63.

Philipp, K. R., and R. T. Field. 2005. "*Phragmites australis* Expansion in Delaware Bay Salt Marshes." *Ecological Engineering* 25:275–91.

Portnoy, J. W., and A. E. Giblin. 1997. "Effects of Historic Tidal Restrictions on Salt Marsh Sediment Chemistry." *Biogeochemistry* 36:275–303.

Ralph C. J., J. R. Sauer, and S. Droege. 1995. *Monitoring Bird Populations by Point Counts.* General Technical Report PSW-GTR-149. Albany, CA: USDA Forest Service.

Raposa, K. B. 2002. "Early Responses of Fishes and Crustaceans to Restoration of a Tidally Restricted New England Salt Marsh." *Restoration Ecology* 10:665–76.

Raposa, K. B., and C. T. Roman. 2003. "Using Gradients of Tidal Restriction to Evaluate Nekton Community Response to Salt Marsh Restoration." *Estuaries* 26:98–105.

Raposa, K. B., C. T. Roman, and J. F. Heltshe. 2003. "Monitoring Nekton as a Bioindicator in Shallow Estuarine Habitats." *Environmental Monitoring and Assessment* 81:239–55.

Roman, C. T., M. J. James-Pirri, and J. F. Heltshe. 2001. *Monitoring Salt Marsh Vegetation: A Protocol for the Long-Term Coastal Ecosystem Monitoring Program at Cape Cod National Seashore.* Wellfleet, MA: Cape Cod National Seashore. www.nature.nps.gov/im/monitor/protocoldb.cfm.

Roman, C. T., K. B. Raposa, S. C. Adamowicz, M. J. James-Pirri, and J. G. Catena. 2002. "Quantifying Vegetation and Nekton Response to Tidal Restoration of a New England Salt Marsh." *Restoration Ecology* 10:450–60.

Rozas, L. 1992. "Bottomless Lift Nets for Quantitatively Sampling Nekton on Intertidal Marshes." *Marine Ecology Progress Series* 89:287–92.

Salafsky, N., R. Margoluis, and K. Redford. 2001. *Adaptive Management: A Tool for Conservation Practitioners.* Washington, DC: Biodiversity Support Program. http://www.fosonline.org/Site_Docs/AdaptiveManagementTool.pdf.

Scatolini, S. R., and J. B. Zedler. 1996. "Epibenthic Invertebrates of Natural and Constructed Marshes of San Diego Bay." *Wetlands* 16:24–37.

Short, F. T., D. M. Burdick, C. A. Short, R. C. Davis, and P. Morgan. 2000. "Developing Success Criteria for Restored Eelgrass, Salt Marsh and Mud Flat Habitats." *Ecological Engineering* 15:239–52.

Simenstad, C. A., and R. M. Thom. 1996. "Functional Equivalency Trajectories of the Restored Gog-Le-Hi-Te Estuarine Wetland." *Ecological Applications* 6:38–56.

Smith, S. M., and R. S. Warren. 2007. "Determining Ground Surface Topography in Tidal Marshes Using Watermarks." *Journal of Coastal Research* 23:265–69.

Society for Ecological Restoration International (SER) International Science and Policy Working Group. 2004. *The SER International Primer on Ecological Restoration.* Tucson, AZ: Society for Ecological Restoration International.

Stewart-Oaten, A., and J. R. Bence. 2001. "Temporal and Spatial Variation in Environmental Impact Assessment." *Ecological Monographs* 71:305–39.

Teal, J. M., and S. B. Peterson. 2005. "Introduction to the Delaware Bay Salt Marsh Restoration." *Ecological Engineering* 25:199–203.

Teal, J. M., and L. Weishar. 2005. "Ecological Engineering, Adaptive Management, and Restoration Management in Delaware Bay Salt Marsh Restoration." *Ecological Engineering* 25:304–14.

Teo, S. L. H., and K. W. Able. 2003. "Habitat Use and Movement of the Mummichog (*Fundulus heteroclitus*) in a Restored Salt Marsh." *Estuaries* 26:720–30.

Thom, R. 2000. "Adaptive Management of Coastal Ecosystem Restoration Projects." *Ecological Engineering* 16:365–72.

Underwood, A. J. 1992. "Beyond BACI: The Detection of Environmental Impacts on Populations in the Real, but Variable, World." *Journal of Experimental Marine Biology and Ecology* 161:145–78.

US Environmental Protection Agency (EPA). 2002a. *Wetland Monitoring and Assessment: A Technical Framework*. EPA-843-F-02-002(h). Washington, DC: Office of Water, US Environmental Protection Agency.

US Environmental Protection Agency (EPA). 2002b. *Methods for Evaluating Wetland Condition: Study Design for Monitoring Wetlands*. EPA-822-R-02-015. Washington, DC: Office of Water, US Environmental Protection Agency.

US Environmental Protection Agency (EPA). 2002c. *Methods for Evaluating Wetland Condition: Using Vegetation to Assess Environmental Conditions in Wetlands*. EPA-822-R-02-020. Washington, DC: Office of Water, US Environmental Protection Agency.

US Environmental Protection Agency (EPA). 2002d. *Methods for Evaluating Wetland Condition: Developing an Invertebrate Index of Biological Integrity for Wetlands*. EPA-822-R-02-019. Washington, DC: Office of Water, US Environmental Protection Agency.

US Environmental Protection Agency (EPA). 2002e. *Methods for Evaluating Wetland Condition: Biological Assessment Methods for Birds*. EPA-822-R-02-023. Washington, DC: Office of Water, US Environmental Protection Agency.

Warren, R. S., P. E. Fell, J. L. Grimsby, E. L. Buck, G. C. Rilling, and R. A. Fertik. 2001. "Rates, Patterns and Impacts of *Phragmites australis* Expansion and Effects of Experimental *Phragmites* Control on Vegetation, Macroinvertebrates, and Fish within Tidelands of the Lower Connecticut River." *Estuaries* 24:90–107.

Warren, R. S., P. E. Fell, R. Rozsa, A. H. Brawley, A. C. Orsted, E. T. Olson, V. Swamy, and W. A. Niering. 2002. "Salt Marsh Restoration in Connecticut: 20 Years of Science and Management." *Restoration Ecology* 10:497–513.

Weis, J., and P. Weis. 2003. "Is the Invasion of the Common Reed, *Phragmites australis*, into Tidal Marshes of the Eastern US an Ecological Disaster?" *Marine Pollution Bulletin* 46:816–20.

Weishar, L., J. M. Teal, and R. Hinkle. 2005. "Designing Large-Scale Wetland Restoration for Delaware Bay." *Ecological Engineering* 25:231–39.

Wigand, C., R. McKinney, M. Chintala, S. Lussier, and J. Heltshe. 2010. "Development of a Reference Coastal Wetland Set in Southern New England (USA)." *Environmental Monitoring and Assessment* 161:583–98.

Williams, B. K., R. C. Szaro, and C. D. Shapiro. 2007. *Adaptive Management: The U.S. Department of the Interior Technical Guide*. Washington, DC: Adaptive Management Working Group, US Department of the Interior.

Wozniak, A. S., C. T. Roman, S. C. Wainright, R. A. McKinney, and M. J. James-Pirri. 2006. "Monitoring Food Web Changes in Tide-Restored Salt Marshes: A Carbon Stable Isotope Approach." *Estuaries and Coasts* 29:568–78.

Zedler, J. B. 2006. "Wetland Restoration." Pp. 348–406 in *Ecology of Freshwater and Estuarine Wetlands*, edited by D. P. Batzer and R. R. Sharitz. Berkeley: University of California Press.

Zedler, J. B., and J. C. Calloway. 1999. "Tracking Wetland Restoration: Do Mitigation Sites Follow Desired Trajectories?" *Restoration Ecology* 7:69–73.

Chapter 15

Recovering Salt Marsh Ecosystem Services through Tidal Restoration

GAIL L. CHMURA, DAVID M. BURDICK, AND GREGG E. MOORE

Some would maintain that conservation and restoration activities are justified on ethical grounds alone (see review by Brennan and Lo 2008). However, demonstration of the economic benefit of ecosystems can help drive social and governmental support for conservation; and restoration and economic limitations could force choices among restoration activities. To aid decision making we need to estimate the values that restored ecosystems will provide for society. But defining these values remains a significant challenge, particularly within the context of restoration in which functions have been impaired and may contribute only incremental services over the varying course of the restoration process. Nonetheless, wetlands have direct and indirect economic value to local communities, and they provide services that benefit society as a whole. The term "ecosystem services" encompasses benefits that have direct economic value and those that have indirect public benefits. Evaluating and quantifying ecosystem services is a challenge regardless of the system status: natural, disturbed, or in various stages of restoration.

The Concept of Ecosystem Services

Mooney and Ehrlich (1997) reviewed the history of the ecosystem services concept, which they suggest existed even before the term "ecosystem" was introduced by Tansley (1935). The concept was first directly articulated by the Study of Critical Environmental Problems (SCEP 1970), which produced a list of nine "environmental services" (pest control, insect pollination, fisheries, climate regulation, soil retention, flood control, soil formation, cycling of organic matter, and composition of the atmosphere). Holdren and Ehrlich (1974) later used the term "public-service functions of the global environment" and added two more (maintenance of soil fertility and of a genetic library). The concept was later referred to as nature's services (Westman 1977) and eventually as ecosystem services by Ehrlich and Ehrlich (1981). Ecosystem services were the focus of a book edited by Daily (1997, 3) who defined ecosystem services as "the conditions and processes through which natural ecosystems, and the species that make them up, sustain and fulfill human life," with a list of fourteen services embracing those mentioned by Holdren and Ehrlich (1974). The term "ecosystem services" has since been in common use, although many have noted that "services" and the ecosystem functions or processes that provide them are frequently confused (e.g., Brander et al. 2006; Fisher et al. 2009). This and a varying list of services and meanings confound attempts to place a dollar value on wetlands.

An effort has been made to reconcile the various ecosystem services applicable to salt marshes (table 15.1). Costanza et al. (1997) listed ecosystem services for tidal marshes and mangroves; the Millennium Ecosystem Assessment (2005) considered estuaries and marshes (including mangroves); and Zedler and Kercher (2005) considered wetlands. These have provided more general terms for services that condense the list in Daily (1997). For instance, "waste treatment" or "waste processing" encompasses Daily's two services "purification of air and water" and "decomposition of wastes." Direct benefits such as production of food, fuel, and raw materials, considered "provisioning services" by the Millennium Ecosystem Assessment (2005), were not included in Daily's list, but were noted in the latter reports. Most recently, Brander et al. (2006) reported ten ecological functions that support sixteen economic goods and services.

Costanza et al. (1997) categorized ecosystem services by biomes and assigned a dollar value to each, including tidal marshes and mangroves. Even though economists showed their economic valuation was flawed (Bockstael et al. 2000), the results were surprising and show why most shallow water habitats have become highly valued by society and are protected in the United States. The combined extent of tidal marshes and mangroves accounted for 0.3 percent of the global area and 5 percent of the annual global value, yielding a per hectare value of $9990. Subsequently, we have learned more about marsh ecosystem services. Atmosphere and climate regulation, cultural and amenity services, and aesthetics should all be recognized as values of tidal marshes (Brander et al. 2006). The next section of this chapter discusses the ecosystem services as they apply to tidal marshes and provides direct and indirect examples of how these services are returned to recovering or restored marshes.

Many attempts to quantify economic values of wetlands, including tidal marshes, have fueled meta-analyses with interesting results (Woodward and Wui 2001; Brander et al. 2006). Both studies found that values range widely ($1 to over $25,000 per hectare), are based on few services (typically one), and tend to be site

TABLE 15.1

Comparison of ecosystem services identified by Daily (1997) to those specified for wetlands or tidal marshes by others

General ecosystems (Daily 1997)	Ecosystem Services			
	Tidal marsh and mangroves (Costanza et al. 1997)	Estuaries and marshes (Millennium Assmnt. 2005)	Wetlands (Zedler & Kircher 2005)	Wetlands (Brander et al. 2006)
Not included	Food production; raw materials	Fiber, timber, fuel	Food production; raw materials	Commercial and recreational fishing and hunting; harvesting of natural materials; energy resources
Maintenance of biodiversity	Habitat/refugia	Biodiversity	Habitat/refugia	Appreciation of species existence
Provision of aesthetic beauty and intellectual stimulation that lift the human spirit	Recreation	Cultural and amenity; aesthetics; recreational	Cultural; recreational	Recreational activities; appreciation of uniqueness to culture/heritage
Protection of coastal shores from erosion by waves.	Disturbance regulation	Flood/storm protection; erosion control	Disturbance regulation	Storm protection flood protection
Protection from ultraviolet rays; partial stabilization of climate; moderation of weather extremes and impacts	Not included	Atmosphere and climate regulation	Gas regulation	Climate stabilization; reduced global warming
Purification of air and water; detoxification and decomposition of wastes	Waste treatment	Waste processing	Waste treatment	Improved water quality; waste disposal
Cycling and movement of nutrients	Not included	Nutrient cycling and fertility	Nutrient cycling	Improved water quality; waste disposal

specific. With all of its flaws, the value proposed by Costanza et al. (1997) ($9990) fell into the high end of the range but is reasonable, since most services were included. Five studies supported by strong data and including four to five services averaged $9900 per hectare in 1990 dollars (Woodward and Wui 2001). Gedan et al. (2009) suggest the current value of services from tidal marshes originally estimated by Costanza et al. (1997) is now $14,397 per hectare annually due to inflation.

Ecosystem Services and Evidence of Losses and Recovery

This section addresses evidence of losses and recovery of ecosystem services that have been documented in studies of tidal marshes of the Gulf of Maine.

Food Production and Raw Material

Costanza et al. (1997, 254) describe the food production service as "that portion of gross *primary* production extractable as food," and give the following example: "production of fish, game, crops, nuts, fruits by hunting, gathering, subsistence farming or fishing." But their wording presents a contradiction, as fish and game are actually secondary production available through habitat provision of an ecosystem. They list the ecosystem function for refugia as "habitat for resident and transient populations" with this example: "nurseries, habitat for migratory species, regional habitats for locally harvested species, or overwintering grounds." The overlap of these two services and functions is particularly apparent in salt marshes, which are widely recognized as critical habitat for migratory waterfowl and refugia for juvenile fish. Here we differentiate the two by limiting food production to primary production or subsistence harvests of secondary production; otherwise the harvest activity is considered recreation.

In the United States and Canada, protective environmental regulations generally prevent individuals or communities from gaining direct economic benefits by using salt marsh vegetation for food production or harvesting raw material. An exception is the harvest of *Spartina patens* (salt meadow cordgrass) where a long cultural history of harvest for livestock fodder and mulch continues to this day, although quite limited in occurrence throughout New England and Atlantic Canada. Florists are known to collect *Limonium carolinianum* (sea lavender) blooms to sell commercially (though this is prohibited in some US states). Acadians eat *Plantago maritima* (seaside plantain), and New Englanders gather *Salicornia maritima* (slender glasswort) for salads. Secondary production in salt marshes of northeastern North America is more important than primary production for subsistence living and recreational pursuits. In particular ducks and geese are hunted in the fall migration. Migratory fish are harvested throughout the year from coastal waters adjacent to marshes and range from anadromous species like shad, river herring, and smelt, to sport fishes like winter flounder and striped bass. In this region such harvests probably have minimal importance as subsistence hunting and fishing; thus fish and waterfowl production are more appropriately covered under habitat and refugia.

Recovery of primary production that supports direct harvests and secondary production may be assessed by direct comparisons of vegetation communities in recovering and reference marshes. To regain the full values associated with primary production we should consider the species composition of the production. Diversity indices in association with totals of primary production can be misleading when one is considering the success of recovery. Reference and restored marshes could have similar levels of primary production but quite different species composition. Increases in species richness in wetlands that typically have low diversity could be driven by establishment of invasive or terrestrial species (Keddy 2000). Neither species richness nor other diversity indices will reveal these differences, and thus are not valuable as functional indicators.

Less salt tolerant and exotic plants such as *Lythrum salicaria* (purple loosestrife) and *Phragmites australis* (common reed) typically invade marshes with restricted tidal regimes (Roman et al. 1984). A Eurasian variety of *P. australis* has invaded tidal marshes in the Northeast (Saltonstall 2002)—often leading to a monoculture—excluding native species and negatively impacting many functions and services of tidal marshes (Chambers et al., chap. 5, this volume). Reintroduction of tides can reduce dominance of invasive plants, especially in marshes where average salinities become greater than 20 parts per thousand. To gauge plant community response to restoration, simple indices such as native plant cover or a ratio of invasive to native plants have been useful (Konisky et al. 2006).

Alternatively, investigators (Thom et al. 2002; Byers and Chmura 2007) have used a modified Sorenson Index to determine similarity of recovering and reference marshes. Byers and Chmura (2007) compared the composition and standing crop of vegetation in two pairs of recovering and reference marshes on the Bay of Fundy. The recovering marshes, Saints Rest, New Brunswick (lower Bay), and John Lusby Marsh, Nova Scotia (upper Bay), had been drained, ditched, and under management for terrestrial agriculture for one hundred and three hundred years, respectively. Despite differences in marsh size, weighted similarities based upon reference marshes revealed that Saints Rest Marsh had 74 percent similarity with respect to plant cover and 90 percent with respect to standing crop. John Lusby had 47 percent similarity in plant cover and 71 percent in standing crop.

Biodiversity, Habitat, and Refugia

Tidal salt marshes are not particularly diverse. Few vascular plants have evolved mechanisms that enable them to tolerate flooded and saline soils. But despite low species richness, salt marshes provide habitat for rare species of flora and fauna, thus contributing to global biodiversity. A number of rare species require salt marsh habitat for at least one stage of their life cycle. In the United States, twenty-six species listed as endangered or threatened by the US Fish and Wildlife Service utilize salt marshes or mangrove swamps, and in Canada, six endangered species use salt marshes for some phase of their life cycle (Chmura et al. in preparation).

These globally rare species may be considered a genetic resource, a value not attributed to salt marshes by Costanza et al. (1997). A long-term research program led by Gallagher (1985) reviews the potential of salt marsh plants for agricultural use, including forage, grain, vegetable crops, and even oil crops such as *Kosteletz-kya virginica* (Virginia saltmarsh mallow) for energy. The salt tolerance of *K. virginica* and other salt marsh species makes them valuable options for culture in salinized soils, which will become more prevalent as sea level rises and climate varies.

The value of tidal marshes in support of secondary production is widely noted (e.g., Turner 1977). Exported primary production becomes part of a detrital food chain where the nutritional value of dead vascular plants is enhanced by microbes (Fenchel and Jorgensen 1977) beginning a trophic relay. Detritivores strip off digestible portions of detritus and are consumed by a variety of small predators like shrimp and fish, which carry the energy seaward to tidal creeks and coastal embayments. Here, a portion is consumed by larger fish and birds, which carry the energy even farther from the marsh (Kneib 2002). Probably just as important are marsh creeks, ponds, and edges in providing refuge to juvenile fish, which feed on soil fauna when they access marsh surfaces during flooding tides.

Studies at multiple sites reveal rapid recovery of nekton populations after restoring tidal flooding to salt marshes (Burdick et al. 1997; Raposa and Talley, chap. 6, this volume). Following the model used for vegetation, Sorensen's similarity index was applied to assess recovery of secondary production at Sachuest Point Marsh in Rhode Island (Roman et al. 2002). Response of nekton in creeks and pools was evaluated after tidal restrictions were removed and 2360 square meters of creek and pool habitat were created using open marsh water management techniques. Similarity between the reference and restored sites was 99 percent in the first year of restoration and 98 percent in the second. Recovery of the nekton use of the vegetated marsh surface was also rapid—89 percent similarity between reference and restored sites during the first year of restoration, with an increase to 98 percent in the second year. Even though recovery of species composition and abundance appears rapid—and growth rate and production of nekton such as *Fundulus heteroclitus* (mummichog) are comparable between natural and tide-restored salt marshes (Teo and Able 2003)—the passage of nekton through new structures installed to reestablish tidal flow can be curtailed by high water velocity interrupting export (i.e., trophic relay; Eberhardt et al. 2011).

Recreational Uses

Most recreational uses are directly derived from habitat and refugia services. The wildlife populations supported by marshes provide indirect economic benefits to the communities that provide supplies and services for those participating in consumptive uses such as waterfowl hunting or nonconsumptive uses like bird-watching.

Both residents and tourists enjoy visiting marshes using trails and boardwalks provided on conservation lands (e.g., wildlife refuges and national, state, and local parks). Marshes with histories of disturbance can provide the same recreational value. Restored marshes have as great, or perhaps greater, value in this regard, as the cultural history and restoration process can make them more interesting. Recreational uses can become an integral component of local economies through enhancing nature tourism activities (Day 2009). In a meta-analysis of wetland values from services; amenity value, biodiversity, water quality, and flood control ranked highest (Brander et al. 2006).

Flood and Storm Protection and Erosion Control

Marshes create surface friction for tidal surges and attenuate wave energy (Barbier et al. 2008; Morgan et al. 2009), thus reducing damage to adjacent personal property and infrastructure during storms. Koch et al. (2009) demonstrated that, although wave attenuation decreases with decreasing distance between the seaward edge of the wetland and the inland property, it is not linear in either space or time. Wave attenuation value increases with plant height and plant density, each varying with the distinct plant community zonation. Tidal restrictions generally do not affect the erosion control service of marshes but are associated with dramatic flooding events that can damage adjacent property (Diers and Richardson, chap. 11, this volume). Conversely, restored hydrology can lead to flooding of property built subsequent to the tidal restriction (Portnoy, Adamowicz and O'Brien, and Reiner, chaps. 18, 19, 21, this volume). Amelioration or exacerbation of flooding with hydrologic restoration of tidal marshes will continue to be an important consideration for restoration design as sea level rises.

Regulation of Atmospheric Gases (Climate Regulation)

Despite their widely recognized value in support of secondary production, much of the primary production of tidal salt marshes is buried in the marsh soil, sequestering the carbon dioxide trapped through photosynthesis. Chmura et al. (2003) estimated that, globally, tidal salt marshes store an average of 210 grams of carbon per square meter per year. Most wetlands are good carbon sinks, but in freshwater wetlands anaerobic decomposition produces methane, a greenhouse gas with twenty-five times the global warming potential of carbon dioxide. The presence of sulfates in salt marsh soils reduces the activity of microbes that produce methane, and emissions of this gas from tidal salt marshes with soil salinity more than 22 parts per thousand is negligible (Bridgham et al. 2006). Methane production should decrease where tidal flows and salt marsh conditions are restored in areas that had become freshwater wetland, but we are unaware of any documenting studies.

If tidal marshes are able to sustain themselves with rising sea level they will continue to accrete and store carbon, in contrast to terrestrial soils, which increase little in volume and reach an equilibrium in their organic matter content (Chmura 2009). Thus, on an area basis, salt marshes and mangroves are some of the world's most valuable natural carbon sinks. Assessment of the recovery of this value requires that the rate of carbon sequestration be determined; that is, the mass of carbon (or, alternatively, organic matter) burial per unit time. Unfortunately, most surveys simply document the percent of organic matter by weight, which can be misleading as soils with high proportions of organic matter and low bulk densities may store less carbon than soils with low proportions of organic matter and high bulk densities (Chmura et al. 2003).

Rates of carbon storage in restored marshes appear to be comparable to those from reference sites. In Connecticut, Anisfeld et al. (1999) determined soil accretion rates of marshes with tidal restrictions, those where restrictions had been removed, and reference sites. Using their records of organic matter and bulk density, we can calculate rates of carbon storage to determine response to restoration. At three undisturbed *Spartina alterniflora* (smooth cordgrass) marshes, average long-term rates of carbon storage (165 ± 35 grams per square meter per year) were more than twice that in the tide-restricted marsh (72 grams per square meter per year), and the two restored sites showed rates comparable to the undisturbed site (181 and 182 grams per square meter per year). In *Spartina patens* marsh the average rate of carbon storage in three undisturbed sites (139 ± 28 grams per square meter per year) was not as distinct from the average at six restricted sites (109 ± 27 grams per square meter per year).

The carbon regained in soils of restored marshes has a market value, albeit extremely variable. During 2009, prices ranged from US$0.10 per 1000 kg of carbon dioxide (CO_2) equivalent in the Chicago Climate Exchange—a voluntary carbon market—to €15 per tonne in the European Union exchange. The latter is presently the world's only regulated carbon market, and carbon price is a function of the demand and supply for credits in the EU cap and trade system. The global average rate of CO_2 sequestered per hectare of salt marsh is 7.7 tonnes (based upon estimates reported by Chmura et al. 2003) giving salt marsh soil carbon a value of US$0.77 to US$138 per hectare per year (assuming €1 = $US1.196).

Waste Treatment

We consider waste treatment to include the trapping of heavy metals in marsh soils, from direct uptake from sewage (Giblin et al. 1983) to sequestration in relatively pristine areas (Hung and Chmura 2007). Halophytes are known to translocate metals from soils through salt excretion glands (Kraus 1988), but the vertical accumulation of the soil and the soil chemistry driven by flooding of saltwater make salt marshes effective traps for heavy metals. Hung and Chmura (2007) found that the extent to which marshes trap trace metals is directly related to the rate of surface sediment deposition. They also showed that mercury accumulation in marsh sediment of the Bay of Fundy comprised 14 percent of the atmospheric deposition to the provinces of Nova Scotia and New Brunswick. They concluded that the higher redox potential of Fundy's macrotidal marshes resulted in reduced translocation of mercury and reduced production of methyl mercury compared with salt marshes in New England.

In their studies of metal sequestration, Hung and Chmura (2006, 2007) examined both undisturbed and recovering marshes. Although their study was not designed to determine the extent to which soils of recovering marshes regained capacity to trap metals, a comparison of metal sequestration rates suggests that this capacity is easily returned along the sediment-rich coast of the Bay of Fundy.

Nutrient Filtering

All wetlands, including tidal marshes, are widely noted as filters for nutrients in coastal waters (e.g., Valiela et al. 2002). Both primary production and decomposition of the vegetation fueling the detrital food web are nitrogen limited (Valiela et al. 1985; Sousa et al. 2008) and provide temporary sinks that remove dissolved nitrogen, reducing effects of eutrophication (Valiela et al. 2002). In addition, denitrification is an important process in salt marsh soils, and as an ecosystem service, Piehler and Smyth (2011) have calculated the value of tidal marshes to be $6128 per hectare based on the North Carolina nutrient offset program.

Some portion of the nitrogen taken up by plants cycles back to the sediments; and gas flux studies have shown that wetlands enriched in nitrogen release nitrous oxide (Moseman-Valtierra et al. 2011), a greenhouse gas with 298 times the global warming potential of carbon dioxide (Forster et al. 2007). Thus the service provided by nutrient regulation may be offset by a reduction in greenhouse gas regulation. The net positive value of the marsh soil as a carbon sink would be maintained only if more than 298 molecules of carbon dioxide were stored in biomass for each molecule of nitrous oxide released.

The value of a salt marsh as a nutrient "filter" also might be at the expense of the sustainability of the salt marsh. Fertilization experiments show that the two dominant grasses of Atlantic salt marshes, S. alterniflora (Darby and Turner 2008) and S. patens (Chmura, unpublished data) increase their aboveground production but decrease their belowground biomass in response to nutrient additions. This could reduce carbon storage and the ability of the marshes to grow vertically and thus keep pace with local rates of sea level rise (Donnelly and Bertness 2001). By exposing marshes to very high nutrient levels, the nutrient filter ecosystem service may compromise marsh sustainability. Therefore, over some threshold the nutrient filter service might occur at the expense of all others performed by a salt marsh and should not be seen as an acceptable compromise to better management of pollution sources from watersheds or human activities (e.g., urban sewage).

Ecosystem Functions That Support Services

The return of ecosystem services requires the return of marsh function and processes that support those services. A major control underlying all tidal salt marsh functions is the frequency and duration of tidal flooding of marsh surfaces. Marsh channels play a key role in this regard, delivering floodwater to marsh surfaces, as well as nekton that breed and feed on marsh surfaces during flood tides. One downside to tidal restoration, although perhaps short term, is the potential loss of nesting habitat for Ammodramus caudacutus (sharp-tailed sparrow) (Adamowicz and O'Brien, chap. 19, this volume; Golet et al., chap. 20, this volume).

A major disturbance to marsh channel networks is the incorporation of systems of ditches to drain marshes for mosquito control or for conversion to terrestrial agriculture (Adamowicz and Roman 2005; Crain et al. 2009). Extensive areas of marsh in the Bay of Fundy were transformed for the purpose of terrestrial agriculture. MacDonald et al. (2010) assessed the extent to which hydrologic systems recovered after tidal flooding was returned to two such agricultural marshes. Although some original channels had been lost, drainage ditches became incorporated into a new hybrid channel network with a comparable density to the reference marsh, but with reduced sinuosity. Pools covered 13 percent of the marsh platform as compared to 5 percent at the reference marsh.

Regardless of the hydrologic path, flooding increases with tidal restoration and reestablishes marsh building processes that allow marshes to maintain elevations relative to sea level under conditions of slow sea level rise, perhaps up to 5 mm per year (Morris et al. 2002). Marsh accretion occurs through sediment deposition and organic matter inputs, notably from belowground production. Greater flooding also leads to soil reduction, which decreases oxidative carbon loss to the atmosphere. Hydrologic restrictions typically lead to marsh oxidation and subsidence as documented by Anisfeld (chap. 3, this volume). At sites where the marsh surface has not subsided below elevations where emergent vegetation can survive, measurements of marsh elevations following restoration show partial yet dramatic recovery (Boumans et al. 2002; Anisfeld, chap. 3, this volume).

Trajectory Model to Assess Recovery of Ecosystem Services and Economic Valuation

Salt marsh creation and restoration efforts in the United States and Canada have resulted in a broad range of outcomes. We have learned that recovery begins after the construction equipment leaves the marsh, and the full range of functions can take over twenty years to develop (Craft et al. 1999; Warren et al. 2002). However, resource managers need to assess restoration performance and enumerate benefits more quickly.

The Gulf of Maine monitoring protocol (Neckles et al. 2002) was developed to standardize assessment of structural and functional indicators and is now commonly used in the region to assess salt marsh restoration (Konisky et al. 2006; Kachmar and Hertz, chap. 12, this volume). The protocol includes four broad categories of indicators (hydrology, soils, vegetation, and wildlife) that are assessed before and after restoration at the managed system and a suitable reference marsh. Building upon the protocol, Moore et al. (2009) designed the restoration performance index (RPI) to document the performance of salt marsh restoration projects by integrating the contribution of several structural and functional indicators simultaneously. The RPI can use any number of variables to assess functional categories associated with many ecosystem services to gauge the progress of a restoration project relative to a reference marsh. Yet, note that some indicators could represent several ecosystem services and more than one indicator could inform us about a single service.

One can compare the structural and functional indicators (rather than estimates of the services themselves) of impacted versus reference marshes to estimate the services lost from the impacted marsh before restoration. Similarly, one can estimate the services gained following the restoration action. This approach assumes the proportional *difference* in indicators equals the proportional *difference* in ecosystem services. For example, a marsh exhibiting 50 percent of indicator levels compared to its reference will only provide 50 percent of the ecosystem services. One way to express restoration benefits is to use an economic value of ecosystem services of reference marshes (e.g., Costanza et al. 1997), and multiply the dollar value by the proportional increase in indicators—and thus services—over the area of the restored marsh.

Consider an example using long-term monitoring data. Tides were restored at Little River Marsh, New Hampshire, where several indicators were used over time to assess benefits (Burdick et al. 2010). Over a hundred years ago, a small tidal inlet was filled and replaced with a new waterway terminating in an undersized culvert. Tides were restored in 2000 by widening the artificial waterway and installing a larger culvert. A reference was established at nearby Awcomin Marsh. Hydrology, soil salinity, vegetation, and nekton were assessed using the Gulf of Maine Protocol before and after restoration (table 15.2).

First, we evaluate impacts to ecosystem functions caused by the tidal restriction at Little River, and equate losses to diminished ecosystem services. To determine the portion of hydrology function lost at the impacted site, we calculated the ratio of the indicator measured at the impacted site to its reference. In our example four variables of hydrology were measured. For the first variable (table 15.2) 45.5 percent of the impacted site was flooded as compared to 98.2 percent of the reference. Dividing 45.5 by 98.2 provides the proportion of hydrologic function remaining for this variable as 0.463. This calculation is repeated for the remaining three variables of hydrology, then all are averaged to derive a single score for hydrologic function remaining at the impacted site: 0.24. We then employ the same comparison to assess the retained functions of soil chemistry (pore water salinity), vegetation, and nekton, averaging all the variables to reveal that the impacted site retained 0.59 of its original ecosystem functions, or overall ecosystem services supplied.

This value (0.59) is used as the baseline for similar calculations derived from monitoring indicators over several years following restoration (table 15.2). With installation of the expanded culvert, hydrologic benefits were realized immediately. Spring tidal range increased over 2.4-fold and approached 75 percent of the reference marsh. The proportion of marsh area flooded during spring tides increased from 45 to 99 percent, and the area flooded during neap tides increased from 0 to 74 percent, much greater than neap tide flooding of the reference area (18 percent; table 15.2). To derive the hydrologic portion of the RPI, we determine how much each indicator approached the reference condition from its original impacted baseline. The average hydrologic score is 0.80 (table 15.2). Soil

TABLE 15.2

Restoration performance index (RPI) for assessing the Little River Marsh restoration

Functional area	Core variable	Year 0 (Pre) Ref.	Year 0 (Pre) Pre-rest.	Year 1 (Post) Ref.	Year 1 (Post) Rest.	Partial RPI	Year 2 Ref.	Year 2 Rest.	Year 3 Ref.	Year 3 Rest.
Hydrology	Area flooded (%) spring tide	98.2	45.5	98.2	99.1	1.00	98.2	99.1	98.2	99.1
	Area flooded (%) neap tide	18.6	0.0	18.6	74.3	1.00	18.6	74.3	18.6	74.3
	Potential range (m) spring tide	1.77	0.57	1.83	1.36	0.63	1.83	1.36	1.83	1.36
	Potential range (m) neap tide	1.26	0.23	1.17	0.77	0.58	1.17	0.77	1.17	0.77
Pore water	Salinity (ppt)	26.2	17.5	25.0	25.7	1.00	29.2	26.6	31.5	31.0
Vegetation	Halophyte cover (%)	65.9	59.9	70.9	39.9	0.00			64.6	34.5
	Invasive cover (%)	0.0	10.2	0.0	2.7	0.59			0.1	2.1
Nekton	Density (#/m²)	20.1	12.7							
	Species richness	7.5	3.1							
RPI			0.59		0.70			0.79		0.71

Functional area	Core variable	Year 4 Ref.	Year 4 Rest.	Year 5 Ref.	Year 5 Rest.	Year 6 Ref.	Year 6 Rest.	Year 7 Ref.	Year 7 Rest.
Hydrology	Area flooded (%) spring tide	98.2	99.1	98.2	99.1	98.2	99.1	98.2	99.1
	Area flooded (%) neap tide	18.6	74.3	18.6	74.3	18.6	74.3	18.6	74.3
	Potential range (m) spring tide	1.83	1.36	1.83	1.36	1.83	1.36	1.83	1.36
	Potential range (m) neap tide	1.17	0.77	1.17	0.77	1.17	0.77	1.17	0.77
Pore water	Salinity (ppt)	31.5	24.9	25.4	19.6	22.7	23.6	28.6	29.4
Vegetation	Halophyte cover (%)			68.0	51.8	63.4	63.6		
	Invasive cover (%)			0.0	1.9	0.0	3.1		
Nekton	Density (#/m²)			35.6	13.8	35.6	38.0		
	Species richness			9.0	4.0	6.00	9.0		
RPI			0.67		0.51		0.91		0.91

Note: As an example, RPI partial scores are calculated in Year 1 for each component (RPI partial = ($Rest_{YR1}$ − Pre-rest) /Ref_{YR1} − Pre-rest) and then averaged using equal weighting. Values are truncated at 1.00 (cannot exceed reference) and 0.00 for each partial score before averaging.

salinity increased in the first year from 17.5 parts per thousand (66 percent of reference salinity) to 25.7 parts per thousand (over 100 percent of reference) so its contribution is 1.0. Vegetation changed from shrub to meadow due to salt water flooding and both invasive and native plants were reduced in cover and vigor (Burdick et al. 2010). Halophyte cover contributed 0.00 and invasive declines contributed 0.59, averaging 0.30 for the vegetative contribution to the RPI. The three indicator groups are then averaged and resulted in Little River achieving RPI of 0.70 relative to the reference marsh after year one (fig. 15.1). That is, the recovering marsh had increased indicator levels (and we assume ecosystem services) from the restricted condition of just 59 percent of reference function to 70 percent. Vegetation declines in large pannes and lower salinity levels due to abundant rainfall led to declines in the RPI in years four and five. Fish collections made in year six indicated populations were increasing and approaching abundances found in the reference marsh, leading to rebound of the RPI to about 90 percent of reference marsh indicators (fig. 15.1).

The economic values attributed to marsh ecosystem services were used to assess the value lost due to impacts or gained as a result of restoration. For illustrative purposes, if we assume the annual ecosystem services derived from tidal marshes are worth $14,535 per hectare ($14,397 from Gedan et al. 2009 plus $138 for carbon storage), then annual losses at Little River constituted 41 percent (since 59 percent were retained) or $6057 per hectare, or just under $424,000 for the 70 hectare marsh.

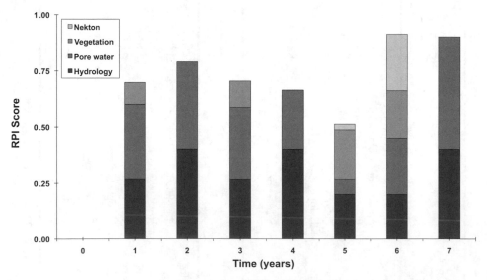

FIGURE 15.1. Restoration performance index (RPI) of Little River Marsh, New Hampshire.

Following restoration, Little River Marsh has accrued considerable ecosystem service benefits over time (fig. 15.1). We can calculate benefits by multiplying the lost values ($6057 per hectare) by the RPI proportion restored that year for each hectare. Restoration benefits were valued at $296,242 in 2001 (year one) with an RPI score of 0.70, and increased to $382,290 in 2007 (RPI = 0.91). Over the seven years of restoration, this effort has accrued $2,200,000 in ecosystem services, outweighing the restoration costs that included construction and monitoring ($1,310,000).

Future Considerations

Restoration actions must be effective and efficient because resources to protect and restore marshes are limited and the threats to marsh sustainability are great. Further research is needed to document the ecosystem services provided by tidal marshes (especially for carbon storage, climate regulation, and amenity services) and development of economic valuations that can be generalized and applied to any tidal marsh. The RPI provides one method to assess restoration performance, offering flexibility and insight into how restoration projects are faring. As coastal resource managers engage in restoration projects, monitoring should be used to document changes and adapt management strategies. Future work may assess ecosystem services in degraded and restored marshes directly, rather than estimating changes from structural and functional indicators. Until that time, economic valuation of ecosystem services based on indicators can be conducted to support ongoing and future restoration investments.

Acknowledgments

Funding was provided by the Natural Resources and Engineering Research Council of Canada (GC) and the NOAA Restoration Center (DB and GM; grant no. NA05OAR4171149). The authors thank P. Gayaldo and NOAA Restoration Center staff as well as local partners: T. Diers and staff of the New Hampshire Coastal Program and M. Dionne and staff of the Wells National Estuarine Research Reserve. We are indebted to students from McGill University: S. Byers, D. Charbonneau, G. Hung, G. MacDonald, P. Noël, and A. Sabourin, and from University of New Hampshire: C. Peter, A. Eberhardt, K. Nelson, and R. Vincent. Jackson Estuarine Laboratory Contribution #506.

REFERENCES

Adamowicz, S. C, and C. T. Roman. 2005. "New England Salt Marsh Pools: A Quantitative Analysis of Geomorphic and Geographic Features." *Wetlands* 25:279–88.

Anisfeld, S. C., M. J. Tobin, and G. Benoit. 1999. "Sedimentation Rates in Flow-Restricted and Restored Salt Marshes in Long Island Sound." *Estuaries* 22:231–44.

Barbier, E. B., E. W. Koch, B. R. Silliman, S. R. Hacker, E. Wolanski, J. Primavera, E. F. Granek, et al. 2008. "Coastal Ecosystem–Based Management with Nonlinear Ecological Functions and Values." *Science* 319:321–23.

Bockstael, N. E., A. M. Freeman, R. J. Kopp, P. R. Portnoy, and V. K. Smith. 2000. "On Measuring Economic Values for Nature." *Environmental Science and Technology* 34:1384–89.

Boumans, R. M. J., D. M. Burdick, and M. Dionne. 2002. "Modeling Habitat Change in Salt Marshes after Tidal Restoration." *Restoration Ecology* 10:543–55.

Brander, L., R. Florax, and J. Vermaat. 2006. "The Empirics of Wetland Valuation: A Comprehensive Summary and a Meta-analysis of the Literature." *Environmental and Resource Economics* 33:223–50.

Brennan, A., and Lo, Y. S. 2008. "Environmental Ethics." In the *Stanford Encyclopedia of Philosophy*, fall 2008 ed., edited by N. Zalta. http://plato.stanford.edu/archives /fall2008/entries/ethics-environmental/.

Bridgham, S. D., J. P. Megonigal, J. K. Keller, N. B. Bliss, and C. Trettin. 2006. "The Carbon Balance of North American Wetlands." *Wetlands* 26:889–916.

Burdick, D. M., M. Dionne, R. M. Boumans, and F. T. Short. 1997. "Ecological Responses to Tidal Restorations of Two Northern New England Salt Marshes." *Wetlands Ecology and Management* 4:129–44.

Burdick, D. M., R. Vincent, and C. R. Peter. 2010. *Evaluation of Post-restoration Conditions at Little River Marsh in North Hampton, New Hampshire*. Final Report. Portsmouth: New Hampshire Coastal Program.

Byers, S. E., and G. L. Chmura. 2007. "Salt Marsh Vegetation Recovery on the Bay of Fundy." *Estuaries and Coasts* 30:869–77.

Chmura, G. L. 2009. "Tidal Salt Marshes." Pp. 5–11 in *The Management of Natural Coastal Carbon Sinks*, edited by D. d'A. Laffoley and G. Grimsditch. Gland, Switzerland: IUCN.

Chmura, G. L., S. C. Anisfeld, D. R. Cahoon, and J. C. Lynch. 2003. "Global Carbon Sequestration in Tidal, Saline Wetland Soils." *Global Biogeochemical Cycles* 17:1–12.

Chmura, G. L., E. Lemieux, and D. Charboneau. In prep. "Dependence of Endangered Species on North American Tidal Salt Marshes."

Costanza, R. , R. d'Arge, R. deGroot, S. Farber, M. Grasso, B. Hannon, K. Limburg, et al. 1997. "The Value of the World's Ecosystem Services and Natural Capital." *Nature* 387:253–60.

Craft, C., J. Reader, J. N. Sacco, and S. W. Broome. 1999. "Twenty-five Years of Ecosystem Development of Constructed *Spartina alterniflora* L. Marshes." *Ecological Applications* 9:1405–19.

Crain, C. M., K. B. Gedham, and M. Dionne. 2009. "Tidal Restrictions and Mosquito Ditching in New England Marshes; Case Studies of the Biotic Evidence, Physical Extent, and Potential for Restoration of Altered Hydrology." Pp. 149–70 in *Human Im-

pacts on Salt Marshes. A Global Perspective, edited by B. R. Silliman, E. D. Grosholz, and M. D. Bertness. Berkeley: University of California Press.

Daily, G. C. 1997. "What Are Ecosystem Services?" Pp. 1–10 in *Nature's Services: Societal Dependence on Natural Ecosystems*, edited by G. C. Daily. Washington, DC: Island Press.

Darby, F. A., and R. E. Turner. 2008. "Effects of Eutrophication on Salt Marsh Root and Rhizome Biomass Accumulation." *Marine Ecology Progress Series* 363:63–70.

Day, O. 2009. *The Impacts of Climate Change on Biodiversity in Caribbean Islands: What We Know, What We Need to Know, and Building Capacity for Effective Adaptation.* Caribbean Natural Resources Institute Technical Report no. 386.

Donnelly, J. P., and M. D. Bertness. 2001. "Rapid Shoreward Encroachment of Salt Marsh Vegetation in Response to Accelerated Sea-Level Rise." *Proceedings of the National Academy of Sciences* 98:14218–223.

Eberhardt, A. L., D. M. Burdick, and M. Dionne. 2011. "The Effects of Road Culverts on Nekton in New England Salt Marshes: Implications for Tidal Restoration." *Restoration Ecology* 19:776–85.

Ehrlich, P. R., and A. H. Ehrlich. 1981. *Extinction: The Causes and Consequences of the Disappearance of Species.* New York: Random House.

Fenchel, T., and B. B. Jorgensen. 1977. "Detritus Food Chains of Aquatic Ecosystems: The Role of Bacteria." Pp. 1–57 in *Advances in Microbial Ecology*, edited by M. Alexander. New York: Plenum Press.

Fisher, B., R. K. Turner, and P. Morling. 2009. "Defining and Classifying Ecosystem Services for Decision Making." *Ecological Economics* 68:643–53.

Forster, P., V. Ramaswamy, P. Artaxo, T. Berntsen, R. Betts, D. W. Fahey, J. Haywood, et al. 2007. "Changes in Atmospheric Constituents and in Radiative Forcing." Pp. 130–234 in *Climate Change 2007: The Physical Science Basis. Contribution of Working Group I to the Fourth Assessment Report of the Intergovernmental Panel on Climate Change*, edited by S. Solomon, D. Qin, M. Manning, Z. Chen, M. Marquis, K. B. Avery, M. Tignor, and H. L. Miller. Cambridge: Cambridge University Press.

Gallagher, J. 1985. "Halophytic Crops for Cultivation at Seawater Salinity." *Plant and Soil* 89:323–36.

Gedan, K. B., B. Silliman, and M. D. Bertness. 2009. "Centuries of Human-Driven Change in Salt Marsh Ecosystems." *Annual Review of Marine Science* 1:117–41.

Giblin, A. E., I. Valiela, and J. M. Teal. 1983. "The Fate of Metals Introduced into a New England Salt Marsh." *Water, Air and Soil Pollution* 20:81–98.

Holdren, J. P., and P. R. Ehrlich. 1974. "Human Population and the Global Environment." *American Scientist* 62:282–92.

Hung, G. A., and G. L. Chmura. 2006. "Mercury Accumulation in Surface Sediments of Salt Marshes of the Bay of Fundy." *Environmental Pollution* 142:418–31.

Hung, G. A., and G. L. Chmura. 2007. "Heavy Metal Accumulation in Surface Sediments of Salt Marshes of the Bay of Fundy." *Estuaries and Coasts* 30:725–34.

Keddy, P. A. 2000. *Wetlands Ecology: Principles and Conservation.* Cambridge: Cambridge University Press.

Kneib, R. T. 2002. "Salt Marsh Ecoscapes and Production Transfers by Estuarine Nekton in the Southeastern United States." Pp. 267–91 in *Concepts and Controversies in Tidal Marsh Ecology*, edited by M. Weinstein and D. Kreeger. Boston: Kluwer Academic.

Koch, E. W., E. B. Barbier, B. R. Silliman, D. J. Reed, G. M. E. Perillo, S. D. Hacker, E. F. Granek, et al. 2009. "Non-linearity in Ecosystem Services: Temporal and Spatial Variability in Coastal Protection." *Frontiers in Ecology and the Environment* 7:29–37.

Konisky, R. A., D. M. Burdick, M. Dionne, and H. A. Neckles. 2006. "A Regional Assessment of Salt Marsh Restoration and Monitoring in the Gulf of Maine." *Restoration Ecology* 14:516–25.

Kraus, M. L. 1988. "Accumulation and Excretion of Five Heavy Metals by the Saltmarsh Cordgrass *Spartina alterniflora*." *Bulletin of the New Jersey Academy of Science* 33:39–43.

MacDonald, G. K., P. E. Noel, D. van Proosdij, and G. L. Chmura. 2010. "The Legacy of Agricultural Reclamation on Surface Hydrology of Two Recovering Salt Marshes of the Bay of Fundy, Canada." *Estuaries and Coasts* 33:151–60.

Millennium Ecosystem Assessment. 2005. *Ecosystems and Human Well-Being: Synthesis*. Washington, DC: Island Press.

Mooney, H. A., and P. R. Ehrlich. 1997. "Ecosystem Services: A Fragmentary History." Pp. 11–19 in *Nature's Services: Societal Dependence on Natural Ecosystems*, edited by G. E. Daily. Washington, DC: Island Press.

Moore, G. E., D. M. Burdick, C. R. Peter, A. Leonard-Duarte, and M. Dionne. 2009. *Regional Assessment of Tidal Marsh Restoration in New England Using the Restoration Performance Index*. Final Report. Gloucester, MA: NOAA Restoration Center.

Morgan, P. A., D. M. Burdick, and F. T. Short. 2009. "The Functions and Values of Fringing Salt Marshes in Northern New England, USA." *Estuaries and Coasts* 32:483–95.

Morris, J. T., P. V. Sundareshwar, C. T. Nietch, B. Kjerfve, and D. R. Cahoon. 2002. "Responses of Coastal Wetlands to Rising Sea Level." *Ecology* 83:2869–77.

Moseman-Valtierra, S., R. Gonzalez, K. D. Kroeger, J. Tang, W. C. Chao, J. Crusius, J. Bratton, A. Green, and J. Shelton. 2011. "Short-Term Nitrogen Additions Can Shift a Coastal Wetland from a Sink to a Source of N_2O." *Atmospheric Environment* 45:4390–97.

Neckles, H. A., M. Dionne, D. M. Burdick, C. T. Roman, R. Buchsbaum, and E. Hutchins. 2002. "A Monitoring Protocol to Assess Tidal Restoration of Salt Marshes on Local and Regional Scales." *Restoration Ecology* 10:556–63.

Piehler, M. F., and A. R. Smyth. 2011. "Habitat-Specific Distinctions in Estuarine Denitrification Affect Both Ecosystem Function and Services." *Ecosphere* 2 (1): art12. doi:10.1890/ES10-00082.1

Roman, C. T., W. A. Niering, and R. S. Warren. 1984. "Salt Marsh Vegetation Changes in Response to Tidal Restriction." *Environmental Management* 8:141–50.

Roman, C. T., K. B. Raposa, S. C. Adamowicz, M. J. James-Pirri, and J. G. Catena. 2002. "Quantifying Vegetation and Nekton Response to Tidal Restoration of a New England Salt Marsh." *Restoration Ecology* 10:450–60.

Saltonstall, K. 2002. "Cryptic Invasion by a Non-native Genotype of the Common Reed, *Phragmites australis*, into North America." *Proceedings of the National Academy of Sciences* 99:2445–49.

Sousa, A. I., A. I. Lillebo, I. Cacador, and M. A. Pardal. 2008. "Contribution of *Spartina maritima* to the Reduction of Eutrophication in Estuarine Systems." *Environmental Pollution* 156:628–35.

Study of Critical Environmental Problems (SCEP), Massachusetts Institute of Technology. 1970. *Man's Impact on the Global Environment: Assessment and Recommendations for Action*. Cambridge, MA: MIT Press.

Tansley, A. G. 1935. "The Use and Abuse of Vegetational Concepts and Terms." *Ecology* 16:284–307.

Teo, S. L. H., and K. W. Able. 2003. "Growth and Production of the Mummichog (*Fundulus heteroclitus*) in a Restored Salt Marsh." *Estuaries* 26:51–63.

Thom, R. M., R. Zeigler, and A. B. Borde. 2002. "Floristic Development Patterns in a Restored Elk River Estuarine Marsh, Grays Harbor, Washington." *Restoration Ecology* 10:487–96.

Turner, R. E. 1977. "Intertidal Vegetation and Commercial Yields of Penaeid Shrimp." *Transactions of the American Fisheries Society* 106:411–16.

Valiela, I., J. M. Teal, S. D. Allen, R. Van Etten, D. Goehringer, and S. Volkman. 1985. "Decomposition in Salt Marsh Ecosystems: The Phases and Major Factors Affecting Disappearance of Above-Ground Organic Matter." *Journal of Experimental Biology and Ecology* 89:29–54.

Valiela, I., M. L. Cole, J. McClelland, J. Hauxwell, J. Cebrian, and S. B. Joye. 2002. "Role of Salt Marshes as Part of Coastal Landscapes." Pp. 23–36 in *Concepts and Controversies in Tidal Marsh Ecology*, edited by M. Weinstein and D. Kreeger. Boston: Kluwer Academic.

Warren, R. S., P. E. Fell, R. Rozsa, A. H. Brawley, A. C. Orsted, E. T. Olson, V. Swamy, and W. A. Niering. 2002. "Salt Marsh Restoration in Connecticut: 20 Years of Science and Management." *Restoration Ecology* 10:497–513.

Westman, W. E. 1977. "How Much Are Nature's Services Worth?" *Science* 197:960–64.

Woodward, R. T., and Y. S. Wui. 2001. "The Economic Value of Wetland Services: A Meta-analysis." *Ecological Economics* 37:257–70.

Zedler, J. B., and S. Kercher. 2005. "Wetland Resources: Status, Trends, Ecosystem Services, and Restorability." *Annual Review of Environment and Resources* 30:39–74.

Chapter 16

Role of Simulation Models in Understanding the Salt Marsh Restoration Process

RAYMOND A. KONISKY

Interest in model-assisted restoration planning is growing among salt marsh resource professionals, and managers increasingly recognize the need to consider factors beyond tidal hydraulics in designing restoration solutions. In fact, practical experiences show that restoration designs based entirely on hydrologic assessment may not result in optimal recovery of marsh habitat, ecological functions, or estuarine biodiversity. As Zedler (2000, 402) noted in a review of wetland restoration progress, "it takes more than water to restore a wetland." It is not possible to anticipate long-term habitat responses without also considering a myriad of interrelated ecological factors. Inability to account for complex ecological interactions, even if tidal exchange is restored, may lead to unintended and less than optimal results when measured against management goals of recovering specific acreage of salt marsh habitat and controlling invasive species.

This chapter provides a synthesis of existing salt marsh process knowledge into a single integrated model for simulating habitat response. It presents a simulation model capable of predicting potential habitat outcomes for changing conditions of tidal flooding, elevation dynamics, and plant community composition. Based on project experiences, the chapter reviews the multiple benefits that simulation models provide to regional restoration managers and planners. Examples show how managers use model outcomes to prioritize sites, select practical design alternatives, improve monitoring plans, and communicate restoration plans and objectives to stakeholders. Simulation models have a critically important role in the continued advancement of salt marsh restoration practices.

Modeling Salt Marsh Processes

Hydraulic engineers design restoration solutions based on expected tidal flood levels, but can we really predict habitat response based solely on this element of the physical environment? Several ecological modelers have taken a step beyond tidal hydraulics to consider biological responses, by linking plant species distribution with ecologically important tide levels (Roman et al. 1995; Boumans et al. 2002; Office of Ocean and Coastal Resource Management 2006). Typically, elevation ranges of dominant salt marsh species are identified by survey at unrestricted reference locations, and important boundaries are determined, such as elevation of the *Spartina alterniflora* (smooth cordgrass)–*Spartina patens* (salt meadow cordgrass) border. Correlative relationships between plant distribution and tide elevations are then used to estimate expected acreage of low or high marsh recovery following restoration.

While correlative approaches have proven to be valid and valuable, salt marsh modelers rarely venture into ecosystem-based simulations of restoration response. Important earlier modeling works from Weigert (1986), Morris and Bowden (1986), and Voiniv et al. (1999), among others, are acknowledged as defining critical linkages among primary productivity, tidal hydrology, and sediment formations, but these efforts did not consider biological and landscape-scale processes that influence marsh responses over longer time horizons. For instance, how will the presence of a competitive dominant species like *Phragmites australis* (common reed) alter restoration outcomes? Or how will changing rates of sediment accretion and sea level rise impact plant community distribution over the long term? This chapter offers an ecosystem perspective on the broad conceptual view that multiple marsh processes and ecological relationships interact to determine marsh habitat response over time. The synthesized model, as described here, exists as a geographic information system (GIS) simulation environment that has been used to assess salt marsh restoration outcomes for many regional projects in New England and Maritime Canada.[1] This chapter provides a summary of the model development, assumptions, and scenarios; Konisky et al. (2003) provide a complete description of the model.

Figure 16.1 shows a conceptual model of key components and interactions, beginning with processes of *tidal hydraulics*. Tidal regime is measured or modeled to determine tidal signal and to compare with marsh elevation for relative flood levels. Maps of current elevation profiles are acquired from field sources. Marsh surface elevations may change over time through processes of sediment accretion, peat biomass production, and subsidence. These *elevation dynamics* processes, including response to sea level rise, are used to track changes in relative marsh elevation over long time horizons. Elevation and tidal regime combine

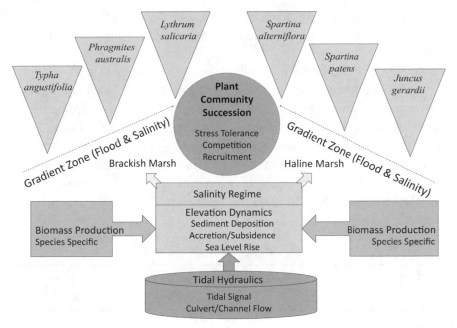

FIGURE 16.1. Conceptual ecosystem model of primary processes, dominant plant species, and ecological interactions considered in salt marsh habitat response to restoration activities.

with salinity levels as major drivers of plant species distribution. Flood and salinity regimes interact to create gradient zones of physical stress that either improve or diminish plant species ability to persist. In addition to physical influences, plants respond to biotic factors of recruitment and interspecific competition from other halophytes and brackish species. Community responses to stress, competition, and recruitment are defined by processes of *plant community succession*.

The model framework is based on responses of six common plant species that aggregate to form salt marsh habitat zones. Halophyte perennials *Spartina alterniflora*, *Spartina patens*, and *Juncus gerardii* (black rush) are dominant plant species of northeastern US salt marshes (Niering and Warren 1980). Native halophytes are commonly replaced by the brackish invasive species *Phragmites australis*, *Typha angustifolia* (narrowleaf cattail), and *Lythrum salicaria* (purple loosestrife) in areas of tidal restriction (Roman et al. 1984; Sinicrope et al. 1990; Burdick et al. 1997). To better understand species response to changing hydrology, a field experiment was conducted to transplant these six species across nine gradient zones of varying flood and salinity conditions. Seasonal measures of transplant survival and growth allowed physical stress tolerance to be quantified for each species, and pairwise arrangements provided measures of relative competitive ranking. Stress tolerance and competition were considered as the primary determinants of habitat response (Bertness and Ellison 1987), although other factors such as seed dispersal, wrack, and surface flow dynamics are certainly important at finer scales (Smith 2007; Smith et al. 2009).

A conceptual model of dominant plant species and physical gradient zones was used as a spatial framework for model application. Existing marsh conditions and restoration alternatives are investigated as modeling scenarios. Applied spatially, the model produces maps that managers can use to quantify the extent and location of expected habitat recovery and to visualize future potential conditions at a site. Model drivers use a combination of experimental results, site-specific field measures, and standardized parameters from the literature. Inputs and sources are summarized in table 16.1. To make the model as transferable as possible, field

TABLE 16.1

Summary of site information needed to configure the marsh restoration model

Parameters	Description	Data sources
Spatial inputs		
Vegetation grid	Raster image of dominant plant zones	Digitized aerial photo interpretation, field survey, or hyperspectral imagery
Elevation grid	Raster image of digital elevation map (DEM)	Digitized aerial photo with photogrammetry, field survey/interpolation, or LIDAR sources
Tidal source	Point coverage identifying locations of high-salinity tidal inflows	GPS waypoints converted to GIS point coverage
Elevation dynamics		
Local rate of sea level rise (mm/yr)	Local/regional annual increase in relative sea level	Published estimates
Rate of sedimentation (mm/yr)	Average annual deposition of sediment on marsh	Feldspar marker-horizons or sediment elevation tables
Vegetation composition		
Plant zone composition	Relative percentage of marsh cover for six dominant plant communities	Field vegetation survey with quadrat plots along transects
Tree/shrub successor	Presence of brackish species (*Phragmites*, *Typha*, or *Lythrum*) likely to colonize restored marsh	Field assessment
Tidal hydrology		
Tide levels (for known datum)	Mean high water Medium spring tide Maximum tide	Observations or models of tidal heights on ocean side and impact side of marsh over lunar cycle

data collection requirements are based on a set of standard measures defined in a regional marsh monitoring protocol (Neckles et al. 2002). The following sections describe the set of model process specifications developed as primary ecosystem components.

Tidal Hydraulics

Tidal hydraulic processes determine the tidal signal resulting from existing or potential tidal sources. To determine water levels, dataloggers are used to simultaneously measure the tidal signal at an impact site and at a nearby reference site (i.e., unrestricted). A minimum two-week sample is collected to capture spring and neap tides, but longer-term datasets are preferable. The tidal signal observed at the reference site represents the potential for hydrologic restoration; the signal at the restricted site is the measure of impact.

FLOOD REGIME

At an impact site, flooding regime is determined by the reference signal, the hydraulic capacity of any water control structure or earthworks, and the drainage characteristics of the site. Existing and potential flood regimes are typically modeled by engineers using HydroCAD (Autodesk, Inc.), the Stormwater Management Model (SWMM; US Environmental Protection Agency), the Hydrologic Engineering Center River Analysis System (HEC-RAS; US Army Corp of Engineers), or other models (MacBroom and Schiff, chap. 2, this volume). Basic engineering formulas for culvert and channel hydraulics seem to work well when calibrated with site-specific flow measures, such as the simple hydraulic model Marsh Response to Hydrological Modifications (MRHM) (Boumans et al. 2002). Using the downstream signal as input, calibration parameters are adjusted iteratively until the highest coefficient of correlation is achieved between simulated results and observed upstream conditions. Once parameters are calibrated, culvert dimensions and/or additional inflow sources are input to simulate design scenarios. MRHM results inform restoration managers of potential options, but engineers are enlisted to develop as-built specifications and site plans.

HYDRAULIC SYNTHESIS

For ecological modeling, tidal series for each considered scenario are synthesized into a set of ecologically significant elevation levels. From long-term averaged time series of observed or modeled water levels, three elevations are determined that delineate upland borders of low/mid/high marsh: (1) Low marsh is delimited by mean high water (MHW), based on commonly observed delineation of S. *alterniflora* and S. *patens* plant zones (Niering and Warren 1980; Bertness and Ellison 1987; McKee and Patrick 1988); (2) midmarsh is delimited by median spring tide (MST) based on observations by Bertness and Ellison (1987) of the boundary between S. *patens* and *J. gerardii*, and Warren et al. (2001) for the boundary between *P. australis* and *Typha* spp. For both studies, the border elevation is roughly the height at which 15 percent of tides regularly flood. This height is approximate with median spring tide as the fourth highest average tide in a lunar series. Other modelers use mean higher high water for a similar purpose (Office of Ocean and Coastal Resource Management 2006); (3) high marsh is delimited by the maximum extent of nonstorm tidal flooding (MAX).

Elevation Dynamics

Elevation dynamics produce subtle changes in marsh surface elevation over time and therefore require highly resolved inputs as digital elevation models (DEMs). Site DEMs are developed using techniques that generate grid cells of varying resolution: (1) 10–25 square meter cell-size resolution from rod-and-laser field surveys with kriging extrapolation (Konisky et al. 2003); (2) 5–10 square meters from aerial photography interpretation (Woodlot Alternatives 2007); and (3) 1 square meter from flight-based Light Detection and Ranging (LIDAR) sensors (Rogers et al. 2007). Changes to DEM values are modeled using sediment dynamics processes defined by Rybczyk et al. (1998), who developed a sample-intensive calibration model from fractionalized soil cores. This approach is simplified by using a single soil column based on (1) organic plant contributions, (2) inorganic sediment deposition, and (3) pore space. Organic and inorganic inputs are simulated at each marsh location as will be described; for pore space, a regionally averaged estimate of 70 percent is used based on Burdick et al. (1999). To account for sea level rise (SLR), marsh elevation results are reduced by the SLR rate to simulate longer-term elevation changes.

ORGANIC SEDIMENT CONTRIBUTIONS

Aboveground leaf and belowground root litter biomass accumulates, degrades, and eventually becomes marsh peat. Because biomass varies by species, a site map of distribution and relative composition for each dominant species zone is first developed. Other plants are combined with the six dominant species based on observed co-occurrences, and woody species are included as a single tree-and-shrub community. Site-averaged cover proportions for each dominant zone are computed. With estimates of plant abundance and distribution, species-specific measures of leaf and root litter from biomass production curves are estimated as modeled by Fitz et al. (1996). Species leaf litter amounts are derived from production curves calibrated to regionally averaged peak aboveground biomass, and root litter is determined from proportions measured by Konisky et al. (2003). Only 20 percent of litter carbon is considered available for sediment formation, with the rest lost to export and oxidation (Chalmers et al. 1985). Litter is allocated to labile (fast decomposing) and refractory (slow decomposing) bins using labile:refractory ratios of 80:20 percent for leaf litter (Valiela et al. 1985) and 20:80 percent for root litter (Hemminga and Buth 1991), and reduced by weekly decomposition rates of 2 percent and 0.2 percent for labile and refractory components, respectively (Valiela et al. 1985). Accumulated sediment carbon is converted to sediment height with a dry weight multiplier of 2.5 grams carbon per gram (Gallagher 1983) and volumetric conversion of 1.14 grams dry weight per cubic centimeter (DeLaune et al. 1983). Derived annual organic contributions to sediment height for each species are: *S. alterniflora* 0.31 millimeters per year; *S. patens* 0.23 millimeters per year; *J. gerardii* 0.23 millimeters per year; *P. australis* 0.27 millimeters per year; *L. salicaria* 0.16 millimeters per year; and *T. angustifolia* 0.35 millimeters per year.

INORGANIC SEDIMENT CONTRIBUTIONS

Estimates of inorganic contributions are based on field measures of sediment deposition using feldspar marker horizons at stations 10 meters landward of the main tidal creek (Cahoon and Turner 1989; Burdick et al. 1999). To estimate accretion away from the creek, the creek-side rate is apportioned as a decreasing function of tidal inundation and elevation (after Stumpf 1983), with marsh locations less than or equal to MHW at 100 percent, MHW-MST at 50 percent, MST-MAX at 25 percent, and greater than MAX at 0 percent. Elevation-adjusted inorganic contributions are reduced by 95 percent to account for mineral-only volumetric components of typical sediment deposits (Gosselink and Hatton 1984; Turner et al. 2000).

ELEVATION DYNAMICS RESULTS

To derive height of the sediment column, contributions from organic and inorganic sources are added and expansion by the pore-space percentage is considered. Heights are then reduced by annual SLR to determine elevation relative to the tidal cycle. Accretion occurs where the marsh tracks with SLR; subsidence where SLR exceeds gains from sediment-building processes. To validate implementation of the Rybczyk et al. (1998) model, elevation dynamics were simulated at four regional marsh sites (two restricted, two unrestricted) and compared results with elevation changes estimated with surface elevation tables (SETs; Burdick et al.1999). Site simulations were configured for a mixed S. *alterniflora–S. patens* community with field measures of sediment deposits from marker horizons, and an average SLR rate of 1.5 millimeters per year (based on Wood et al. 1989; Anisfeld et al. 1999; Nerem 1999; Donnelly and Bertness 2001).The model results concurred with field measures of less than 3 millimeters per year subsidence at the two restricted sites, and greater than 1 millimeter per year accretion at the two unrestricted sites, although modeled rates of change were conservative relative to SET estimates. Given a high degree of variability in SET elevations, it is concluded that this conservative model of elevation dynamics adequately simulates marsh elevation change over restoration-scale time frames.

Plant Community Succession

Plant succession processes tie together habitat responses to physical conditions of saltwater flooding and biological conditions of competition and recruitment. This section describes an allocation scheme for simulating plant succession processes based on model drivers of species-specific parameters developed through field experiments. Measures of stress tolerance and competitive strengths were derived by transplanting 252 individual shoots of each of the six dominant species into open-bottom containers across a 3 meter by 3 meter gradient of tidal flooding (high flood less than MHW; midflood greater than MHW, less than MST; low flood greater than MST) and salinity regime (high salinity greater than 22 parts per thousand [ppt]; medium salinity 14–22 ppt; low salinity less than 14 ppt). For each species at each gradient zone (fig. 16.2), we measured transplant survival, and growth was measured and transplants were arranged into pairwise combinations to develop relative competitive strength ranks. The biomass metric, relative aboveground biomass growth (RABG) is ending live aboveground biomass divided by starting live aboveground biomass. A tolerance factor (TF) was computed for each species at each gradient location as RABG multiplied by the proportion of surviving transplants, expressed as a proportion of the maximum value observed for the species across all locations. A pairwise competition factor (CF) is derived for species combinations as the overall mean of all pairings RABG divided by the mean species RABG in each respective pairing zone. Table 16.2 shows TF and CF values from Konisky and Burdick (2004).

Gradient Zone Determinants

Simulating restoration by increasing salinity and flooding, and restriction by reducing salinity and flooding, is determined by scenario hydraulics and gradient zone

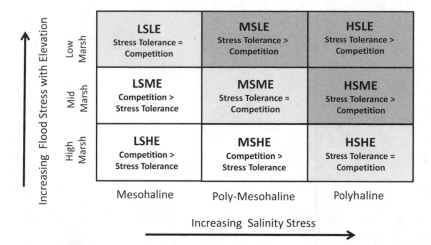

FIGURE 16.2. Schematic diagram of salt marsh gradient zones with three salinity regimes (LS: low salinity, MS: midsalinity, HS: high salinity) and three flood regimes (LE: low elevation, ME: midelevation, HE: high elevation). Shading shows model assumption of gradient locations with similar stress conditions, and relative weighting of succession factors for stress tolerance and competitive strength.

distribution. Flood zones are delineated by elevation relative to the tidal cycle, with high flooding less than MHW, midflooding greater than MHW and less than MST, and low flooding greater than MST. Salinity zones are simulated based on general observations that salinity levels diminish due to dilution with increased distance from the open-ocean or tidal source (Odum et al. 1984; Pearlstine et al. 1993; Gardner et al. 2002; Warren et al. 2002) and also due to interactions of upland elevation slope, substrate hydraulics, tidal signal, and the tidal pressure wave with distance from the tidal creek (Harvey et al. 1987; Pearlstine et al. 1993; Gardner et al. 2002). Spatial relationships to determine zones are developed based on (1) overall site maximum distance between tidal source and the furthest marsh edge, *DistanceMaxExtent*; (2) distance to the nearest open water, *DistanceOpenWater*; (3) distance to the nearest tidal source, *DistanceTidalSource*; and (4) distance to the nearest upland edge, *DistanceUpland*. Cell position relative to the maximum tidal source extent is then derived:

RS: *RelativeSource = DistanceTidalSource / DistanceMaxExtent*

and open water/upland edge,

RC: *RelativeCreek = DistanceOpenWater / (DistanceOpenWater + DistanceUpland)*.

Resulting values are organized into proportional categories of "furthest" (greater than or equal to .67), "mid" (between .67 and .33), and "closest" (less than or

TABLE 16.2

Tolerance and competition factors for dominant halophyte and brackish plant species from transplant experiments
(Konisky and Burdick 2004)

| | Dominant plant species | | | | | |
| | Halophyte Species | | | Brackish Invasive Species | | |
Tolerance factor by gradient location	Spartina alterniflora	Spartina patens	Juncus gerardii	Phragmites australis	Lythrum salicaria	Typha angustifolia
Low salinity						
Low elevation	0.76	0.04	0	0.18	0	0.39
Mid elevation	0.84	0.3	0.15	0.4	0	0.93
High elevation	0.39	0.47	0.79	1	0.94	1
Mid salinity						
Low elevation	1	0.72	0	0.2	0	0.15
Mid elevation	0.78	0.77	0.11	0.3	0	0.3
High elevation	0.23	0.81	0.29	0.13	1	0.08
High salinity						
Low elevation	0.5	0.47	0.32	0.36	0	0
Mid elevation	0.03	1	0.98	0.12	0	0
High elevation	0.07	0.67	1	0.03	0.14	0.01
Competition factor by competitor						
Spartina alterniflora	–	0.81	1.2	0.53	No data	1.08
Spartina patens	0.8	–	1.14	1.02	1.36	1.07
Juncus gerardii	0.7	1.04	–	0.87	1.06	1.05
Phragmites australis	1.18	1.13	1.51	–	1.96	0.59
Lythrum salicaria	no data	0.34	1.94	0.63	–	No data
Typha angustifolia	0.56	0.71	1.42	1.29	No data	–

Note: Tolerance factors (TF) show relative biomass (from 0 to 1) at a gradient location compared to species top biomass across all locations. Competition factors (CF) show proportion of average seasonal growth achieved for a species versus each competitor.

equal to .33), and grouped into a matrix that fits the model of low-mid-high salinity zones. Low salinity zones occupy areas furthest from the tidal source and open water, and nearest the uplands. High salinity zones are closest to the tidal source and open water, and furthest from the uplands. Midsalinity zones are in between. All areas above the tidal extent (greater than MAX) are assigned to the low salinity zone. The model checks annually for potential zone changes. For tree and shrub areas, succession only occurs in areas that change from the low flood/low salinity zone. When tree/shrub areas are flooded, the model switches dominance to the most common brackish species at a site, and succession proceeds with the "known" dominant species.

SUCCESSION SCHEME

A plant cover reallocation scheme was devised that allows zone-tolerant and competitively advantaged species to flourish at the expense of others. This general approach of "preemptive advantage" is based on the succession traits of established wetland plants identified by Grace (1987). The probability of species persistence at a location is assessed by allocating a portion of percent cover to be "at-risk" for annual replacement by others as percent cover percentage multiplied by 1 minus the species tolerance factor at the gradient location, with a minimum of 5 percent for random chance (at-risk$_{species}$; equation 16.1). The at-risk portion is then reallocated according to factors of neighbor recruitment potential, relative competitive ability, and tolerance of physical conditions, with tolerance most important in the stressful low marsh and competition most important in the benign high marsh (Bertness and Ellison 1987). TF/CF is weighted 80/10 percent for the three highest-stress zones, 45/45 percent for midstress, and 10/80 percent for lowest stress zones, with recruitment considered less influential and weighted at a constant 10 percent across zones (fig. 16.2).

For recruitment reallocation, the at-risk portion is multiplied by the neighbor cover values and summed across species, with the product multiplied by the recruitment weighting factor (*recruit_wf*; equation 16.2). Lastly, each interspecific combination in the cell is reallocated as the difference in species competition factors (CFs) multiplied by the product of the at-risk pool and the competition weight factor at the location (*comp_wf*), plus the difference in tolerance factors (TFs) multiplied by the product of the at-risk pool and the tolerance weight factor at the location (*tol_wf*; equation 16.3). Pairwise reallocations are prorated by the relative cover percentage of each species to not exceed 100 percent. The model processes succession allocations annually. The scheme requires one adjustment for pairwise reallocations between *Typha* and *Phragmites* in the low salinity–high elevation zone (TF = 1 for both species). In tidally flooded locations, *Phragmites* is given an advantage by reducing *Typha* TF from 1 to 0.5, based on observations that *Phragmites* is more likely than *Typha* to occupy salt marsh areas with regular tidal flooding (Warren et al. 2001; Konisky et al. 2003).

$$\text{At-risk}_{\text{species i–j}} = \text{cover}_{\text{species i–j}} \times \text{MAX}(.05,(1 - \text{TF}_{\text{species i–j at location}})) \tag{16.1}$$

$$\begin{aligned}\text{Recruit reallocation}_{\text{species i–j}} = \textit{recruit\_wf} &\times \Sigma(\text{at-risk}_{\text{species i–j}} \\ &\times \text{ neighbor cover}_{\text{species i–j}})\end{aligned} \tag{16.2}$$

$$\begin{aligned}\text{Pairwise reallocation}_{\text{species i from j}} = &((\text{CF}_{\text{species i on j}} - \text{CF}_{\text{species j on i}}) \\ &\times \text{At-risk}_{\text{species i–j}} \times \textit{comp\_wf}_{\text{location}})) \\ &+ ((\text{TF}_{\text{species i at location}} - \text{TF}_{\text{species j at location}}) \\ &\times \text{At-risk}_{\text{species i–j}} \times \textit{tol\_wf}_{\text{location}}\end{aligned} \tag{16.3}$$

SUCCESSION RESULTS

The succession model was tested at two past restoration sites by comparing model simulations to field observations of current habitat conditions. For Drakes Island Marsh in Wells, Maine (partial restoration), and Mill Brook Marsh in Stratham, New Hampshire (full restoration), historic cover maps were created from limited field data collected prior to hydrologic change (Burdick et al. 1999). Prerestoration cover for Drakes Island was 100 percent *Typha* with initial species percentages set at 50 percent *Typha* and 10 percent for each other species. At Mill Brook, the prerestoration marsh was 40 percent *Lythrum*, 50 percent *Typha*, 5 percent *Phragmites*, 3 percent *S. patens*, and 2 percent *Juncus*, with initial percentages also set at 50/10 percent for dominant/other species. Lacking prerestoration elevation data, site DEMs were used from recent elevation surveys, and re-created starting elevations were derived by running the sediment dynamics model in reverse back to the time of restoration (nine and thirteen years, respectively, for Drakes Island and Mill Brook). As described earlier, model validation of elevation change was done as a standalone process. For both test sites, observed measures of current (postrestoration) tidal signal were applied to trigger the habitat response.

Model simulations allowed a comparison of predicted versus actual area of marsh cover for aggregated halophytes (*S. alterniflora, S. patens, J. gerardii*) and brackish species (*P. australis, L. salicaria, T. angustifolia*). For Drakes Island, starting with 100 percent brackish cover, the model predicted a postrestoration community of 39 percent halophyte and 61 percent brackish species. This compares with actual habitat observations of 38 percent halophyte and 62 percent brackish (1 percent error). For Mill Brook, the initial mix of 95 percent brackish and 5 percent halophyte converted to a community of 55 percent halophyte and 45 percent brackish cover following restoration. These results compare to observations of 50 percent halophyte and 50 percent brackish (9 percent error). Combining results to minimize site-specific influences, the overall composite model error was 4 percent. Therefore, a model accuracy of 96 percent was achieved for predicting total halophyte and brackish areas at two sites following hydrologic restoration.

Spatial accuracy was also addressed to determine how well the model predicts where plant communities develop. For both sites combined, error was 27 percent for the halophyte community and 23 percent for the brackish plants, with an overall error of 25 percent. The model correctly chose the location of a halophyte or brackish cell three out of four times. Spatial accuracy is especially dependent on starting vegetation maps, which in the test cases were developed from very limited field datasets. Still, the overall validation results provide confidence in model outcomes, especially for predictions of acreage totals for halophyte and brackish areas. These aggregate habitat measures are one of the most frequent metrics used by regional managers to assess restoration project success (Konisky et al. 2006).

GIS Spatial Model Environment

The model delivery platform is ArcView 9.x (ESRI or Environmental Systems Research Institute). As an ESRI Arc extension, the model supports GIS professionals with access to very high resolution spatial datasets (>500,000 pixels/cell). The GIS extension Salt Marsh Assessment and Restoration Tool (SMART) was developed as part of a project to investigate LIDAR and hyperspectral imagery as remote sensing data sources for salt marsh restoration assessments (Rogers et al. 2007). ESRI ArcObjects and Spatial Analyst functions were utilized to manipulate and analyze spatial datasets and provide a user panel to specify spatial inputs (vegetation, elevation, and tidal sources) and site-specific parameters prior to model runs of one to five years (table 16.1).

Model operations use spatial functions that, for each vegetated cell, generate measures of (1) distance to open water, (2) distance to upland, (3) distance to nearest tidal source, and (4) the dominant cover type in the nearest neighborhood. Model inputs are merged into a single mosaic raster with encoded pixel values for cover type, elevation, and tidal source points, and processed one cell at a time to execute the simulation. The spatial datasets were used to determine gradient zones and apply succession model algorithms to reallocate species cover percentages and assess cell cover types. A habitat map layer is generated for each year of model operation and added to the GIS display (fig. 16.3). Annual cover totals are tabulated and stored in a text file for scenario analysis and reporting. Year-by-year series of habitat maps are also exported as standard jpeg files and sequenced with freeware software to create animations of succession over time.

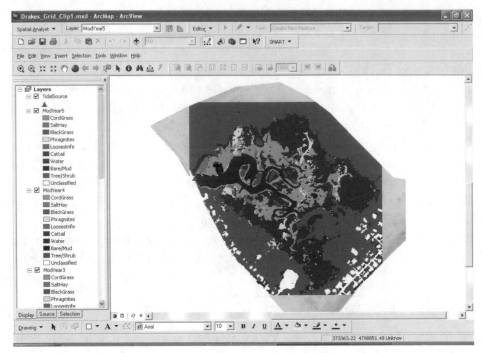

FIGURE 16.3. Sample model outcomes from Drakes Island Marsh simulation showing computer-generated GIS map layers of dominant cover types for multiple model years.

Using Simulation Outcomes for Restoration Design and Management

Model simulations have been conducted for eleven tidally restricted marsh sites around the Gulf of Maine, often as a centerpiece of restoration assessment planning (table 16.3). The assessments, funded by project proponents seeking support of design decisions and public approvals, are typically completed within about one month of effort, including field data collection, model development, and reporting. A wide range of local, state, and federal partners have been involved, and many uses for model deliverables have been discovered as summarized here.

Measuring Site Degradation Trajectory

Projects are often prioritized by resource managers according to the size of the impacted area, the practicality of restoring tidal hydrology, the likelihood of funding, and the level of public acceptance. With modeling, a new assessment dimension was added by quantifying the site trajectory of degradation (i.e., likely outcome of "no action"). Expansion of brackish habitat over time is the model outcome most useful as a predictor of degradation trends at an impacted site.

TABLE 16.3

Summary of modeled sites, project factors, significant model outcomes, and current restoration status

Project site	Impact area hectares	Issue	Lead proponent	Key model outcomes	Status
Oak Knoll (Rowley, MA)	15	Multiple culverts	Massachusetts Audubon	Brackish increase of ~1.5 percent yr.$^{-1}$; low potential for halophyte increase	No restoration action
North Pool (Plum Island, MA)	66	Water control structure	US Fish and Wildlife Service	Open structure allows 10 percent halophyte marsh; removal recovers 56 percent for halophytes	Expanded water control structure under review
Fresh Creek (Dover, NH)	40	Perched culvert	NH Coastal Program	Restoration would recover 3.5 hectares of low marsh	Pending engineering feasibility
Little River (North Hampton, NH)	70	Culvert/blocked channel	NH Coastal Program	Restoration returns 72 percent of marsh to halophytes from 44 percent prerestoration, with *Phragmites* < 10 percent[1]	Full restoration with twin box culverts and new cut channels
Mill Brook (Stratham, NH)	6	Culvert	NH Coastal Program	Restoration returns 65 percent of marsh to halophytes from 5 percent prerestoration[1]	Full restoration with elliptical culvert
Wheeler Refuge (York, ME)	7	Culvert	NOAA, York Rivers Association	Removal of berm and creation of new channel increases halophyte area by 20 percent	Partial restoration with berm grading and new channels
Drakes Island (Wells, ME)	35	Culvert	NOAA, Wells National Estuarine Research Reserve	Self-regulating tidegate (SRT) settings capable of maintaining 52 percent of marsh as halophyte, from 38 percent pre-SRT, with *Phragmites* < 10 percent	Partial restoration with box culvert and self-regulating tide gate; settings under review

267

TABLE 16.3
Continued

Project site	Impact area hectares	Issue	Lead proponent	Key model outcomes	Status
New Salt and Randall Marsh (Old Orchard, ME)	2	Multiple culverts	Town of Old Orchard Beach	Halophyte area of 40 percent would increase to 80 percent with full restoration, but only partial restoration is practical due to nearby structures	Adjustable tide gate reset to allow more inflow; settings under review
New Meadows (Brunswick, ME)	34	Culvert	Town of Brunswick	Halophyte area of 20 percent would increase to 62 percent with full removal, but culvert expansion could increase halophytes to 58 percent	Project under review
Pemaquid Marsh (South Bristol, ME)	2	Failed culvert	NOAA, Gulf of Maine Council	Impoundment due to failed culvert causing > 50 percent marsh loss; new culvert increases salt marsh habitat to 75 percent of area	Full restoration with box culvert
Cheverie Creek (Cheverie, Nova Scotia, Canada)	30	Culvert	NOAA, Ecology Action Centre	Marsh halophyte areas increased from 25 to 75 percent with full restoration	Full restoration with bridge span

[1] Models developed during or after design phase of project as retrospective analysis.

Degradation outcomes are especially valuable for those sites in early stages of brackish invasion. For instance, the threat of rapid *Phragmites* expansion was a primary management concern at Oak Knoll (Rowley, MA), Little River (North Hampton, NH), Drakes Island (Wells, ME), and New Salt Road (Old Orchard, ME). For these sites, model predictions of annual rates of *Phragmites* expansion were used to quantify the severity and scope of the invasive threat. At Drakes Island and New Salt Road, model animations were presented at public hearings to show neighbors the expected outcome of no action, with year-by-year frames showing the extent and expected rate of expansion for brackish species. These series are especially effective for communicating with marsh neighbors concerned about losing their views to encroaching *Phragmites*, and the moving pictures generate considerable dialogue about marsh processes and ecology (e.g., why is this section of the marsh expected to change while others are not?).

For resource managers, the invasive threat heightens awareness and adds urgency to action plans. At Little River, *Phragmites* occupied only 14 percent of the marsh before restoration, but a simulation of continued brackish conditions predicted rapid expansion of 3 percent marsh area per year. This pressing concern, together with flood control issues, helped elevate regional prioritization and funding for restoration work. Other sites showed less immediate invasive threats. At Oak Knoll, for example, higher salinities and fewer *Phragmites* patches contributed to a model outcome that predicted fairly slow expansion of less than 1.5 percent per year.

Assessing Restoration Potential

One of the driving questions for restoration prioritization and decision making is, What does full tidal restoration look like at a site? An understanding of full restoration potential is always a major management consideration. Even though it may not be practical in the short term to reintroduce full flooding at a site, estimates of maximum recoverable habitat are important factors for regional cost–benefit analyses.

At some sites, even under full tidal flow scenarios, marsh topography is a limiting factor. At the Wheeler Refuge (York, ME), the marsh was used for years as a dredge spoil dump, and the elevated marsh surface was thought to be a near-complete barrier to tidal flows. Modeling showed, however, that enhanced tidal flows to portions of the marsh following berm removal and channel creation would increase salt marsh cover to about 20 percent of the site. This expectation was sufficient to convince local partners to proceed with the project. At New Meadows Lake (Bath, ME), a large tidal impoundment exists above a perched tidal culvert, surrounded by the steep topography of lake shore. Restoration proponents believed that salt marsh potential might be limited to small fringing areas, although a low-elevation expanse in the upper reaches of the lake appeared within the range of natural tides. The model was applied to simulate an expanded culvert at a much lower invert elevation and confirmed that fully restored flows would in fact extend to the upper lake reaches, converting an estimated 60 percent of that brackish area to salt marsh. A similar early-stage assessment for an impoundment above a perched culvert at Fresh Creek (Dover, NH) showed the potential for low marsh to form as flooded areas reemerged following culvert replacement.

Several project sites experience near-complete restriction of tidal flows, and for these highly degraded marshes, models show dramatic potential for restoration. North Pool (Plum Island, MA), has a perched water control structure and waterfowl impoundment built in salt marsh that had filled with brackish invasive plants. Model results showed that, even if the structure were left open permanently, only 10 percent of the fringing marsh would become salt marsh. Complete removal and regrading of the structure, however, would recover 56 percent to salt marsh. This scenario was received favorably at public hearing, but concerns about avian habitat loss ultimately stalled management action. At Pemaquid Marsh (South Bristol, ME), an undersized, silted-in culvert had impounded a back-barrier creek and converted salt marsh habitat to freshwater. Model assessment, also presented at public hearing, showed that native salt marsh would quickly reestablish and expand to 75 percent of the area with implementation of a properly sized box culvert. With help from private funding sources, the Pemaquid project was approved and the marsh was restored within a single year of planning.

Evaluating Partial or Incremental Restoration Designs

Given practical considerations of costs, encroaching residential structures, and degradation state of the impact area, partial or incremental restoration is sometimes the best possible outcome. In these cases, the model can be used to investigate marsh responses to a range of possible flood levels. Scenario results are used by managers to find an engineering design that best meets acceptable flow conditions. Outcomes often represent a balance between flood control and ecological benefits.

At two sites with self-regulating tide gates (SRTs), the model was used to recommend new closure settings based on the ecological health of the marsh. The marsh at New Salt Road (Old Orchard, ME) is surrounded by low-elevation homes and roads that limit the amount of allowable flooding, resulting in very restrictive tide-gate settings. A public presentation showed that the closure setting could be incrementally increased with little flooding risk. The Public Works Department agreed to raise the setting a few inches and reassess each year. This adaptive management approach has a much longer history at the Drakes Island Marsh in Wells, ME (see also Adamowicz and O'Brien, chap. 19, this volume). The history of "managed" tidal flows there has evolved over time, from no-flow prior to 1988, to an undersized culvert 1988–2001, and most recently to a box culvert and adjustable SRT. This site has been modeled extensively, and initial assessments were instrumental in gaining public approval for the current box and SRT structure. But like New Salt Road, flood control settings on the SRT are, in some views, excessively restrictive for long-term ecological health of the marsh. The site was recently remodeled with high-resolution elevation data (LIDAR) and a sophisticated hydraulic model to make the case for increased flood allowances (Rogers et al. 2007). The new model suggests that the SRT may not be necessary as protection from nonstorm tidal flooding. It remains to be seen what additional flood allowances will be made there.

Improving Site Monitoring Plans

Integral to adaptive management is the need for postrestoration monitoring, and all federally funded projects now have a mandate to track marsh responses for at least several years following restoration activities. Monitoring plans almost always include physical measures of hydrologic and salinity changes, and biological measures of vegetation responses, but details regarding monitoring duration, scope of variables, location, and methods are highly variable (Konisky et al. 2006).

Through modeling assessments, marsh areas of concern can be identified to help managers develop more focused and effective monitoring programs. At Drakes Island Marsh, the first model predicted that a very large area of cattail would likely convert to *S. alterniflora*. However, the inaccessibility of this cattail area resulted in poor spatial coverage of elevation measures and therefore greater uncertainty in model predictions. Further, these backlands appeared to be at an elevation very near peak projected flood levels for the initial SRT setting. From continued work with local resource managers on the project, it became evident that this section of the marsh required greater monitoring focus. Plans were adjusted accordingly to include additional salinity wells and vegetation transects, and field results were reported annually for management team review of SRT settings. Recognizing that monitoring is labor-intensive and expensive, any assistance in developing a targeted program can be a considerable benefit.

Communicating Plans to Stakeholders

Positive responses about the value of model outcomes are consistently received as aids for envisioning and better understanding restoration project objectives.

Model simulations seem particularly beneficial for nontechnical people, such as local residents, who connect quickly with easily recognized marsh features, upland borders, and areas of concern. Spatial model outcomes help assure residents that proposed hydrologic changes and habitat responses will not impact their properties, even at peak tides. Year-to-year time series animations are especially effective to demonstrate that expected changes may be subtle, preserving marsh aesthetics into the future. Model animations are particularly useful for public hearing presentations.

Technical managers also find model outcomes useful for communicating the need for action and for justifying fund expenditures. In several cases, models were shared with private and public funders prior to securing financial resources. As experience grows, further uses and benefits of simulations will be realized in areas of consensus building, public approval, permitting, funding, and communication among diverse stakeholders.

Benefits of Ecosystem Simulation Models

A synthesized ecosystem model brings together knowledge from decades of field observations, theoretical studies, and experimentation. As with all models, high-quality inputs are of critical importance (e.g., engineer-grade tidal hydraulics, high-resolution digital elevation models, and fine-scale vegetation maps). Based on sufficient spatial inputs, the model deliverables are credible, understandable, and consistent with observed patterns of plant community response to tidal reintroduction. Scenario images and animated results of expected habitat responses provide new ways for managers to assess potential outcomes and to communicate, inform, and reassure stakeholders. In the final analysis, the real value of models is determined by those people directly involved in identifying, planning, and implementing improvements to degraded salt marshes. Uncertainties and limitations remain, but model simulation technology is a useful and practical tool that advances the future science and management of salt marsh restoration.

Acknowledgments

David Burdick provided much of the inspiration for our restoration model and advised in all aspects of initial development and field experimentation. Roelof Boumans and Fred Short contributed significantly to conceptual design and technical implementation. We thank the Cooperative Institute for Coastal and Estuarine Environmental Technology (CICEET) for support of model development and technology transfer.

NOTE

1. The complete version of the simulation model that is presented in this chapter is contained in a technical report by Konisky et al. (2003). Visit http://rfp.ciceet.unh.edu /projects/search.php to view a copy of this technical report. Enter "spatial modeling" in the title box, then click "search."

REFERENCES

Anisfeld, S. C., M. J. Tobin, and G. Benoit. 1999. "Sedimentation Rates in Flow-Restricted and Restored Salt Marshes in Long Island Sound." *Estuaries* 22:231–44.

Bertness, M. D., and A. M. Ellison. 1987. "Determinants of Pattern in a New England Salt Marsh Plant Community." *Ecological Monographs* 57:129–47.

Boumans, R. M. J., D. M. Burdick, and M. Dionne. 2002. "Modeling Habitat Change in Salt Marshes after Tidal Restoration." *Restoration Ecology* 10:543–55.

Burdick, D. M., M. Dionne, R. M. Boumans, and F. T. Short. 1997. "Ecological Responses to Tidal Restoration of Two New England Salt Marshes." *Wetlands Ecology and Management* 4:129–44.

Burdick, D. M., R. M. Boumans, M. Dionne, and F. T. Short. 1999. *Impacts to Salt Marshes from Tidal Restrictions and Ecological Responses to Tidal Restoration.* Final Report to National Oceanic and Atmospheric Administration (NOAA), grant no. NA570R0343.

Cahoon, D. R., and R. E. Turner. 1989. "Accretion and Canal Impacts in a Rapidly Subsiding Wetland, II: Feldspar Marker Horizon Technique." *Estuaries* 12:260–68.

Chalmers, A. G., R. G. Wiegert, and P. L. Wolf. 1985. "Carbon Balance in a Salt Marsh: Interactions of Diffusive Export, Tidal Deposition and Rainfall-Caused Erosion." *Estuarine, Coastal Shelf Science* 21:757–71.

DeLaune, R. D., R. H. Baumann, and J. G. Gosselink. 1983. "Relationship among Vertical Accretion, Coastal Submergence, and Erosion in a Louisiana Gulf Coast Marsh." *Journal of Sedimentary Petrology* 53:147–57.

Donnelly, J. P., and M. D. Bertness. 2001. "Rapid Shoreward Encroachment of Salt Marsh Cordgrass in Response to Accelerated Sea-Level Rise." *Proceedings of the National Academy of Sciences (PNAS)* 98:14218–23.

Fitz, C., E. DeBellevue, R. Costanza, R. M. J. Boumans, T. Maxwell, L. Wainger, and F. H. Sklar. 1996. "Development of a Generic Transforming Ecological Model (GEM) for a Range of Scales and Ecosystems." *Ecological Modeling* 88:263–95.

Gallagher, J. L. 1983. "Seasonal Patterns of Recoverable Underground Reserves in *Spartina alterniflora* Loisel." *American Journal of Botany* 70:212–15.

Gardner, L. R., H. W. Reeves, and P. M. Thibodeau. 2002. "Groundwater Dynamics along Forest-Marsh Transects in a Southeastern Salt Marsh, USA: Description, Interpretation and Challenges for Numerical Modeling." *Wetlands Ecology and Management* 10:145–59.

Gosselink, J. G., and R. S. Hatton. 1984. "Relationship of Organic Carbon and Mineral Content to Bulk Density in Louisiana Soils." *Soil Science* 137:177–80.

Grace, J. B. 1987. "The Impact of Preemption on the Zonation of Two *Typha* Species along Lakeshores." *Ecological Monographs* 57:283–303.

Harvey, J. W., P. F. Germann, and W. E. Odum. 1987. "Geomorphological Control of Subsurface Hydrology in the Creekbank Zone of Tidal Marshes." *Estuarine, Coastal, and Shelf Science* 25:677–91.

Hemminga, M. A., and G. J. C. Buth. 1991. "Decomposition in Salt Marsh Ecosystems of the S.W. Netherlands: The Effects of Biotic and Abiotic Factors." *Vegetation* 92:73–83.

Konisky, R. A., and D. M. Burdick. 2004. "Effects of Stressors on Invasive and Halophytic Plants of New England Salt Marshes: A Framework for Predicting Response to Tidal Restoration." *Wetlands* 24:434–47.

Konisky, R. A., D. M. Burdick, F. T. Short, and R. M. Boumans. 2003. *Spatial Modeling and Visualization of Habitat Response to Hydrologic Restoration in New England Salt Marshes*. Final report to Cooperative Institute for Coastal and Estuarine Environmental Technology (CICEET), University of New Hampshire, Durham.

Konisky, R. A., D. M. Burdick, M. Dionne, and H. A. Neckles. 2006. "A Regional Assessment of Salt Marsh Restoration and Monitoring in the Gulf of Maine." *Restoration Ecology* 14:516–25.

McKee, K. L., and W. H. Patrick. 1988. "The Relationship of Smooth Cordgrass (*Spartina alterniflora*) to Tidal Datums: A Review." *Estuaries* 22:143–51.

Morris, J. T., and W. B. Bowden. 1986. "A Mechanistic, Numerical Model of Sedimentation, Mineralization, and Decomposition for Marsh Sediments." *Soil Science Society of America Journal* 50:96–105.

Neckles, H. A., M. Dionne, D. M. Burdick, C. T. Roman, R. Buchsbaum, and E. Hutchins. 2002. "A Monitoring Protocol to Assess Tidal Restoration of Salt Marshes on Local and Regional Scales." *Restoration Ecology* 10:556–63.

Nerem, R. S. 1999. "Measuring Very Low Frequency Sea Level Variations Using Satellite Altimeter Data." *Global and Planetary Change* 20:157–71.

Niering, W. A., and R. S. Warren. 1980. "Vegetation Patterns and Processes of New England Salt Marshes." *BioScience* 30:301–7.

Odum, W. E., T. J. Smith III, J. K. Hoover, and C. C. McIvor. 1984. *The Ecology of Tidal Freshwater Marshes of the United States East Coast: A Community Profile*. FWS/OBS-87/17. Washington, DC: US Fish and Wildlife Service.

Office of Ocean and Coastal Resource Management (OCRM). 2006. *MAPTITE: Integrating Land Elevations and Water Levels for Habitat Restoration*. Bethesda, MD: National Oceanic and Atmospheric Administration (NOAA).

Pearlstine, L. G., W. M. Kitchens, P. J. Latham, and R. D. Bartleson. 1993. "Tide Gate Influences on a Tidal Marsh." Pp. 433–40 in *Proceedings from Symposium on Geographic Information Systems and Water Resource, American Water Resources Association, Mobile, Alabama*.

Rogers, J., R. Konisky, and J. Mustard. 2007. *Salt Marsh Assessment and Restoration Tool (SMART): An Evaluation of Hyperspectral, LIDAR and SWMM for Producing Accurate Habitat Restoration Predictions*. Final report to Cooperative Institute for Coastal and Estuarine Environmental Technology (CICEET), University of New Hampshire, Durham.

Roman, C. T., W. A. Niering, and R. S. Warren. 1984. "Salt Marsh Vegetation Change in Response to Tidal Restrictions." *Environmental Management* 8:141–49.

Roman, C. T., R. W. Garvine, and J. W. Portnoy. 1995. "Hydrologic Modeling as a Predictive Basis for the Ecological Restoration of Salt Marshes." *Environmental Management* 19:559–66.

Rybczyk, J. M., J. C. Callaway, and J. W. Day. 1998. "A Relative Elevation Model for a Subsiding Coastal Forested Wetland Receiving Wastewater Effluent." *Ecological Modeling* 112:23–44.

Sinicrope, T. L., P. G. Hine, R. S. Warren, and W. A. Niering. 1990. "Restoration of an Impounded Salt Marsh in New England." *Estuaries* 13:25–30.

Smith, S. M. 2007. "Removal of Salt-Killed Vegetation during Tidal Restoration of a New England Salt Marsh: Effects on Wrack Movement and the Establishment of Native Halophytes." *Ecological Restoration* 25:268–73.

Smith, S. M., C. T. Roman, M. J. James-Pirri, K. Chapman, J. Portnoy, and E. Gwilliam. 2009. "Responses of Plant Communities to Incremental Hydrologic Restoration of a Tide-Restricted Salt Marsh in Southern New England (Massachusetts, U.S.A.)." *Restoration Ecology* 17:606–18.

Stumpf, R. P. 1983. "The Process of Sedimentation on the Surface of a Salt Marsh." *Estuarine, Coastal, and Shelf Science* 17:495–508.

Turner, R. E., E. M. Swenson, and C. S. Milan. 2000. "Organic and Inorganic Contributions to Vertical Accretion in Salt Marsh Sediments." Pp. 583–95 in *Concepts and Controversies in Tidal Marsh Ecology*, edited by M. P. Weinstein and D. A. Kreeger. Dordrecht, Netherlands: Kluwer Academic.

Valiela, I., J. M. Teal, S. D. Allen, R. Van Etten, D. Goehringer, and S. Volkman. 1985. "Decomposition in Salt Marsh Ecosystems: The Phases and Major Factors Affecting Disappearance of Above-Ground Organic Matter." *Journal of Experimental Biology and Ecology* 89:29–54.

Voinov, A. A., H. Voinov, and R. Costanza. 1999. "Surface Water Flow in Landscape Models, II: Patuxent Watershed Case Study." *Ecological Modeling* 119:211–30.

Warren, R. S., P. E. Fell, J. L. Grimsby, E. L. Buck, C. G. Rilling, and R. A. Fertek. 2001. "Rates, Patterns, and Impacts of *Phragmites australis* Expansion and Effects of Experimental *Phragmites* Control on Vegetation, Macroinvertebrates, and Fish within Tidelands of the Lower Connecticut River." *Estuaries* 24:90–107.

Warren, R. S., P. E. Fell, R. Rozsa, A. H. Brawley, A. C. Orsted, E. T. Olson, V. Swamy, and W. A. Niering. 2002. "Salt Marsh Restoration in Connecticut: 20 Years of Science and Management." *Restoration Ecology* 10:497–514.

Weigert, R. E. 1986. "Modeling Spatial and Temporal Variability in a Salt Marsh: Sensitivity to Rates of Primary Production, Tidal Migration, and Microbial Degradation." Pp. 405–26 in *Estuarine Variability*, edited by D. A. Wolfe. London: Academic.

Wood, M. E., J. T. Kelley, and D. F. Belknap. 1989. "Patterns of Sediment Accumulation in the Tidal Marshes of Maine." *Estuaries* 12:237–46.

Woodlot Alternatives Inc. 2007. *New Meadows Lake, Brunswick and West Bath, Maine: Model Analysis of Expected Plant Communities Response to Potential Tidal Restoration Conditions.* Prepared for Town of Brunswick, 28 Federal Street, Brunswick, ME.

Zedler, J. B. 2000. "Progress in Wetland Restoration Ecology." *TREE* 15:402–7.

Chapter 17

Incorporating Innovative Engineering Solutions into Tidal Restoration Studies

WILLIAM C. GLAMORE

Restoring tidal flows to salt marsh wetlands is typically undertaken for environmental, social, or compensatory (i.e., compensating for loss elsewhere) purposes. In most circumstances the objective of the project is to create or restore an environmental setting that historically existed on-site or nearby. During the process of restoring the tidal environment, significant attention is given to the desired environmental outcomes with specific interest on the notable flora and fauna. To achieve these end goals attention has become increasingly focused on the importance of reestablishing a desired hydrologic and/or hydrodynamic regime. The process of restoring the site's hydrology, however, is often complicated by a range of factors, and, in many cases, it is not possible to restore the site back to some pre-existing "natural" state. In these circumstances, engineering solutions are required to achieve the desired hydrologic and/or hydrodynamic outcome.

This chapter presents a variety of engineering principles and solutions related to salt marsh restoration projects. The chapter begins by discussing design concepts and issues to consider before modifying on-ground structures to restore tidal flushing. With this background information, various state-of-the-art engineering concepts and innovative designs are detailed. Specific examples related to modified floodgates and site monitoring are provided. While these examples have been tailored to the Australian landscape, the on-ground solutions are relevant to the United States and other coastal locations worldwide.

Engineering and Design Concepts in Tidal Restoration Projects

Previous chapters in this book have outlined various tools available to generate a hydrologic model (conceptual or numerical) of the site. This information is typically used to formulate site-specific restoration plans. Applying these plans to the site often involves determining the most appropriate on-ground engineering practices. Of particular concern are the selection, design, construction, and testing of on-ground structures. These structures are often the greatest expense to the project, requiring a large investment of resources, and typically control the hydrology of the restored site. Therefore, this chapter presents key guidelines that should be considered during the design/planning phase, prior to extensive and costly on-ground works.

Based on water quality and adjacent landowner flooding concerns (determined via site data collection and numerical modeling exercises), and working within the infrastructure and hydraulic constraints present at any study site, modifications of on-ground structures should comply with a range of predetermined operational guidelines. For this purpose, the following list provides general guidelines that should be satisfied for all new or modified structures prior to construction. Failure to comply with these guidelines can be rationalized for a particular site; however, consideration of all the criteria is warranted.

- Operability
  - Control of infrastructure to allow adjustments during critical events such as flooding and/or gate obstruction (e.g., debris removal)
  - Sufficient flexibility to trial several management strategies including optimizing water levels and seasonal variations
  - Ease of transport and installation
- Durability
  - Low maintenance
  - Long-lasting materials
- Safety
  - Safe to operate and maintain
  - Resistant to vandalism
  - Designed to function during extreme rainfall and discharge events
- Applicability
  - Can be implemented within existing infrastructure
  - Is suitably designed to minimize headloss, hydrostatic forces, and water velocity
  - Considers flood mitigation effects for adjacent landholders
  - Has potential for widespread application
  - Can adapt to include effects of climate change and varying sea levels
- Connectivity
  - Will permit passage of desired flora and fauna upstream of the structure

- Cost
  - Reasonably priced
  - Inexpensive to maintain

These guidelines are designed to help avoid common pitfalls associated with the design and implementation of on-ground structures during salt marsh restoration projects. Of particular concern is the ability of a structure to operate during extreme environmental events such as large flooding events. If the proposed structure is to be retrofitted to existing infrastructure additional concern should be taken as most existing infrastructure in salt marsh environments was originally designed as one-way flow. Modification of these structures to allow for bidirectional tidal flows may require water to flow upslope or against designed hydrostatic forces. Failure to take these guidelines into account could result in significant miscalculations of flood levels behind infrastructure or catastrophic failure during critical events (e.g., floods, spring tides, etc.).

In many restoration projects adjoining private landholders may have concerns related to flooding or ponding of saltwater on their properties. As such, this must also be considered during the design of a structure. In several cases this concern results in the installation of complex structures that can control the tidal volume, tidal amplitude, or hydroperiod within the salt marsh. While these structures may be effective at limiting the flooding of adjacent properties, the land manager must also consider how the structure will operate under conditions likely to be experienced due to climate change (e.g., rising sea levels or increased storm frequency/duration).

Despite the foregoing guidelines, a paucity of information is available related to general wetland restoration/engineering and design techniques. In an attempt to fill this deficiency there have been numerous requests for a comprehensive salt marsh restoration engineering design guideline. The production of a wetland engineering guideline focused on coastal issues would be a vital step toward collating the range of relevant engineering design procedures while providing an important source of information for future restoration studies.

Innovative On-Ground Engineering Structures for Restoration Projects

Once the initial concepts have been considered and approved, then on-ground works would commence. The on-ground practices of primary importance to a tidal restoration project typically include (1) removing preexisting flow-retarding structures, (2) installing flow-diverting structures (e.g., culverts, levees, and weirs to contain/divert the restored tidal flows), (3) undertaking earthworks to modify the landscape (e.g., creating channel sinuosity and ponds, removing contaminated soils, preparing the site for planting, etc.), (4) establishing pre- and postmonitoring sites, and (5) installing tidal control devices. As the majority of these tasks are standard engineering practices that are undertaken at many field sites (e.g., construction of levees or installation of culverts), the focus here is on the installation of modified tidal gate structures specifically related to salt marsh restoration projects. These devices are used in situations where the hydroperiod or tidal prism within the wetland must be manipulated to achieve the desired outcomes and can either be installed on preexisting headworks or form part of a new engineering design.

Tidal Floodgate Hydraulics

In coastal settings throughout the world salt marsh restoration projects are occurring in areas that were previously drained, modified, and engineered to eliminate or dramatically reduce tidal flushing. In many of these modified environments, structures such as top-hinged tidal floodgates (or flap gates) are installed to restrict the tide and maintain the upstream water levels at a low tide mark. These tidal floodgates (fig. 17.1) range from simple pipe culvert structures to sophisticated multicompartment box culvert systems. Restoring tidal flows to these sites typically involves modifying the existing floodgate structures to allow either full or muted (i.e., partial) tidal flushing. Therefore, following a brief discussion of typical tidal floodgates, a range of innovative modifications currently being applied to these existing structures to reinstate tidal flushing are presented next. Due to the wide range of structures currently in practice, the structures highlighted are commonly used devices (at Australian sites); however, other devices are available.

Though radial and sluice gates can be found in selected flood mitigation systems, the majority of existing flood mitigation structures found in estuarine environments are top-hinged flap gates. The working objective of flap gates is to permit drainage upstream of the floodgate structure and prevent flood and tidal flows from downstream waterways into the drained area. Top-hinged floodgates operate under two design criteria. First, when the downstream water elevation is higher than the upstream levels, hydrostatic pressure closes the gate, and reverse flow cannot occur. Second, when upstream water levels are higher than downstream levels, hydrostatic pressure opens the gate, and water is discharged from the drain. Assuming these conditions remain, water will continue to discharge until the drain invert or the sill level is reached. Naturally, when the water level is equal on both sides, the gate remains closed and acts as a nonreturn valve. Slanting the headwall enhances gate closure, and a compression seal between the flap face and the headwall structure decreases leakage.

FIGURE 17.1. Typical top-hinged tidal floodgates. Ebb tide flow is indicated by the white arrow. The floodgate flap remains closed when downstream water levels are higher (due to tides or flooding) than upstream water levels. When upstream water levels are higher (due to low tides, large inflows, or backswamp flooding), hydrostatic pressure forces the floodgate flap to open.

These operating conditions imply that flap gates "automatically" drain the wetland to the lowest tidal level. However, drainage will not occur unless there is a significant difference between upstream and downstream water levels; therefore, the rate of fall upstream of the gate cannot be greater than the rate of fall downstream. Nonetheless, drainage times are site specific and related to a number of factors, including the following (Patterson and Smith 2000):

• The relationship between water height and volume stored in a particular area
• The size of the floodgate structure
• The design of the floodgate
• The size of the drain
• The hydraulic efficiency of the drain

Limited information is available on the hydraulic considerations and stage discharge relationship of top-hinged flap gates. The majority of available information

(Pethick and Harrison 1981; Lewin 1995) concerns the discharge of water through multiple gates and its impact on flow interference. Specifically, when a series of flap gates are in close proximity, discharge under and around the gate causes flow interferences and results in hydraulic loss. To maximize discharge and reduce bed scour from flow interferences, training walls and concrete aprons are often installed. With this in mind, any major structural modifications to the floodgate infrastructure must consider the interactions between adjoining floodgates and hence compensate for reduced discharge.

Modifying a floodgate to allow for tidal restoration (i.e., two-way flow) requires a good understanding of tidal hydraulics. This is particularly important if numerical modeling is to be undertaken to simulate tidal restoration under a range of different floodgate modifications and an approximation of discharge is required for development of stage–discharge curves. Due to cost and site restraints, modifications are typically undertaken within the existing floodgate headworks and usually involve the modification of the one-way tidal floodgate flap to allow for two-way (or bidirectional) flow. Modifications differ in design and range from simple lifting mechanisms to complex automated gates. A number of modification styles and related methods to approximate tidal volumes are described here, but only in brief. The reader is advised to consult with a specialist in fluid mechanics to gain a better understanding of the entire range of related hydraulic issues, including supercritical flows, prior to modifying tidal structures.

Potentially the simplest floodgate modification is to cut a hole in the floodgate to allow for tidal waters to pass through. The hole is typically cut toward the bottom of the gate; thus the aperture is below the waterline. In these circumstances the amount of water that may flow into the wetland can be estimated by applying an equation for flow through an orifice (equation 17.1).

$$Q = C_d A \sqrt{2g(HW - B/2)} \qquad (17.1)$$

Where, Q is discharge in meters-cubed per second, C_d varies with head, culvert type, and geometry (0.6 can be used as an initial guide); A is the cross-sectional area of the orifice, HW is the water height, g is the gravitational constant, and B is the flow height (from top of culvert).

For planning purposes the size of the aperture (i.e., orifice) can be manipulated in equation 17.1 to determine the optimal cross-sectional area (i.e., the area that provides the optimal tidal flushing with limited head loss). This method may also be applied to estimate flow through a modified gate when it is vertically lifted to allow tidal flushing (i.e., say by hand winches similar to an undershot weir). This method may also be applied when estimating flow through a culvert; however, it is only applicable when the flow is inlet controlled and the culvert is relatively short, being less than 2 meters long.

Orifice-type modifications that allow for tidal restoration are useful as they typically provide a high degree of water-level control. This style of modification is practical in small wetland systems (1–10 hectares) or those with a relatively modest tidal range when only a small volume of tidal water is required to restore wetland processes. In larger systems it is important to consider both the discharge (as part of the stage–discharge relationship) and the resultant velocity. Water velocity (i.e., discharge per cross-sectional area [Q/A]) is important as high velocities (and turbulence) can have a deleterious effect on estuarine biota, may cause excessive scouring, and can result in large unwanted forces being applied to the structure. Headloss is an exponential function of velocity, and, as shown in equation 17.2, a small increase in velocity will result in a large increase in headloss.

$$HL \approx V^2/2g \tag{17.2}$$

Where, HL is an approximation of headloss in meters, V is velocity (in meters per second), and g is the gravitational constant.

In most wetlands with limited tidal range the six-hour tidal period may limit flow velocities. However, in sites with greater tidal range high (jetlike) velocities could occur, resulting in excessive erosion/scouring unless a larger aperture size is provided. In these cases a general rule is to try to maintain velocities below 0.6 meters per second by manipulating the cross-sectional area.

In environments where the orifice approach is not desired, it may be preferable to use weir structures to manipulate the flows into the restored tidal wetland. This is a common practice in Australia where the emphasis is on maintaining high groundwater tables (to combat acid sulphate soils) by withholding a portion of the tidal prism upstream of the weir each tide. Flow over a sharp-crested weir can be estimated using equation 17.3, whereas flow over a broad crested weir (i.e., similar to a levee bank or low-set dyke) is given in equation 17.4. Additional weir types, such as V-notch weirs, can also be employed. A full description of weir structures and design guidelines can be found in Olin et al. (2000). It is worth noting that although weirs can provide adequate water-level control they may restrict fish passage and increase stagnation.

$$Q = 0.66 \times cB * (2g)^{0.66} \times H^{1.5} \tag{17.3}$$

$$Q = 0.35 \times B * (2g)^{0.5} \times H^{1.5} \tag{17.4}$$

Where, Q is the discharge (meters-cubed per second), c is a discharge coefficient (average value of 0.62), B is the width of the weir (in meters), and H is the height of the water above the weir (measured 4 times H upstream).

In circumstances where it is preferable to remove all structures (e.g., removing culverts, surrounding levees, and existing floodgates), yet maintain a muted tidal amplitude, a number of methods can be employed to calculate the dimensions of the entrance channel and the resultant tidal signature. In these cases friction is the dominant force that reduces the tidal energy. As will be detailed here, the majority of methods currently available are derived from estuary inlet stability theory and assume an idealized estuary and wetland/basin. As only a brief discussion is provided here, the reader is encouraged to visit the source documents for a complete synthesis of empirical and analytical formulations.

O'Brien (1931) established that the tidal prism could be used to determine a stable inlet cross-sectional area. This work was significantly expanded upon by Kuelegan (1967), who proposed relationships between the tidal phase lag, the ratio between the wetland to ocean tidal amplitudes, and a term referred to as the repletion coefficient. Brunn et al. (1978) further expanded upon the theory and provided a method for simply calculating water levels within a tidal wetland based on downstream tidal characteristics.

Recent methods to determine water levels and flux have been proposed by Roman et al. (1995) and Williams et al. (2002). Roman et al. (1995) employed a modified Manning's equation and a continuity equation to calculate the resultant wetland tidal heights. In their study the volume of water being discharged into the restored wetland is proportional to the square root of the water height difference between water levels upstream and downstream of a dike. Conversely, Williams et al. (2002) used field data collected from a range of restored wetlands in San Francisco Bay to develop relationships between entrance channel dimensions and tidal prism or marsh area.

Modification of any engineered structure should be based on the previously provided design guidelines and discharge calculations outlined earlier. The intricacy of a design is typically correlated with the degree of upstream water-level control required, with manual gates having the least control, buoyant gates having moderate control (i.e., 0.1 meters), and intricate automated gates having the greatest control of upstream water levels (and typically the greatest installation expense). A range of innovative modified floodgate structures from simple winch lifting gates to complex SmartGate systems are presented next. Commonly used water control devices such as sluice gates and stop logs are not discussed here, but further information on these and other basic structures can be found at Olin et al. (2000).

Modified Gates: Winch Driven

The simplest forms of modified floodgate structures are winch-driven lifting gates. Of these, two designs are most apparent: vertical lifting and overhead cantilever lifting. The overhead cantilever lifting design, as shown in figure 17.2, is designed

FIGURE 17.2. Example of modified cantilever floodgates.

to raise the floodgate away from the face of the headworks, lifting the gate above the waterline. The advantage of this style of modified floodgate is the ability to completely raise the floodgate from the water and thus eliminate drag forces. However, this limits the amount of upstream water-level control as the gate is either fully open or fully closed. Further, if the gate is not fully open and is thus subject to the drag force of the incoming tidal waters, uneven drag forces can cause excessive oscillating strain on the gate's support arms and hinges. As such, this style of modified gate is best suited to systems that are to be left fully open or fully closed and are not to be operated during large flow events.

In contrast, vertical lifting winch gates can be safely controlled in large flow events as the hinge mechanism is incorporated into the modified design. As shown in figure 17.3, a vertical lifting winch gate consists of two I-beams welded to a base plate with holes drilled into the plate to align with the existing headwall bolts. Two load-compensating arms are installed within the gate with one inserted through holes drilled within the I-beams, and the other bolted onto the arms

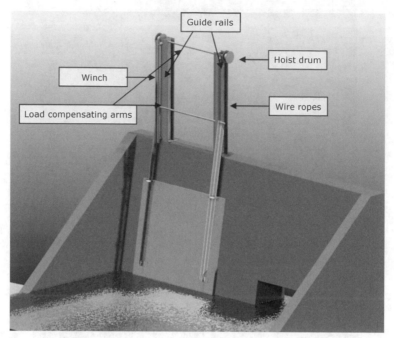

FIGURE 17.3. Schematic of modified vertical lifting floodgates.

of the flap gate. A hoist drum is then welded onto one end of the top load-compensating arm, and a winch plate is welded onto the I-beam below the hoist drum. Stainless steel wire ropes (safety factor of 10) are then run between the winch and the hoist drum and between the load-compensating arms. The wire ropes are designed so that if one rope fails the other rope will continue to operate.

Unlike other lifting gates, in the vertical lifting gate the mass of the flap is not overbalanced, and the main drive effort is in opening or lifting the gate. To enhance this "fail-safe" design, a series of guides are installed along the exterior of the I-beams that enclose the lower load-compensating arm. These guides allow the gate to rise within a predetermined slot, while maintaining the gate pivot point along the lower load-compensating arm. The bottom bolts on the guides also stop the gate from being excessively lowered, which could damage the flap or the sill, and act as a catch in the unlikely case of dual wire rope failure. Further, while hydraulic oscillating forces are important in high-headed cantilever lifting gates, they are not a concern with vertical lifting gates because under open channel conditions hydraulic downpull forces are negligible and can be absorbed by the margin of hoisting force provided in the gate installation. Vertical lifting gates can also provide some upstream water level control as the gates can be opened at preset intervals (i.e., less than fully opened) to allow the ingress of a muted tide.

Modified Gates: Buoyant or Self-Regulating Tide Gates

The modified gates already discussed are typically preset in a known position and are manually manipulated on-site when required. As such, the upstream water level control provided by these structures is minimal. When greater control is required, other styles of modified gates should be employed. To improve upstream water level control, buoyant tide gates (also called self-regulating tide gates) are commonly used.

As shown in figure 17.4, self-regulating tide gates are attached to an aperture within a floodgate flap and consist of a hinged door mechanism that is opened and closed by the rise and fall of a buoyant cylinder. As the cylinder falls, the door on the aperture opens, allowing tidal water upstream of the floodgate. Conversely when the cylinder rises with the tide to a preset level, the door closes and no tidal water is allowed upstream. Adjustments to the cylinder height and buoyancy (by adding fill such as sand to the cylinder) can be made to fine-tune the opening and closing heights of the gate mechanism to regulate upstream water levels.

FIGURE 17.4. Example of modified self-regulating tide gates.

Self-regulating tide gates provide adequate upstream water level control on small drainage systems. In small drainage networks the gates provide upstream water level control in the order of ± 0.25 meters and can be an effective long-term, low-maintenance tool for restoring tidal flows. Indeed, hundreds of self-regulating tide gates have been installed throughout small to medium (1–10 hectares) tidal restoration sites in coastal Australia. Project planners should be sure to include sufficient capacity to store interior runoff above tidally induced flood heights.

In larger wetland systems the volume of water that discharges through a self-regulating tide gate may lead to the gate's operating ineffectively if the culvert is not designed properly. When the number of openings is insufficient or the designed gates are too small, a large volume of water is forced through a relatively small gate aperture. This results in excessive velocities and thus drag forces on the hinged door mechanism. If these forces exceed those of the buoyant lifting force of the cylinder, the gate will close regardless of the water level height. Due to tidal inequities and the design of the system, this may occur during spring tides, in systems where the culvert is too small for the volume of tidal water required, or in regions with large (meso- or macro-) tidal ranges. As such, prior to installing self-regulating tide gates, calculations should be undertaken to estimate the flow velocities of the tidal water as it goes through the gate aperture (and subsequent calculations of drag forces), and these should be compared against the buoyant lifting force of the cylinder. Failure to consider these competing forces may result in the gate closing sooner than desired and hence a failure of the required objectives. This is of particular concern with self-regulating gates as shown in figure 17.4, although alternative designs used in the United States appear to be more efficient and/or reliable at high flows

Modified Gates: Automatic Gates

Though the foregoing styles of modified gates are useful on a wide majority of restoration systems, in some select locations additional upstream water level control, security, or water level manipulation may be required. In these locations an alternative design, called the SmartGate system, has been developed, tested, and installed at several sites (fig. 17.5).

The SmartGate system was designed to allow a specific volume of tidal water into an environment based on real-time water quality variables. The system consists of a sliding gate mechanism attached to a predetermined aperture within the main gate. The sliding gate movements are based on a number of preset variables that are coded into a data logger. The data logger unit can receive external measurements from a range of sensors, including pressure sensors, rainfall buckets, water quality measurement units, or flow gauges. The parameters measured by

FIGURE 17.5. Example of multiple SmartGate units. Components of the SmartGate are indicated (dashed arrows). The solid white arrow indicates ebb tide flow direction.

these sensors are uploaded into the data logger and, based on trigger values, the position of the sliding gate is determined. A direct current geared motor is automatically activated when a gate movement is required. The entire system is solar powered and can be remotely operated via telemetry.

An example of a typical gate operation is explained as follows:

1. Gate is open to allow tidal flushing.
2. Data logger measurements are uploaded to the control unit at preset intervals (e.g., every 10 seconds).
3. Data logger readings exceed the preset trigger levels and the control system activates the close gate protocols.
4. Once the gate(s) are closed, the system continues to monitor environmental variables.
5. When acceptable levels return, the preset trigger level is breached and the gate reopens.

SmartGates control the volume of water upstream of the floodgate (i.e., within the wetland) and optimize tidal flushing. To accomplish these objectives, a range of user-defined triggers based on real-time data logger readings are layered into

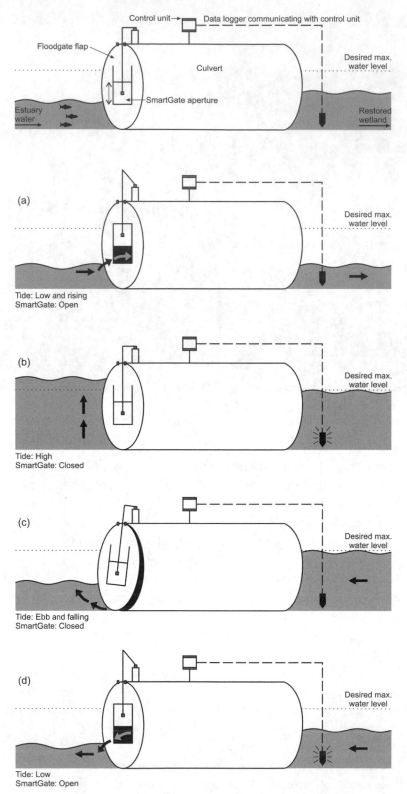

Figure 17.6. Operating modes of typical SmartGate unit installed on a culvert.

the system program. A schematic of a typical SmartGate operation over a full tidal cycle is shown in figure 17.6.

To increase robustness and enable remote control of the system, various "intelligent" designs are built into the SmartGate program. Via telemetry a user can view current conditions, download previously obtained data, change trigger levels, and override automatic functions. This ability is particularly important during flooding periods when access to the site may be restricted. Second, the control unit may also dial out to alert the site manager to breaches in preset alarms. These safety protocols are designed to alert the user to specific problems regarding operational issues such as low battery, jammed gate, energy surge, or various environmental water quality variables (e.g., low pH, high salinity, high flow) via a Short Message Service (SMS) text message. This function can also be used to alert a floodplain manager to activate an automatic sampler based on real-time drain water parameters or to commence an alert warning system. As a fail-safe, the control unit is equipped with manual override switches that can be operated on-site.

In comparison with other buoyancy-driven "automatic" gates driven solely by hydrostatic pressure, the SmartGate's critical advantage is that it can provide precise upstream water level control using multiple parameters, including pH, electrical conductivity, water temperature, dissolved oxygen, or bidirectional flows. This is significant because improved water quality, and not water elevation, is the key factor in restoring tidal flushing in many cases. Advanced water level control provides security that upland landholders will not be impacted by the tidal restoration project.

SmartGates can be constructed as individual or multiple units. Multiple units can be used to restore large salt marsh environments where a range of complex water level controls are required. As shown in figure 17.5, four SmartGate systems (each with a 1 square meter opening) installed at a single site can operate either in tandem or individually to control upstream tidal levels within a large restored salt marsh (greater than 300 hectares). Each SmartGate has its own preset trigger levels, although the entire system is operated from one control unit.

SmartGates are required at the site depicted in figure 17.5 because the primary aim of the study was to create salt marsh for migratory wading birds. Excessive drainage over the past fifty years at this site has caused subsidence of the soil profile; therefore, restoration of natural tidal flows would not restore the site to a functional salt marsh ecosystem. To overcome this limitation and ensure that upland landholders were not impacted by the tidal waters, the SmartGates at this site operate (i.e., open and close) within a predetermined range of water level values. These SmartGates are also being used in conjunction with regional floodplain management to ensure that the restored salt marsh site can function as a storage detention basin during large floods. The alarms and telemetry system within the control unit ensure that the system operates effectively under all conditions despite its remote location.

Though the SmartGates are designed to require minimal maintenance, any automated structure located in an estuarine environment will require ongoing care. Due to the remote nature of most installations vandalism is the greatest concern. As such, most SmartGates components are enclosed within waterproof locked boxes above predetermined flood levels and, where feasible, are located behind security fencing. In addition, telemetry alarms are used to alert the site manager if a component is tampered with. For the remaining parts a quarterly maintenance program is undertaken. Based on these requirements, SmartGate installations (and indeed all floodgate modification projects) have been most successful in locations where maintenance programs are part of the standard working procedures, such as in national parks or within local government authorities.

Incorporating Engineering into Monitoring of Tidal Restoration Projects

Field studies throughout the world commonly employ environmental monitoring as a core part of the project deliverables. As field monitoring data are often used to determine the success of the on-ground works, this section focuses on key data requirements that either incorporate engineering techniques or provide innovative engineering solutions. Other reference texts should be consulted in regard to establishing water quality or ecological monitoring programs.

Field monitoring of restored salt marsh sites is typically based on the before, after, control impact (BACI) monitoring model with individual constituents selected to represent the status of the wetland before and after restoration works. A monitoring program is typically developed to determine concentrations of constituents such as pH, salinity, and related selected ions at a range of locations across temporal domains.

Though relatively easy to gather, a collection of spot point water quality data is not very useful unless discharge (i.e., flow) is simultaneously collected. In combination with discharge data, the concentration data can be used to determine the net flux of various constituents across a site. Determination of net flux (i.e., concentration multiplied by discharge) is particularly important across salt marsh restoration sites as flows may be very small in some locations and larger in others. As such, regions with high pollutant concentrations but low flows may be of lesser concern than regions of moderate concentration and high flows.

Calculating discharge in salt marsh wetlands can be a difficult task due to the shallow waters experienced on-site, the continually changing nature of the channel cross-sectional area, and the low flows that limit traditional methods such as propeller meters. Recent advances in Doppler technology now permit discharge measurement in regions with very low velocities or in flow environments with shallow water. As these units become increasingly more affordable, techniques that are commonly applied in oceanic studies (such as examining multidirectional flows in predetermined depth bins) can be applied in restored estuaries and wetlands.

In locations where modern measurement techniques are not feasible, alternative techniques can be applied. In recent times, imaging techniques have been used in restored salt marsh environments to examine a range of restoration goals, including flow measurements. In these locations a high-resolution digital video camera is placed on a large pole or stand at a height of 15 meters or greater. Equipped with telemetry the digital camera can provide images of the restored site at preset time intervals. By rectifying these images to known coordinates, a plan image of the site can be created. Once a series of rectified photos are available, a grid can be placed over the images to divide the image into distinct quadrants (e.g., dividing the image into 1 meter by 1 meter squares). Using available particle tracking software, the floating debris in the images can be "tracked" to determine flow velocities over different time periods such as the ebb or flood tides. In combination with topographic information and water level data, this information can be used to calculate the flux of water (or contaminants) into or out of a restored site.

In southeastern Australia, the installation of digital video cameras in remote sites has greatly assisted with on-site monitoring and determining if the objectives of a study have been met. In addition to assisting in calculating the flux across the site, the high-resolution cameras can provide information on the following:

- *The evolution of the dendritic channel network* As the tide is restored to a low-lying site, the sinuosity, depth, width, and nature of the existing channel network evolve. The evolution of this channel network under the new flushing regime, as documented in the hourly images, provides useful information regarding the prediction of channel networks elsewhere.
- *The evolution of salt marsh growth under varying flushing regimes* At sites where SmartGates have been installed, different flushing regimes are being tried to determine their effectiveness at promoting the growth of salt marsh species. Since the salt marsh species in this region appear red in the images, a red color filter can be applied to the images to quickly quantify the spatial abundance of the salt marsh.
- *The use of the site by migratory wading birds* At locations where the primary aim of the salt marsh restoration works is to provide nighttime roosting habitat for migratory wading birds, image analysis has assisted with on-ground

bird counts. This analysis assists in determining when the site is most used and what hydrologic conditions attract the greatest abundance of birds.

- *Increased reporting and security* As the images from the camera can be remotely accessed at any time, the pictures have also been used to determine the impact of flooding across large remote wetlands to assist in determining when SmartGates, installed at the site, should be remotely closed or opened. Further, the use of the cameras on-site has assisted in reducing vandalism.

Overall, in the majority of restored salt marsh environments the main priority is to measure water quality variables or other similar biological variables that can be directly linked to the flora and fauna of the site. While these variables provide useful information it is important to link these spot point variables with flow into and out of the system to ensure that the total mass flux is calculated. In addition, other physical variables such as channel sinuosity, channel network evolution, and appropriate hydroperiods should be determined. This information can be easily obtained from a variety of methods, including remote high-resolution digital cameras, and is vital in understanding how a salt marsh system functions and evolves. Further information on an existing salt marsh video camera can be found at http://www.wrl.unsw.edu.au/site/resources/projects/tomago-wetland -remote-monitoring/.

Conclusion: Cross-Disciplinary Design Needs

This chapter has focused on the importance of incorporating engineering concepts into salt marsh restoration projects. On-ground engineering issues largely related to floodgate modifications have been addressed and current state-of-the-art techniques suggested. With the current advancements in the field of engineering and salt marsh restoration it is acknowledged that many of these innovative solutions may become commonplace; however, future innovations should continue to involve cross-disciplinary ideas that integrate ecological and engineering disciplines.

REFERENCES

Brunn, P., A. Mehta, and I. Johnsson. 1978. *Stability of Tidal Inlets: Theory and Engineering*. Amsterdam: Elsevier.

Kuelegan, G. H. 1967. *Tidal Flow in Entrances: Water Level Fluctuations of Basins in Communications with Seas*. Technical Bulletin No. 14. Vicksburg, MS: Committee on Tidal Hydraulics, US Army Engineers Waterways Experiment Station.

Lewin, J. 1995. *Hydraulic Gates and Valves in Free Surface Flow and Submerged Outlets*. London: Thomas Telford.

O'Brien, M. P. 1931. "Estuary Tidal Prism Related to Entrance Areas." *Civil Engineering* 1:8.

Olin, T. J., C. Fischenich, M. R. Palermo, and D. F. Hayes. 2000. *Wetlands Engineering Handbook*. Vicksburg, MS: Wetlands Research Program, Engineer Research and Development Center, US Army Corps of Engineers.

Patterson, C., and R. Smith. 2000. *Acid Sulfate Soils Program: Design Improvements— Rural Drainage Schemes*. Sydney, Australia: Patterson Consultants Pty Limited.

Pethick, R. J., and A. J. W. Harrison. 1981. "The Theoretical Treatment of the Hydraulics of Rectangular Flap Gates." Paper no. 12, subject B (c) in *19th International Association of Hydraulic Research Congress*, Karlsruhe.

Roman, C. T., R. W. Garvine, and J. W. Portnoy. 1995. "Hydrologic Modeling as a Predictive Basis for Ecological Restoration of Salt Marshes." *Environmental Management* 19:559–66.

Williams, P. B., M. K. Orr, and N. J. Garrity. 2002. "Hydraulic Geometry: A Geomorphic Design Tool for Tidal Marsh Channel Evolution in Wetland Restoration Projects." *Restoration Ecology* 10:577–90.

PART V

Communicating Restoration Science

Tidal restoration projects are often accompanied by social and ecological concerns related to property flooding, aesthetics from the die-off of brackish and freshwater vegetation, uncertainty regarding transport of fine sediments from the tide-restricted channels, changes in mosquito production, and effects on wildlife populations that have adapted to the tide-restricted habitats, among others. This part of the book demonstrates how interdisciplinary science, monitoring, and modeling are essential to addressing these concerns, alleviating some of the uncertainty associated with predicting responses to restoration, and communicating with the public, regulatory, and stakeholder audiences. Four case studies are presented, each detailing the process of marsh restoration, from project planning and goal setting to implementation, with specific comments on the role of science communication in the restoration process. Portnoy (chap. 18) and Adamowicz and O'Brien (chap. 19) offer interesting accounts on the full range of societal and ecological concerns that are addressed during the restoration process at sites on Cape Cod and in Maine, respectively. For a restoration effort in Rhode Island, Golet and others (chap. 20) focus on the role of science and effective partnerships in tidal restoration. Reiner (chap. 21) discusses the challenges of reintroducing tidal flow within a complex urban estuary with the installation of eleven self-regulating tide gates to balance flood control needs and habitat restoration goals.

Chapter 18

Salt Marsh Restoration at Cape Cod National Seashore, Massachusetts

The Role of Science in Addressing Societal Concerns

JOHN W. PORTNOY

Cape Cod has a 350-year history of coastal wetland loss due to tide restrictions, including 1400 hectares of original salt marsh estuaries that are still diked today (Justus 2001). Salt marshes were diked for various reasons: to ease foot, wagon, and later automobile and train passage across the many salt marshes that wove throughout the outer Cape's upland farms and villages; to favor salt hay farming by reducing tidal flooding and thereby encouraging the growth of high-marsh grasses such as *Spartina patens* (salt meadow cordgrass); to allow the cultivation of salt-sensitive crops in the organic-rich wetland soils; and to eliminate, through drainage, habitat for floodwater-breeding mosquitoes. Typically an earthen dike was built across an inlet or narrow reach of tidal creek or marsh to an elevation that blocked all but the highest storm tides. This structure effectively blocked seawater flow into upstream wetlands but also tended to impound freshwater that normally discharged to the sea during low tides. Thus, to allow discharge, all dikes were fitted with culverts, albeit with one-way valves on their seaward ends to prevent saltwater inflow during flood tides. To further wetland drainage, diked salt marshes were subsequently ditched and creeks channelized to expedite freshwater discharge. Virtually all of the Cape's diked marshes have been treated in this way.

Major portions of five diked Cape estuaries (fig. 18.1) were incorporated in 1961 within the boundaries of Cape Cod National Seashore, a unit of the US National Park Service (NPS) with the mission of preserving native ecosystems for public enjoyment and education. Diked coastal marshes within Seashore lands presently comprise over 850 hectares, or 42 percent of the native salt marsh habitat present at the time of European settlement. Importantly, although the National Seashore received stewardship responsibility for most of the marshlands, both diked and natural, within park boundaries, ownership and control of the

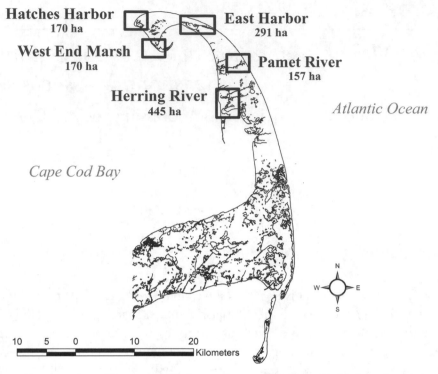

FIGURE 18.1. Location of five major diked estuaries on outer Cape Cod (Massachusetts).

structures (dikes, culverts, tide gates, and weirs) that profoundly affect the salinity, hydroperiod, biogeochemistry, and biota of the extensive wetlands landward of these structures were retained by the local towns and the state.

In addition, infrastructure development within and adjacent to the diked flood plains has been far from static both before and since Seashore establishment. By 1961, diked outer Cape coastal flood plains included a municipal airport and major portions of a golf course. Aside from the airport, these tidally restricted marshes escaped the filling that effectively eliminated many once-expansive marshlands elsewhere along the northeastern US coast. This was probably because of the remoteness and lack of urban and industrial development on outer Cape Cod throughout the twentieth century. Research leading to the recognition of salt marsh values and legal protection in the 1970s ended most direct wetland loss through new diking or filling projects.

Documenting the Impacts of Diking and Drainage

Beginning in the early 1980s, the NPS began a multidisciplinary program of research and impact assessment in the Seashore's large diked estuaries. This work was prompted by redigging of drainage ditches and channelization of original tidal creeks by the regional mosquito control authority (a program begun in the 1930s), and concurrent observations of fish kills. Water quality monitoring showed that decades of wetland drainage had caused the oxidation of sulfide-rich wetland sediments, creating acid sulfate soils and releasing both acidity and aluminum, acid-leached from native aluminosilicate clays (Soukup and Portnoy 1986; Portnoy and Giblin 1997a; Anisfeld, chap. 3, this volume). This outcome of diking and draining sulfur-rich coastal sediments has in fact been observed throughout the world (Breemen 1982; Kittrick et al. 1982; Dent 1986; Melville et al. 1999). Even after up to a hundred years of diking, large reserves of reduced sulfur remained in diked marshes (Portnoy and Giblin 1997a). Thus repeated channelization further oxidized these normally anaerobic sediments, leading to extremely low pH (2.5–4) and consequent fish kills in now fresh and low-alkalinity receiving waters.

A program of comprehensive water quality monitoring revealed additional problems consequent with tidal restrictions. Despite decades of diking and drainage, extensive deposits of wetland peat rich in organic matter persist behind the dikes. During summer, dissolved and particulate organic matter and reduced chemical species are released by these wetlands and comprise a high oxygen demand. In the absence of semidiurnal flushing with oxygen-saturated Cape Cod Bay seawater, this demand puts a severe strain on the system's oxygen budget. This has been manifested in chronic dissolved oxygen depletions and fish kills (Portnoy 1991). Importantly, these kills included high-profile migratory species like *Alosa pseudoharengus* (alewife), *A. estivalis* (blueback herring), and *Anguila rostrata* (American eel) that, in addition to being ecologically important, have long been a social focus for outer Cape townspeople.

About 1984 the Commonwealth of Massachusetts intensified fecal coliform (FC) monitoring of shellfish-growing waters, and subsequently closed vast inter- and subtidal areas to shellfishing statewide, including oyster beds seaward of the Herring River dike (Moles 2005; fig. 18.2) at great financial loss to local shellfishermen. Surveys by National Seashore scientists showed that FC, probably from wildlife given the low density of development and human sources, were concentrated just above tidal restrictions, due to the lack of flushing (Portnoy and Allen 2006). In addition, it is well documented (Bordallo et al. 2002) that the survival time of enteric bacteria is prolonged by low salinity, as in the diked Herring River. Thus observed high FC in ebb flows from tide-restricted wetlands could have been expected: diking blocks the semidiurnal infusion of high-salinity, low-FC Cape Cod Bay water that would otherwise greatly dilute and reduce the survival time of enteric bacteria.

Finally, studies addressing the long-term physical and biogeochemical effects

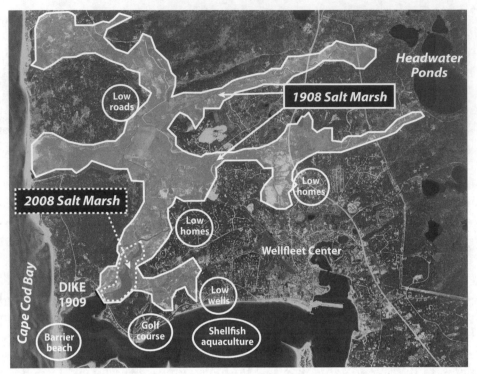

FIGURE 18.2. The extent of salt marsh in 1908 and 2008, and general locations of social concerns (circled), in the Herring River floodplain, Wellfleet, Massachusetts.

of diking and drainage (Portnoy and Giblin 1997a) revealed wetland-surface subsidence of up to 80 centimeters due to pore space collapse, organic matter loss, and the blockage of sediment normally supplied from the marine environment (Portnoy 1999). Clearly, diked salt marshes were in a particularly vulnerable state in the face of accelerating sea-level rise with predicted global climate change.

Developing a Restoration Strategy

The aforementioned adverse impacts have compelled the NPS and other agencies with management responsibility to undertake additional research and planning for tidal restoration. The critical conceptual framework is one of "adaptive management" (Walters 1986) in which restoration is undertaken in small steps accompanied by regular environmental monitoring that informs and refines each subsequent step. Adaptive management acknowledges the uncertainty in predicting the effects of tidal restoration on highly complicated and historically altered natural systems, and therefore strives continually to incorporate new data and analyses throughout project planning and implementation.

An important scientific tool to guide the restoration process is hydrodynamic modeling of (1) existing tide-height and salinity conditions and (2) the effects of proposed physical restoration alternatives (Roman et al. 1995). Modelers use copious field observations of tidal forcing and bathymetry to produce numerical models, which they then calibrate with tide-height and salinity distribution under existing conditions. Once calibrated, and validated with a completely different set of field data, numerical models can be used to test the physical effects of various alternatives for restoring tidal flow. Within the adaptive management context, actual observations of system response can be used iteratively throughout the restoration process to recalibrate the models for improved predictions of system response.

Addressing Social Concerns

Hydrodynamic modeling and an adaptive management approach are crucial for responsibly addressing the many social concerns that have arisen on the outer Cape in relation to diked wetland remediation, and for gaining scientific, management, and public confidence in a project's outcome. The remainder of the chapter focuses primarily on these social concerns and how science has been used to address them.

Tidal Flooding of Public Structures and Private Properties

Despite promulgation of wetland protection legislation and establishment of the 18,000 hectare National Seashore several decades ago, the region's diked wetlands are not without public and private infrastructure that could be damaged by tidal restoration. Since the first major dikes were built in the late nineteenth century, many private and public projects were undertaken with the assumption that dikes would always be in place to block tidal flooding. To this date, federal flood insurance rate maps (FEMA 1992) show the hundred-year flood elevation behind dikes as only a few feet above mean sea level, compared to 3–4 meters just seaward of the structure. This has only added to the misperception that historic flood plains, which may include public and private property and structures, are forever protected from tidal flooding or, even worse, no longer even defined as wetlands with attendant legal protection.

Hatches Harbor, Provincetown, Massachusetts (fig. 18.1), is a good example. In 1930 the coastal floodplain was bisected by an earthen dike in an attempt to control floodwater mosquito breeding. Shortly thereafter, a small aircraft-landing field was built on the diked wetland and adjacent low dunes. By the 1950s, the airfield had been expanded on fill and upgraded to become the town's municipal

airport. Amid the emergency of a dike breach in 1987, the NPS and Massachusetts Coastal Zone Management office recommended to federal, state, and local
airport officials that current diking of seawater for airport flood protection was excessive and could be scaled down to allow partial salt marsh restoration landward
of the dike. An agreement was reached whereby the Federal Aviation Administration would ascertain the airport's critical flood threshold, while NPS would develop a hydrodynamic model to prescribe an optimum opening through the dike
(i.e., a culvert design that maximizes the area of tidal restoration while protecting
the airport from tidal flooding). Based on a one-dimensional hydrodynamic
model, a culvert 7 meters wide would deliver enough tidal volume to restore
about 36 hectares of tidal marsh; meanwhile, restricting culvert height to 1 meter
would keep spring and storm tides from exceeding the airport threshold (Roman
et al. 1995). The NPS installed new culverts in winter 1998–99 and opened them
incrementally over the next five years while monitoring tide-height as well as ecological variables. The system has performed as predicted, with an extensive area of
salt marsh restoration and no adverse effect on airport operations (Portnoy et al.
2003; Smith et al. 2009).

In other situations, structures cannot be protected in situ from tidal flooding
and, again based on careful field surveys and modeling (Spaulding and Grilli
2001), must be moved or raised in elevation. The 445 hectare Herring River
(Wellfleet, MA) floodplain, for example, includes nearly 7000 meters of low-lying
and regularly traveled roads, six domestic wells, and several fairways of a golf
course (fig. 18.2). Planning and environmental review are still under way at this
writing, but full restoration would require relocating or elevating all of this infrastructure. The role of science in these cases is to define precisely the extent and
degree of the problem, to recommend alternative solutions, and to study their impacts with respect to both ecological objectives and property protection.

Sediment Transport and Barrier Beaches

Sediment transport and the stability of barrier beaches are issues of concern for
the restoration of tidal exchange of both back-barrier and riverine salt marshes on
Cape Cod. Perhaps counterintuitively, diking can destabilize barrier beaches by
reducing diked-wetland water levels and thereby increasing the hydraulic head
between diked marsh and open ocean (Bruun 1978). Diking also cuts off the marine source of sediment that allows marshes to accrete (Thom 1992) and protect
adjacent properties from storm surges. For example, in the high wave energy environment of the Atlantic shore, the barrier beach at the eastern end of the Pamet
River estuary (fig. 18.1) is more prone to storm overwash because of the restricted
regular tidal exchange with Cape Cod Bay to the west. Elsewhere, some shellfish

aquaculturists in Wellfleet Harbor have expressed concern that increased tidal exchange into Herring River could cause the barrier beach near the river's mouth to breach (fig. 18.3a), change the harbor's salinity and temperature regime, and thereby degrade shellfish-growing conditions. However, NPS-supported geomorphological research (Dougherty 2004) has shown that the paleodevelopment of

(a)

(b)

Figure 18.3. Coastal geomorphology and sediment transport are important considerations in tidal restoration projects. (a) At Herring River, science addressed concerns that restored tidal exchange could breach a barrier beach, degrading the salinity and temperature regime for shellfish aquaculture in Wellfleet Harbor. (b) At East Harbor the question is how to restore a permanent opening through the barrier beach, for restoration of tides in the back-barrier lagoon, without interrupting alongshore sediment transport, starving down-gradient beaches of sand and thereby endangering beachfront cottages. (Photo source: Cape Cod National Seashore).

the barrier beach actually cut off the river from direct discharge to Cape Cod Bay, and that even before diking, unfettered estuarine tidal exchange was too weak to breach the barrier beach.

Sediment transport is of both ecological and socioeconomic concern at East Harbor (Truro; fig. 18.3b). This 291 hectare back-barrier lagoon and salt marsh was historically connected to the bay via a 300 meter wide inlet, filled for railway construction about 1870. The system has received some tidal exchange through a small culvert since 2002, but continued poor flushing and eutrophication have prompted planning for a larger opening through the barrier beach. However, cottages built along the beach depend on littoral sand transport to sustain their beaches. An artificial inlet, constrained in width by adjacent development, will interrupt this transport; therefore, planning is under way to study and model sediment transport.

Changes in sedimentation as a result of restored tides have also been a cause for concern among shellfish aquaculture seaward of the Herring River dike. Some have argued that fine sediments that have accumulated just upstream of the dike structure might be carried downstream, via increased tidal volume, to cover their *Mercenaria mercenaria* (hard clam) beds in the river mouth (fig. 18.2). The problem was studied in two ways: (1) by using historic charts and aerial photographs to assess sedimentation patterns in the river mouth both before and after the system was originally diked and (2) by using the results of prior hydrodynamic modeling (Spaulding and Grilli 2001) to predict restored flow velocities and to compare them to velocities required to resuspend river sediments (Dougherty 2004). This work showed that sedimentation has not changed significantly since the mid-1800s (before diking), and that restored tides would not increase channel velocities enough to resuspend sediment. Further, the model predicted that flood tide velocities will, as today, exceed ebb tide flows; therefore, the net flow of any resuspended sediment must be in an upstream direction. Finally, observations from a period in the mid-1970s during and just after the dike breached indicate that sedimentation on shellfish beds was not a problem until after the dike was repaired. These findings are consistent with the current paradigm for the net inflow of sediment into coastal marshes (Thom 1992), contributing to their accretion and, at least for studied Cape Cod sites, keeping them above rising sea level over the past several thousand years (Redfield 1972; Roman et al. 1997).

Groundwater Quality

Saltwater intrusion into coastal freshwater aquifers is a global concern heightened by predictions of accelerating sea level rise (Barlow 2003). The issue is serious for unconfined coastal aquifers composed of highly permeable sands separating fresh groundwater from oceanic seawater, as is typical of outer Cape Cod. Proposals to restore seawater flooding to wetlands that have been diked for many decades naturally raises fears that adjacent private wells, most of which have been drilled long after tides were blocked, may be contaminated with seawater following tidal restoration.

This potential problem was addressed at Herring River in several ways. First, geophysical soundings, ground truthed with deep observation wells, described the depth to the freshwater–saltwater interface, and thus the thickness of the freshwater lens in the developed upland surrounding the diked floodplain (Fitterman and Dennehy 1991). The authors found a 20 meter thick freshwater lens even at the edge of the diked wetland and concluded that the predicted 0.5 meter increase in mean estuarine seawater levels could only affect wells located in the floodplain proper and not in the upland.

Second, hydrodynamic model predictions of changes in the estuary's water level and salinity distribution with tidal restoration (Spaulding and Grilli 2001) were coupled with a recently developed US Geological Survey numerical groundwater model for outer Cape Cod (Masterson 2004). The groundwater model predicted, perhaps counterintuitively, that tidal restoration will cause the freshwater lens to thicken, making it less vulnerable to saltwater intrusion than under tide-restricted conditions (Masterson and Garabedian 2007). This is because diking lowers the mean water level in upstream wetlands, causing increased groundwater discharge and a lower water level in the adjacent groundwater aquifer; in contrast, mean estuarine water levels increase with restoration, with a model-predicted increase in adjacent aquifer thickness.

Third, Martin (2004, 2007) surveyed a large number of domestic wells placed very close to saline embayments along the Cape Cod Bay shoreline, including the Herring River, to assess the effects of well depth and proximity to saline surface water on well water quality. Most wells, even those drilled right along the shoreline, showed very low salt content. Exceptions were wells drilled deep enough into the relatively shallow coastal aquifer to reach the freshwater–saltwater interface.

In summary, these studies concluded that tidal restoration would only affect those wells placed within original tidal wetlands and subject to actual surface flooding by seawater. They nevertheless recommended periodic monitoring of both deep observation wells and shoreline domestic wells around restoration sites as part of the adaptive management approach.

Receiving Water Quality

Public education about the water quality problems caused by salt marsh diking (i.e., acidified and hypoxic surface waters and high fecal coliform bacteria) has had unexpected reactions at Herring River. Most prominent has been the perception that tidal restoration could extend these problems into coastal receiving waters that presently enjoy good water quality.

However, regarding acid sulfate soils and acidified surface waters, microcosm experiments showed that the reflooding of these soils with seawater re-creates reducing conditions, generates copious alkalinity, and eliminates acidity and metals mobilization within a few months (Portnoy and Giblin 1997b). In addition, hydrodynamic modeling (Spaulding and Grilli 2005) indicated that a sufficiently wide opening for tidal restoration would increase flushing with normally oxygen-saturated seawater, improving aeration to the extent of seawater incursion throughout the floodplain. Similarly, restored tidal volume would reduce the current fecal coliform problem through simple dilution; in addition, increased salinity and osmotic stress would reduce coliform survival time in the open estuary (Portnoy and Allen 2006). Despite the science, presenting these findings in a way that counters the lay public's intuitive response has proved challenging.

Vegetation Change

The most noticeable effects of diking are changes in vegetation. *Typha angustifolia* (narrowleaf cattail), *Phragmites australis* (common reed), and *Lythrum salicaria* (purple loosestrife) characteristically spread onto tide-restricted marshes in the northeastern United States, shading out and displacing native halophytic grasses (Roman et al. 1984). At higher elevations that are consequently more deeply drained by diking, a large variety of freshwater wetland and even upland plants have invaded diked floodplains once the stresses of salt and waterlogging (low redox potential, toxic sulfides, etc.) are removed. Thus hundreds of hectares of original salt marsh habitat have been converted to freshwater wetlands or even upland habitat within Hatches Harbor, East Harbor, Pamet River, and Herring River on outer Cape Cod (fig. 18.1). Although vegetation change has been extensive and profound, succession has been slow, proceeding over many decades, and therefore little appreciated by even the local public, few of whom experienced the original tidal wetlands. With this background, it is understandable that many object to the aesthetic disruption inevitable in converting what have become shrub thickets back to herbaceous marshlands.

The public perception of, and aversion to, a radical die-off of woody vegetation over an expansive diked floodplain has become an important consideration in planning and implementing tidal restoration at the Seashore. The approach at Hatches Harbor, and the likely approach at Herring River, is to restore tides in small increments, consistent with adaptive management, and to monitor vegetation effects (Smith et al. 2009). In the latter system, cutting and prescribed fire will probably be needed for the removal of about 240 hectares of salt-sensitive shrubs and trees and for the removal of well-established, and somewhat salt-tolerant, *Phragmites*.

Mosquito Production

A major impetus and rationale for salt marsh diking on Cape Cod through the mid-twentieth century was concern about salt marsh mosquito breeding (Whitman and Howard 1906). Thus early proposals to restore tides were met with opposition from individuals and the local mosquito control agency who believed that this action could restore breeding habitat and resurrect a public nuisance. Of note, public health officials have not considered mosquitoes a public health threat on Cape Cod because of the rarity of eastern equine encephalitis or West Nile virus in adult mosquitoes that have been collected and analyzed. This, in turn, is probably because of the lack of habitat for the bird-biting mosquito species that increase virus frequency in local birds. At the same time, nuisance mosquito production after diking was (and still is) a serious problem, evidenced by escalating expenditures for mosquito control drainage shortly *after* salt marshes were diked (Portnoy and Reynolds 1997).

In response to the seeming conflict between insect control and habitat restoration goals, NPS studied mosquito breeding ecology in Seashore wetlands in the early 1980s (Portnoy 1984). Breeding of the dominant human biter (*Ochlerotatus cantator*) was most intensive in the park's diked and degraded wetlands, where predation by fish, normally a major source of natural control (Palanichamy et al. 1983; Blaustein and Byard 1993; Mogi et al. 1995) was obviated by poor water quality. In addition, much breeding occurred in stagnant pools that formed after rain on the subsided wetland surface, which were inaccessible to fish due to both poor water quality and sustained low water levels.

Besides the research already mentioned indicating major improvements in estuarine water quality with restoration (Portnoy 1991; Portnoy and Giblin 1997b), hydrodynamic modeling (Spaulding and Grilli 2001, 2005) showed that much breeding habitat could be eliminated, or at least made accessible to predatory fish, by reestablishing the naturally high tidal range. Restored tides, and estuarine fish, would regularly reach wetland surfaces and mosquito breeding sites. Restoration would also cause lower low tides, enhance low-tide drainage, and reduce standing-water mosquito breeding sites in depressions (i.e., subsided peat and abandoned ditches) on the marsh surface. Local mosquito control officials now support tidal restoration for floodwater mosquito control.

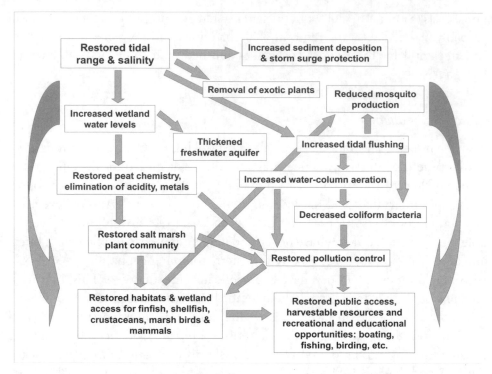

FIGURE 18.4. Processes by which tidal restoration can lead to net public benefits.

Identifying Benefits of Tidal Restoration

The Cape Cod studies described here and concurrent research at other sites argue for the net environmental and social benefits of tidal restoration. The processes by which these benefits can be realized are summarized following here and schematically in figure 18.4. Public understanding of the potential benefits and public appreciation of the peer-reviewed scientific bases are critical to acceptance of what is often a significant change in the perceived landscape.

Restored tidal range leads to higher sediment transport and deposition onto the wetland surface, as sediment-carrying flood tides again flow over creek banks and onto the marsh platform (Thom 1992; Anisfeld et al. 1999). For most Cape systems, this surface has subsided over many decades; therefore, restored sedimentation can allow the wetland surface to rise, thereby increasing storm-surge protection for roads and other structures at the edge of the floodplain.

Restored tidal range (i.e., higher high tides, lower low tides) and, thus, increased intertidal volume, produces greatly increased tidal flushing (Spaulding and Grilli 2001, 2005). Better flushing can reduce floodwater mosquito breeding on the wetland surface (Portnoy 1984), dilute the presently high fecal coliform counts that have closed river-mouth shellfish beds (Portnoy and Allen 2006), and improve water column aeration by flooding the wetland twice each day with oxygen-rich Cape Cod Bay water (Portnoy 1991).

Tidal restoration also produces a higher average water level in the estuary's wetlands, with additional benefits. Groundwater modeling predicts that the fresh-water lens, the source of drinking water for all properties surrounding the flood-plain, will thicken with a higher mean water level in the estuary (Masterson and Garabedian 2007). Within the wetland proper, higher water levels will resaturate wetland soils that have been drained by diking and ditch drainage for decades, and, based on mesocosm experiments (Portnoy 1999), reverse the chemistry that has given rise to high acidity, toxic metals, and fish kills in receiving waters. Along with higher high tides and increased mean water level, low tides will actually be lower with tidal restoration, improving low tide drainage of mosquito breeding sites on the wetland surface. Improved water quality will likely also reduce mos-quito production by enhancing aquatic habitat for their major predators, estuarine fish (Portnoy 1984).

Restored salinity will kill many of the salt-sensitive exotic plants that have in-vaded the floodplain, and will enable recolonization by the native salt marsh plants, which have a large competitive advantage once salinity is restored. Higher salinity will also reduce the survival time of coliform bacteria, thereby adding to the aforementioned dilution effect of increased tidal flushing to further depress fe-cal coliform counts, and remediating a chronic problem in downstream shellfish beds (Portnoy and Allen 2006).

The reestablishment of tidal range, salinity, overall water quality, and the salt marsh plant community will reclaim hundreds of acres of wetland habitats, im-proving physical access to those habitats for finfish, shellfish, marsh birds, and mammals. For people, this means better boat access on higher tides across an open marsh in place of presently dense *Phragmites* or drained shrub thicket, with fewer mosquitoes. More important, it also means more extensive, abundant, and diverse marine resources for observation, education, and harvest, both within the estuary and in nearby coastal waters.

Conclusions

Over the past several decades, science has played critical roles in the preservation of coastal wetlands (1) by describing their functions and values to coastal commu-nities, (2) by identifying how past human alterations have compromised those functions and social values, (3) by developing strategies for ecosystem restoration, (4) by assessing restoration outcomes, and (5) by communicating these findings to the public to engender support for preservation and restoration. On Cape Cod, the final task has been most time consuming and challenging, but crucial to progress in restoring tidal exchange and estuarine habitats.

Environmental amnesia (Diamond 2005), whereby each successive generation has a different baseline as to what is normal or natural in their environment, has been a major obstacle in raising public awareness of damage, albeit severe, done over several human generations. It can be difficult to explain to people gazing over a diked floodplain dominated by *Phragmites*, shrubs, and trees that, without past human disturbance, they would be looking across a broad and open tidal marsh. An effective cure to this environmental amnesia has at times been as simple as the display and discussion of historic maps and photographs showing the undisturbed, and to most eyes beautiful, native estuaries. To the growing number of people who now appreciate the environmental values of salt marshes but are unaware of their local losses, these images can be a revelation. These same people often become active, highly credible, and effective project proponents within their community.

Regardless, reaching public and official consensus to change hundreds to thousands of acres of a familiar landscape has been and probably must be a slow and incremental process. Progress on tidal restoration on Cape Cod requires a long-term (years to decades) commitment to (1) addressing all public concerns with scientific scrutiny and (2) persistently bringing the science to the public and their elected officials at both local and state levels.

REFERENCES

Anisfeld, S. C., M. J. Tobin, and G. Benoit. 1999. "Sedimentation Rates in Flow-Restricted and Restored Salt Marshes in Long Island Sound." *Estuaries* 22:231–44.

Barlow, P. M. 2003. *Ground Water in Freshwater–Saltwater Environments of the Atlantic Coast.* US Geological Survey Circular 1262. Reston, VA: US Geological Survey.

Blaustein, L., and R. Byard. 1993. "Predation by a Cyprinodontid Fish, *Aphanius mento*, on *Culex pipiens*: Effects of Alternative Prey and Vegetation." *Journal of the American Mosquito Control Association* 9:356–58.

Bordalo, A., R. Onrassami, and C. Dechsakulwatana. 2002. "Survival of Faecal Indicator Bacteria in Tropical Estuarine Waters (Ba River, Thailand)." *Journal of Applied Microbiology* 93:864–71.

Breemen, N. V. 1982. "Genesis, Morphology, and Classification of Acid Sulfate Soils in Coastal Plains." Pp. 95–108 in *Acid Sulfate Weathering*, edited by J. A. Kittrick, D. S. Fanning, and L. R. Hossner. Madison: Soil Science Society of America Special Publication no. 10.

Bruun, P. 1978. *Stability of Tidal Inlets: Theory and Engineering.* Developments in Geotechnical Engineering vol. 23. New York: Elsevier.

Dent, D. 1986. *Acid Sulfate Soils: A Baseline for Research and Development.* Wageningen, Netherlands: International Institute for Land Reclamation and Improvement.

Diamond, J. 2005. *Collapse: How Societies Choose to Fail or Succeed*. New York: Viking.

Dougherty, A. J. 2004. *Sedimentation Concerns Associated with the Proposed Restoration of Herring River Marsh, Wellfleet, MA*. Report to Cape Cod National Seashore, Town of Wellfleet and Association of Women Geoscientists.

Federal Emergency Management Agency (FEMA). 1992. *Flood Insurance Rate Map*. Towns of Provincetown, Truro, and Wellfleet: National Flood Insurance Program.

Fitterman, D. V., and K. F. Dennehy. 1991. *Verification of Geophysically Determined Depths to Saltwater Near the Herring River (Cape Cod National Seashore), Wellfleet, Massachusetts*. US Geological Survey Open-File Report 91–321.

Justus, S. 2001. *Cape Cod Atlas of Tidally Restricted Salt Marshes: Cape Cod Massachusetts*. Barnstable, MA: Cape Cod Commission.

Kittrick, I. A., D. S. Fanning, and L. R. Hossner. 1982. *Acid Sulfate Weathering*. Soil Science Society of America Special Publication no. 10. Madison, WI: Soil Science Society of America.

Martin, L. 2004. *Salt Marsh Restoration at Herring River: An Assessment of Potential Salt Water Intrusion in Areas Adjacent to the Herring River and Mill Creek, Cape Cod National Seashore*. Technical Report NPS/NRWRD/NRTR-2004/319. Fort Collins, CO: National Park Service Water Resources Division.

Martin, L. 2007. *Identification of Private Domestic Wells Adjacent to the Herring River Flood Plain That Could Be Affected by Restoration of Tidal Flow*. Natural Resource Report NPS/NRPC/WRD/NRTR—2007/370. Fort Collins, CO: National Park Service Water Resources Division.

Masterson, J. P. 2004. *Simulated Interaction between Freshwater and Saltwater and Effects of Ground-Water Pumping and Sea-Level Change, Lower Cape Cod Aquifer System, Massachusetts*. US Geological Survey Scientific Investigations Report 2004-5014. Washington, DC: US Geological Survey.

Masterson, J. P., and S. P. Garabedian. 2007. "Effects of Sea-Level Rise on Ground Water Flow in a Coastal Aquifer System." *Groundwater* 45:209–17.

Melville, M. D., C. Lin, J. Sammut, I. White, X. Yang, and B. P. Wilson. 1999. "Acid Sulfate Soils (ASS): Their Impacts on Water Quality and Estuarine Aquatic Organisms with Special Reference to East Australia and South China Coasts." *Asian Fisheries Science* 12:249–56.

Mogi, M., V. Nemah, I. Miyagi, T. Toma, and D. T. Sembel. 1995. "Mosquito (Diptera: Culicidae) and Predator Abundance in Irrigated and Rain-Fed Rice Fields in North Sulawesi, Indonesia." *Journal of Medical Entomology* 32:361–67.

Moles, J. B. 2005. *Annual Evaluation of Herring River CCB:12 in the Town of Wellfleet*. Pocasset: Massachusetts Division of Marine Fisheries.

Palanichamy, S., M. P. Balasubramanian, and M. Selvam. 1983. "Influence of Body Weight, Prey Density, Food Quality and Prey Deprivation on Mosquito Predation of Two Larvivorous Fishes." *Environmental Ecology* 1:179–83.

Portnoy, J., and M. Reynolds. 1997. "Wellfleet's Herring River: The Case for Habitat Restoration." *Environment Cape Cod* 1:35–43.

Portnoy, J., C. Roman, S. Smith, and E. Gwilliam. 2003. "Estuarine Habitat Restoration at Cape Cod National Seashore: The Hatches Harbor Prototype." *Park Science* 22:51–58.

Portnoy, J. W. 1984. "Salt Marsh Diking and Nuisance Mosquito Production on Cape Cod, Massachusetts." *Journal of the American Mosquito Control Association* 44:560–64.

Portnoy, J. W. 1991. "Summer Oxygen Depletion in a Diked New England Estuary." *Estuaries* 14:122–29.

Portnoy, J. W. 1999. "Salt Marsh Diking and Restoration: Biogeochemical Implications of Altered Wetland Hydrology." *Environmental Management* 24:111–20.

Portnoy, J. W., and J. R. Allen. 2006. "Effects of Tidal Restrictions and Potential Benefits of Tidal Restoration on Fecal Coliform and Shellfish-Water Quality." *Journal of Shellfish Research* 25:609–17.

Portnoy, J. W., and A. E. Giblin. 1997a. "Effects of Historic Tidal Restrictions on Salt Marsh Sediment Chemistry." *Biogeochemistry* 36:275–303.

Portnoy, J. W., and A. E. Giblin. 1997b. "Biogeochemical Effects of Seawater Restoration to Diked Salt Marshes." *Ecological Applications* 7:1054–63.

Redfield, A. C. 1972. "Development of a New England Salt Marsh." *Ecological Monographs* 42:201–37.

Roman, C. T., W. A. Niering, and R. S. Warren. 1984. "Salt Marsh Vegetation Change in Response to Tidal Restriction." *Environmental Management* 8:141–50.

Roman, C. T., R. W. Garvine, and J. W. Portnoy. 1995. "Hydrologic Modeling as a Predictive Basis for Ecological Restoration of Salt Marshes." *Environmental Management* 19:559–66.

Roman, C. T., J. A. Peck, J. R. Allen, J. W. King, and P. G. Appleby. 1997. "Accretion of a New England (U.S.A.) Salt Marsh in Response to Inlet Migration, Storms, and Sea-Level Rise." *Estuarine Coastal and Shelf Science* 45:717–27.

Smith, S. M., C. T. Roman, M. J. James-Pirri, K. Chapman, J. Portnoy, and E. Gwilliam. 2009. "Responses of Plant Communities to Incremental Hydrologic Restoration of a Tide-Restricted Salt Marsh in Southern New England (Massachusetts, USA)." *Restoration Ecology* 17:606–18.

Soukup, M. A., and J. W. Portnoy. 1986. "Impacts from Mosquito Control–Induced Sulfur Mobilization in a Cape Cod Estuary." *Environmental Conservation* 13:47–50.

Spaulding, M. L., and A. Grilli. 2001. *Hydrodynamic and Salinity Modeling for Estuarine Habitat Restoration at Herring River, Wellfleet, Massachusetts*. Report to the National Park Service, Cape Cod National Seashore, Wellfleet, MA.

Spaulding, M. L., and A. Grilli. 2005. *Simulations of Wide Sluice Gate Restoration Option for Herring River*. Report to the National Park Service, Cape Cod National Seashore, Wellfleet, MA.

Thom, R. M. 1992. "Accretion Rates of Low Intertidal Salt Marshes in the Pacific Northwest." *Wetlands* 12:147–56.

Walters, C. 1986. *Adaptive Management of Renewable Resources*. Caldwell, NJ: Blackburn Press.

Whitman, H. T., and C. Howard. 1906. *Report on Proposed Dike at Herring River, Wellfleet, Massachusetts*. Boston, MA: Whitman and Howard, Civil Engineers.

Chapter 19

Drakes Island Tidal Restoration
Science, Community, and Compromise

SUSAN C. ADAMOWICZ AND KATHLEEN M. O'BRIEN

Shortly after the landing of Pilgrims at Plymouth Rock, Massachusetts, settlers arrived at the southern portion of the Province of Mayne, now the State of Maine (Butler 2005). While the area remained largely undeveloped for nearly two hundred years, by the mid- and late 1800s the then well-settled Town of Wells, Maine, embarked on a series of coastal improvements for the sake of agriculture and vacationing summer residents. Dikes and roads caused tidal restrictions in salt marshes that resulted in changes in vegetation, encroachment of invasive plant species, and elevation subsidence. Tidal restoration to one such area, Drakes Island, has provided a series of social challenges in addition to the standard ecological and engineering concerns. Although restoration construction occurred during late spring 2005, in 2010 the marsh and town continued to be in a period of adjustment. Scientific field investigations, engineering designs, and computer modeling have and continue to be the currency of dialogue between restoration partners and the local community. Homeowner perceptions, particularly about slow stormwater drainage and potential tidal flooding, nonetheless continue to drive management of the system.

Setting and History

While Maine perhaps is best known for its rocky coastline, southern Maine is typified by sandy beaches with a few large back-barrier salt marsh complexes. Beginning their formation four to five thousand years ago, peat deposits now are 3 to 5 meters thick (Kelley et al. 1988). One such system in the Town of Wells, Maine, has an extensive salt marsh complex that developed behind the barrier beach at the mouth of the Webhannet River.

An island at the northern extent of the marsh complex was named Drakes Island ("Drakes") after Thomas Drake, a trader and voyager. Prior to major human alterations, a small tidal channel, Nancy's Creek, cut off Drakes from the mainland and connected the Webhannet and Little Rivers (fig. 19.1; Kelly 1978). Nancy's Creek, the marsh and lands surrounding it, as well as the human communities of Drakes Island and Wells, Maine, are the focus of this chapter.

The earliest record of marsh use dates to 1645, when Stephen Batson was granted 4.1 hectares of marsh on the western end of Drakes Island. The area remained sparsely settled until the mid-1800s (Spencer 1973) when relatively large

FIGURE 19.1. Nancy's Creek marsh and Drakes Island, Wells, Maine. The digital orthophotographic base image was obtained from the Maine Office of Geographic Information Systems online catalog.

infrastructure projects appeared, including an agricultural dike built across Nancy's Creek circa 1848. This drained and likely freshened (as per Portnoy 1999) the marsh, making it suitable for salt haying (Shelley 1997) or pasture (Boumans et al. 2002). Originally, the dike ran about 370 meters (it has since completely eroded at the channel crossing) and currently stands approximately 1.5 meters above the marsh surface.

Marshes are also tied intimately to their surroundings. On the mainland, to the west of Nancy's Creek, a large hillside tract was worked by Henry Boade in 1643. Purchased later by the Lord family and known as Laudholm Farms, it remained actively farmed until 1952 (Butler 2005). Drakes Island itself was purchased in its entirety by Joseph Eaton in 1883, who then petitioned the Town in 1890 to build a road that traversed the dike at Nancy's Creek. Dyke Road provided Eaton with the ability to transport goods and passengers to a series of crofts (cottages) built in the late 1890s and early 1900s. By 1910, Drakes Island had grown to twenty-five dwellings, including the Eaton and Lord crofts. Thus began the gradual transition of Wells from an agrarian and shipbuilding community to a residential and tourism-based town.

During the summer of 1915, the Drakes dike failed and "destroyed the flora on the marshes." The bridge over Nancy's Creek was like a "turn table at every high tide, sometimes turning a half way round." Drakes' residents petitioned for a new road that fall (Kelly 1978).

The replacement road, Drakes Island Road, was built parallel to the dike (fig. 19.1). From the 1920s to the 1950s, it had a box culvert and water control structure. During the 1950s, a 107 centimeter concrete culvert with a flap gate was installed. At some later point, a 91 centimeter corrugated metal pipe was used to line sections where the concrete culvert had failed (Boumans et al. 2002).

Other changes were occurring in the neighborhood too. Historic US Geological Survey maps (1891; 1979) show that the community on Drakes Island expanded dramatically over time. Some of the new homes were built at low elevation with backyards extending into the Nancy Creek marsh system. The Rachel Carson National Wildlife Refuge was established in 1966; part of its holdings include the salt marsh downstream of Drakes Island Road and a portion of the Nancy's Creek marsh. And finally, in 1984 the Laudholm Trust, in partnership with the National Oceanographic and Atmospheric Administration, the State of Maine, and the Town of Wells, purchased Laudholm Farms and created the Wells National Estuarine Research Reserve (Wells Reserve), which today encompasses 911 hectares of uplands and wetlands. This close-knit landscape of property owners means that managing Nancy's Creek marsh involves a diverse group of stakeholders: the Town, Wells Reserve, Rachel Carson National Wildlife Refuge, and local residents.

As time passed, the flap gate on the Drakes Island Road culvert broke and finally fell off in March 1988. Vegetation and fish composition then changed as a result of the new, though limited, tidal exchange (Burdick et al. 1997; Dionne et al. 1999). Drakes Island was reverting to a salt marsh, albeit slowly and incompletely. Residents were unhappy and approached Town and State officials to have the flap gate reinstalled due to concerns over saltwater flooding, loss of the freshwater marsh, and damage to their homes. They were met, however, with a denial by state regulators in 1991.

Community Concerns and Involvement

The Drakes Island Improvement Association, formed circa 1914, laid the foundation for active community involvement in infrastructure needs. Decades later a local newsletter noted that home building increased that year and recorded concerns with parking, one-way streets, and drainage ditches (Clarkson 1950). As the number of houses on low-lying lots increased, so did flooding problems. In the past twenty years, citizen letters and petitions voiced several concerns, including a preference for the freshwater marsh (pre-1988 conditions), a desire to preserve the large area of ponded water in the scour pool upstream of the culvert, a desire to reduce low-tide odors, and a need to increase stormwater drainage and protect homes from coastal storm flooding.

The Path toward Restoration

As part of a research project supported through the Wells Reserve, Boumans et al. (2002) modeled six different hydrologic scenarios for restoring the Nancy's Creek marsh. The installation of an additional culvert, identical in size to the existing one, was predicted to increase upstream marsh elevation and to promote salt marsh vegetation. Other scenarios, including the "no action" alternative, resulted in degraded salt marsh and/or conversion to freshwater marsh or mudflat.

In August 2003, the Wells town manager called a public meeting for Drakes residents to discuss culvert replacement and salt marsh restoration—by that point water was flowing around the culvert pipe and the road had visible cracks. A critical component of the meeting was presentation of a model predicting a 3.8 hectare expansion of *Phragmites australis* (common reed) within 20 years given no change in hydrology (Konisky and Burdick 2003; Konisky, chap. 16, this volume). An alternative model predicted a reduction in both invasive *Phragmites* and *Typha angustifolia* (narrowleaf cattail), by replacing the existing culvert with a larger box culvert and a self-regulating tide gate (SRT).

During 2004, the Town pursued additional engineering studies and designs as well as environmental permits required to install a new box culvert with an SRT (TRC 2005). At the same time, stakeholders including the Wells Reserve, Rachel Carson National Wildlife Refuge, town officials, and area residents drafted an operations and maintenance plan, a monitoring plan, and a framework for a consensus-based process to deal with neighbor concerns, salt marsh restoration goals, and emergency situations. A variety of issues, including construction timing to avoid summer traffic, emergency vehicle access, and even the aesthetics of a safety fence were processed successfully by the stakeholders.

Ecological Conditions and Concerns

Nancy's Creek biota responded in a variety of ways to changes in the surrounding environment. Tidal hydrology is a major driving factor that influences vegetation, nekton, and birds.

Vegetation

Boumans et al. (2002) presumed that, prior to the original diking in 1848, the marsh had been similar to high marsh on the downstream side of the road. Kelly (1978) noted that the 1848 dike caused narrowing of Nancy's Creek so that it no longer connected through to the Little River and that the breach in 1915 caused a massive vegetation die-off of freshwater plants.

The construction of Drakes Island Road after 1915 reestablished a freshwater regime. Prior to 1988, the upstream marsh was dominated by *Typha latifolia* (broadleaf cattail) and *Spartina pectinata* (prairie cordgrass) and also supported *Schoenoplectus robustus* (sturdy bulrush) and *Triglochin maritimum* (seaside arrow grass). The 1988 loss of the flap gate caused yet another tidal reintroduction resulting in the death of *Typha* and *S. pectinata* near the gate (Burdick et al. 1997). By 2003 the two major plant communities were *Typha–Phragmites* (62 percent) and *Spartina alterniflora–Spartina patens* (smooth cordgrass–salt meadow cordgrass) (38 percent) (Konisky and Burdick 2003).

Both drainage and reintroduction of saltwater may lead to subsidence due to oxidation of organic matter (Portnoy 1999). The issue of subsidence is important for two reasons. First, *S. alterniflora*, the lowest-growing salt marsh grass, can tolerate limited inundation (Redfield 1972; Fragosa and Spencer 2008). Tidally restricted, subsided marshes, however, often have elevations too low to support *S. alterniflora* survival under full tidal flow conditions. In these cases, complete removal of a tidal barrier may result in marsh loss and thus may not be desirable.

Second, sea level rise is a stressor even for unrestricted marshes, occurring at 1–2 millimeters per year at Wells, Maine (Kelley et al. 1995; Smith and Warren, chap. 4, this volume). Subsided, tidally restricted marshes are then at a double disadvantage. Once tidal flows are restored, they have to "make up" the elevation lost by subsidence plus keep pace with accelerated sea level rise (Turner 2004). The 0.6–0.9 meter loss in elevation (Dionne et al. 1999) would convert portions of the Nancy's Creek marsh to mudflat in the advent of full tidal flow. Therefore, if the marsh is to be maintained, only partial flow restoration is feasible, at least initially (Boumans et al. 2002).

Nekton

Salt marsh fish and decapod crustaceans (collectively called nekton) are important food web components providing energy to salt marsh consumers and the larger downstream estuary (Deegan et al. 2000). Prior to the culvert–SRT installation, there were more nekton species in open water habitats on the unrestricted downstream marsh than on the restricted Nancy's Creek side: tidal channels had ten nekton species on the unrestricted side versus seven on the restricted side (Eberhardt et al. 2011), whereas salt marsh pools yielded nine nekton species on the unrestricted side and only two on the restricted side (Adamowicz 2004, unpublished data).

Sharp-Tailed Sparrows

Concern over sharp-tailed sparrows has been neglected in most tidal flow restoration projects (Golet et al., chap. 20, this volume). This is perhaps because they are cryptic and difficult to identify, and their nests are hard to find. *Ammodramus nelsoni subvirgatus* (Nelson's sparrow) and *A. caudacutus caudacutus* (saltmarsh sparrow) both breed on the marshes of southern Maine. They build nests in the high marsh, selecting areas with *Spartina patens* and deep thatch (Shriver et al. 2007). But because the nests are only a few centimeters above the marsh surface, they are susceptible to flooding during spring and storm tides. The birds have adapted to nest loss from flooding by quickly re-nesting, synchronizing nesting cycles with tides (Gjerdrum et al. 2005), and, in the case of Nelson's sparrows ("Nelson's"), building their nests higher (Greenlaw and Rising 1994; Shriver et al. 2007). Habitat limitations and frequent nest failures, among other reasons, have ranked these birds as highest priority species for conservation on the Atlantic coast (Rich et al. 2004; US Fish and Wildlife Service 2008). Saltmarsh sparrows ("saltmarsh sparrows") also are classified as globally vulnerable to extinction (IUCN Red List criteria: BirdLife International 2012).

Since these sparrows rely on appropriate vegetation to site nests, tidal flow restoration projects, such as the one at Drakes Island, raise a dilemma. In this case, final designs called for an average increase in mean high water of 15 centimeters (Konisky and Burdick 2003). Unfortunately, vegetation does not respond immediately to alterations in hydrology, and it may take up to 15 years for "restored" sites to develop proper high marsh vegetation cues at appropriate elevations (Warren et al. 2002). In the intervening years, while vegetation distribution adjusts to the altered tidal regime, birds may continue to build nests in inappropriate locations or abandon the site entirely.

Such was the case in the Galilee Bird Sanctuary (Narragansett, RI) where tidal flow was restored to a salt marsh and 91 percent of all nesting attempts were lost due to flooding and only 5 percent successfully fledged (DiQuinzio et al. 2002). In fact, a 2004 resurvey of Galilee seven years after tidal restoration showed that salt marsh sparrows still had not recovered to pre-restoration levels (Golet et al., chap. 20, this volume). The sparrows, however, are expected to recolonize once restoration is complete. In the meantime, tidal flow restoration projects may represent a long-term stress in locations where suitable habitat is already limited.

As for the Rachel Carson National Wildlife Refuge, two survey points within the Nancy's Creek marsh actually have relatively high sharp-tailed sparrow densities (O'Brien et al., unpublished data). So the predicted increase of mean high water levels and reduction of *S. patens* habitat (Konisky and Burdick 2003) was a definite concern. As a result, a proposal was made by the authors to test whether the SRT could be managed to provide pre-restoration hydrologic regimes during the first three summers and full restoration flows during the remainder of the year in order to protect nesting sparrows while also facilitating restoration of salt marsh vegetation.

Technical Solution

Given these social and environmental concerns and based on several engineering studies, the Town of Wells decided to install a 1.5 by 1.2 meter box culvert with flashboards on the upstream side and an SRT on the downstream side. The SRT was necessary in order to "prevent potential adverse effects on surrounding property during heavy precipitation runoff that coincides with high tides and predicted coastal flooding events" (TRC 2005, 2). The flashboards were intended for fine-tuning upstream water levels at low tides.

As construction proceeded in May/June 2005, these plans were met with a series of unfortunate events. The SRT supplier, Waterman Engineering, filed for bankruptcy, although the Town managed to secure an SRT from them. A lawsuit prevented the SRT designer from critical oversight of the SRT's installation and settings. The culvert itself was installed higher than the existing culvert and with a greater pitch than designed. Finally, to the dismay of local residents, initial drainage of impounded waters resulted in a die-off of a blue mussel bed and release of foul odors.

Ecosystem Monitoring and 2004 Conditions

The Rachel Carson National Wildlife Refuge conducted before, after, control, impact (BACI) monitoring as recommended for management projects (Stewart-Oaten et al. 1986). Vegetation, groundwater levels, nekton, and sharp-tailed sparrows were chosen as survey variables because of their responsiveness to changes in tidal conditions and because they reflect important attributes of salt marsh ecosystems (Neckles et al. 2002). Surveys before restoration were conducted in 2004 and after culvert–SRT installation from 2005 to 2007. Methods and detailed results have been reported elsewhere (Adamowicz et al. 2009).

Statistical analyses of univariate data (e.g., groundwater level, salinity, nekton and bird density) used analysis of variance. A significant interaction term (site × year) indicated a change likely due to the culvert replacement. BACI comparisons used multivariate nonparametric tests for changes in vegetation and nekton community over time (Clarke and Warwick 2001; Roman et al. 2002; James-Pirri et al. 2007).

Rachel Carson National Wildlife Refuge staff monitored relative water levels upstream of the roadway before (May 2005) and after culvert–SRT installation (August 2005). Water elevation on the tide-restricted marsh was compared to predicted (unrestricted) levels at Kennebunkport, Maine (figs. 19.2a and 19.2b). The unrestricted tide shows the neap and spring sequence common to US East Coast shores (fig. 19.2a). Water elevations of the restricted marsh also reveal the neap and spring sequence, but during the spring series, flooding of the marsh overwhelms the drainage capacity of the undersized culvert, and water accumulates upstream (fig. 19.2b). The result is visible in figure 19.1—the restricted channel is filled with water (dark signature) while mudflats are exposed downstream of the dike (light signature).

Vegetation

Vegetation plots at Nancy's Creek marsh were dominated by *Spartina alterniflora*, *S. pectinata*, and *S. patens* with average percent cover (± standard error) of 33 ± 9, 31 ± 8, and 24 ± 7, respectively. An invasive form of *Phragmites* was present on the impact site, but, at least initially, none occurred in the survey plots. *Phragmites* at the unrestricted marsh was of the native genotype (Saltonstall 2002).

Groundwater and pore water salinity levels were measured at each permanent

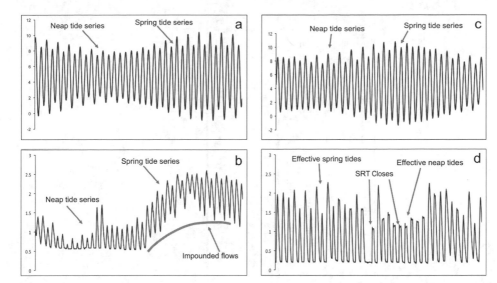

FIGURE 19.2. Predicted and actual tides. (a) Predicted tides for Kennebunkport, May 10–28, 2005. Tide height is given in feet above mean high water (Nobeltec 2004). (b) Actual tides for Nancy's Creek marsh under tide-restricted conditions, May 10–28, 2005, as recorded by a Global Water WL15 water level logger (relative height in feet, not corrected to mean high water). (c) Predicted tides for Kennebunkport, August 10–28, 2005. Tide height is given in feet above mean high water (Nobeltec 2004). (d) Actual tides for Nancy's Creek marsh, following installation of the self-regulating tide gate, August 10–28, 2005, as recorded by a Global Water WL15 water level logger (relative height given in feet, not corrected to mean high water).

vegetation plot using perforated PVC wells and marsh sippers (Roman et al. 2001), respectively. In 2004, prior to installation of new tide gates, there was no significant difference in depth to groundwater between the control and impact sites (table 19.1). Pore water salinities at the control and impact sites were significantly different in 2004 (both depths, $p < 0.001$) with the control site approximately 20 parts per thousand more saline (table 19.1).

Approximately twenty pools in control and impact marshes were sampled for nekton four times each year using a 1 square meter throw trap. Fish densities at impact pools were significantly lower than at the control site in 2004 (table 19.2). This and the lack of crustaceans indicated that the nekton community at Nancy's Creek marsh is incomplete and is unlikely to support the coastal food web as well as the control marsh.

Sharp-tailed sparrow surveys began in 1997, but only the consecutive years of 2000–2007 were examined here. The control and impact sites each had four survey points (Hodgman et al. 2002). At the impact site, densities of these sparrows were higher for most years (table 19.2), emphasizing the importance of this area to the species.

TABLE 19.1

Physical parameter monitoring data before (2004) and after (2005–2007) installation of a self-regulating tide gate, Drakes Island restoration project, Wells, Maine

| Year | Groundwater levels (cm below surface) | | Pore water salinity (ppt) | | | | Pool depth (cm) | | Pool salinity (ppt) | |
| | | | 30 cm depth | | 45 cm depth | | | | | |
	Control	Impact	Control[1]	Impact[1]	Control[2]	Impact[2]	Control[3]	Impact[3]	Control[4]	Impact[4]
2004	11.6±8.4	9.1±9.0	28.6±9.2	8.5±7.0	29.2±10.5	9.6±8.1	33.2±7.8	16.8±5.5	28.5±2.3	6.0±6.8
2005	10.7±8.1	6.2±7.1	28.1±8.8	10.4±7.7	28.5±9.8	10.2±7.4	38.4±8.3	19.0±6.9	26.0±1.9	20.1±4.9
2006	11.9±9.1	12.4±9.5	26.5±10.0	8.4±6.9	27.7±10.7	8.8±7.0	40.2±9.7	13.9±7.8	26.6±2.4	7.4±8.0
2007	11.4±10.9	14.4±8.4	27.9±9.0	9.0±6.1	28.5±10.2	9.0±6.2	38.3±9.5	10.2±6.4	28.9±6.8	10.8±8.8

[1] $F_{7,1} = 274.0, p < 0.001$ (site)
[2] $F_{7,1} = 240.2, p < 0.001$ (site)
[3] $F_{7,3} = 5.138, p = 0.0002$ (site × year)
[4] $F_{7,3} = 16.077, p < 0.001$ (site × year)
Note: Means ± standard errors are shown.
Source: USFWS, unpublished.

TABLE 19.2

Biotic parameter monitoring data before (2003–2004) and after (2005–2007) installation of a self-regulating tide gate, Drakes Island restoration project, Wells, Maine

Year	Percent pools with fish		Fish density (#/m²)		Crustacean density (#/m²)		Nekton community (3 most common species)[4]		Sharp-tailed sparrow density (#/survey point)[3]	
	Control	Impact	Control[1]	Impact[1]	Control[2]	Impact[2]	Control	Impact	Control[3]	Impact[3]
2003									0.49 ± 1.0	2.89 ± 0.49
2004	70	31	12.1 ± 2.0	6.0 ± 1.9	2.3 ± 2.0	0.0 ± 0.0	F. heteroclitus, P. pugio, G. aculeatus	F. heteroclitus, P. pungitius	0.64 ± 0.64	1.44 + 0.74
2005	59	39	13.4 ± 1.5	17.5 ± 7.2	5.9 ± 1.2	0.2 ± 0.2	F. heteroclitus, P. pugio, G. aculeatus	F. heteroclitus, P. pungitius, H. sanguineus	1.21 ± 0.36	2.25 ± 0.36
2006	76	14	13.7 ± 5.9	3.9 ± 1.6	20.6 ± 5.9	0.0 ± 0.0	P. pugio, F. heteroclitus, C. maenas	F. heteroclitus, A. rostrata, P. pungitius	1.44 ± 1.0	1.00 ± 0.49
2007	72	21	11.7 ± 1.7	8.6 ± 3.7	8.0 ± 1.5	0.1 ± 0.2	F. heteroclitus, P. pugio, P. pungitius	F. heteroclitus, P. pungitius, A. quadracus	1.0 ± 0.49	2.25 ± 1.0

[1] $F_{7,1} = 4.368$, $p = 0.039$ (site)

[2] $F_{7,1} = 6.573$, $p < 0.001$ (site × year)

[3] $F_{3,1} = 5.645$, $p < 0.021$ (site)

[4] *Fundulus heteroclitus, Palaemonetes pugio, Gasterosteus aculeatus, Pungitius pungitius, Hemigraspus sanguineus, Carcinus maenas, Anguilla rostrata, Apeltes quadracus.*

Note: Means ± standard errors are shown.

Source: USFWS, unpublished.

Postconstruction Conditions, Community Reactions, and Adaptive Management

Salt marsh restoration requires a variety of skilled disciplines, but just as crucial as science or engineering is community support.

2005

After installation of the new culvert–SRT in June 2005, actual spring and neap tides "switched" periods at Nancy's Creek compared to downstream unrestricted flows (fig. 19.2c and d). The switched tidal periods mean that the SRT was allowing more water to pass through the culvert during a neap sequence than when the SRT closed prior to peak tides of a spring sequence. Other results of the new culvert–SRT were an increase in daily tidal range (approximately 15 centimeters), a decrease of spring high tide levels (10 centimeters), and a further drainage of low tide waters by 10 centimeters (fig. 19.2d). This last condition exposed mudflats and a mussel bed—producing strong low-tide odors. Upset neighbors demanded installation of flashboards to reflood the scour pool and abate objectionable odors. These accommodations were implemented, and additional adjustments to the SRT were made in an attempt to obtain desired water levels and tidal exchange.

Despite best efforts, however, the impact marsh impounded water as noted by a 3 centimeter increase in annual average groundwater levels (table 19.1). Portions of the marsh showed signs of waterlogging stress (weak, discolored vegetation and a large area of continually saturated peat) so the flashboards were removed in the autumn when odors were less of a concern. On the positive side, marsh pool habitat in the Nancy's Creek marsh became more similar to the downstream unrestricted marsh. Pool salinities increased by 14 parts per thousand (table 19.1), and there was an 8 percent increase, compared to 2004, in the number of pools with fish (table 19.2). Fish density nearly doubled, and nekton species increased from two in 2004 to six in 2005.

2006

By spring, neighbor concerns switched to tidal flooding, and they mandated the SRT be adjusted to reduce tidal exchange. The Town complied, the last adjustment was made in April 2006, and the SRT was left alone during the growing season. There were no odor complaints and residents, Wells Reserve, and Rachel Carson National Wildlife Refuge agreed that the low tide water level in the upstream scour pool was appropriate. High tide levels, however, were lower relative to the previous year and appeared to set back marsh restoration.

Vegetation responded with a nearly significant change in vegetation community at the impact site (analysis of similarities [ANOSIM]; $p = 0.067$) through an increase of *Spartina alterniflora, S. patens, Typha angustifolia*, and *Thelypteris palustris* (marsh fern) and a decline in *Spartina pectinata* and open water. *Phragmites* stands on the impact site noticeably expanded with rhizomes over 3 meters long under these drier conditions.

Hydrologic changes included a drop in groundwater levels (table 19.1). Similarly, salt marsh pools became shallower, many dried up, and few contained fish (down from 39 to 14 percent; table 19.2). Because of the negative ecosystem changes documented in 2006, the Technical Advisory Committee called for a 30 centimeter increase in tidal water levels for 2007 in order for the impact marsh to be inundated as frequently as the control site.

2007

With Town approval, five incremental adjustments were made to the SRT with a goal of increasing tidal flow by 20–30 centimeters in order to sustain salt marsh habitat. However, tide levels at the impact site exceeded high marsh elevations only twice from September 19 to November 19, 2007. The Wells Reserve calculated that all the 2007 adjustments resulted in only a 13 centimeter increase in tidal level. At the same time, the northwestern edge of cattail seemed to be advancing over the marsh. With a final SRT adjustment in December 2007, total tidal level increase was estimated at 17 centimeters relative to spring 2007 levels. The Technical Committee recommended to the Town that the SRT settings not be changed until an evaluation of summer 2008 conditions could be made.

Monitoring results from the 2007 growing season showed no change in vegetation from the previous year. Invasive *Phragmites* continued to expand, however, and finally was recorded in one impact survey plot. Control and impact vegetation continued to differ from one another despite expectations that SRT adjustments would flood Nancy's Creek marsh and lead to more salt tolerant plants. Since adjustments did not result in greater flooding of the marsh surface until December, the 2007 growing season supported expansion of invasive plants rather than native halophytes.

Hydrologic and aquatic parameters also reflected the absence of increased tidal flooding in 2007. Pore water salinity levels at Nancy's Creek marsh were about 19 parts per thousand less saline than the control marsh (table 19.1). Impact site pools also became even shallower in 2007—only a fourth as deep as control pools (table 19.1). Impact pool salinities were about half those of 2005, and only 21 percent of pools had any fish at all—compared to 72 percent of control pools.

Fish and crustacean densities remained the same from 2006 to 2007 within both the impact and the control sites, although they still differed from one another significantly (table 19.2). *Palaemonetes pugio* (daggerblade grass shrimp), an important component of salt marsh pools at the control site, were first sampled at the impact site in 2007, a promising sign even though the nekton levels in pool habitat never regained the levels measured in 2005.

Although sharp-tailed sparrow densities were higher at the impact site during most years (table 19.2) there were no significant changes attributable to the new culvert–SRT. In fact, sharp-tailed sparrow densities exhibited a declining trend at the impact site since 2003 versus an increasing trend at the control site. These unexplained changes remain a concern for managers trying to enhance population numbers.

As a final note for 2007, a storm in April (the Patriot's Day Storm) produced floodwaters that surged over Laudholm Farm Road at the north end of Nancy's Creek marsh temporarily reconnecting it with the Little River estuary. The SRT prevented floodwaters from entering on the Drakes Island Road side, making it one of the few roads in Wells that was not damaged by the storm surge.

2008

Qualitative observations indicate the marsh was responding positively to the SRT adjustments made in late 2007. A few tides were just reaching high marsh areas during the highest flooding events, and some plant growth was recovering from the induced dry conditions. But the Technical Advisory Committee determined additional spring tide flooding would be required in order to maintain the site as salt marsh.

Neighbor concerns about poor drainage, however, continued. In response, the Town decided not to increase tidal flows by an additional 10 centimeters as recommended by the Technical Advisory Committee. It is likely that repeatedly flooded yards and basements were due to an exceptionally rainy July rather than the tides. Additionally, it is not yet clear whether the homeowners with drainage concerns were even located near the marsh. It may well be that neighborhood stormwater runoff is more an issue of island soil characteristics (Flewelling and Lisante 1982), unmaintained upland drainage ditches, and building within wetlands.

In the meantime, as of February 2009, tidal flow levels upstream of the culvert–SRT remain up to 13 centimeters below the goal established in 2007. This means that, according to best professional judgment, the Nancy's Creek marsh is not receiving sufficient tidal flows to maintain healthy salt marsh habitat.

Challenges to Long-Term Restoration Success

The Drakes Island project revealed the difficulties of not having unified community commitment over time to support common goals. Residents' concerns often conflicted with previously agreed-upon goals and one another; moreover, they changed rapidly. The result was a dysfunctional management process where goals were more readily changed by political pressure than by sound science, with the consequence that restoration may never be fully realized.

Use of an SRT at Drakes Island has had mixed results. Because the SRT can be adjusted, different parties argued for different settings to benefit individual preferences. In responding to individuals, the Town has further complicated resource management of the marsh because SRT adjustments do not translate into changes in tidal flow that can be quantified in advance, and negative impacts to marsh restoration cannot be evaluated for many months.

The difficulty in adjusting the SRT also made it impossible to manage flows on a fine enough scale to protect nesting sharp-tailed sparrows. As a result, the birds have been subjected to a highly altered flow regime. At this point, vegetation distribution is slowly responding to the latest hydrologic settings and may not adequately cue the birds to safe nesting locations. It is also unclear whether the reversed spring and neap tidal cycles upstream of the SRT impede the ability of the sharp-tailed sparrows to re-nest in a timely fashion. If the birds are tied to lunar events, re-nesting could be delayed by as much as two weeks.

Despite the many concerns, it is widely acknowledged that the new culvert–SRT saved Drakes Island Road from washout during severe spring storms of 2006 and 2007. The culvert–SRT also prevented floodwater impoundment during these events and permitted a more rapid return to preexisting water levels compared to similar road crossings.

Well-documented scientific monitoring has been the most persuasive means of addressing local concerns and setting management goals in an adaptive framework. However, because it takes time and resources to produce quality science, it has not always been possible to supply data to match shifting community demands. Therefore, it is highly recommended that partners in future projects give thoughtful consideration to goals and limitations well before construction plans are ever drafted. For example, the goal of abating stormwater runoff has been transformed by residents into a major issue involving portions of the island far outside the project area.

Drakes Island has a long history of landscape alteration dating back to the mid-1600s. Each era has stressed the Nancy's Creek salt marsh, and it remains noticeably impaired compared to the salt marsh just downstream. The new culvert–SRT is a compromise between a variety of social and ecological concerns originally mediated by stakeholder consensus. The current pull of individual interests and the Town's response to them, however, seriously threaten the ability to achieve previously agreed-upon ecological and social goals.

Acknowledgments

This work was supported through funds from the US Fish and Wildlife Service. The new culvert–SRT installation was made possible by funding from the National Oceanographic and Atmospheric Administration Restoration Center, US Fish and Wildlife Service, Wells National Estuarine Research Reserve, Town of Wells, Gulf of Maine Council on the Marine Environment, Conservation Law Foundation, and Restore America's Estuaries. L. Wagner, K. Springer, N. Williams, A. Chessey, J. Panaccione, and C. Guindon are thanked for their dedicated fieldwork. The authors are indebted to their partners at the Wells National Estuarine Research Reserve and Town of Wells. This chapter was improved through the comments of anonymous reviewers and the editors.

REFERENCES

Adamowicz, S. C., K. M. O'Brien, and E. Bonebakker. 2009. *Monitoring Results of Tidal Restoration at Drakes Island*. Final Report to US Fish and Wildlife Service, Wells, ME.

BirdLife International. 2012. "Species Factsheet: *Ammodramus caudacutus*." http://www .birdlife.org/datazone/speciesfactsheet.php?id=8992.

Boumans, R. M. J., D. M. Burdick, and M. Dionne. 2002. "Modeling Habitat Change in Salt Marshes after Tidal Restoration." *Restoration Ecology* 10:543–55.

Burdick, D. M., M. Dionne, R. M. Boumans, and F. T. Short. 1997. "Ecological Responses to Tidal Restorations of Two Northern New England Salt Marshes." *Wetlands Ecology and Management* 4:129–44.

Butler, J. 2005. *Laudholm: The History of a Celebrated Maine Saltwater Farm, 1642–1986*. Wells, ME: Wells Reserve and Laudholm Trust.

Clarke, K. R., and R. M. Warwick. 2001. *Change in Marine Communities: An Approach to Statistical Analysis and Interpretation*. 2nd ed. Plymouth, UK: PRIMER-E.

Clarkson, F. B. 1950. "Report of Wells Town Meeting." In *Drakes' Island Driftwood* 4 (2).

Deegan, L. A., J. E. Hughes, and R. A. Rountree. 2000. "Salt Marsh Ecosystem Support of Marine Transient Species." Pp. 333–68 in *Concepts and Controversies in Tidal Marsh Ecology*, edited by M. Weinstein and D. Kreeger. Dordrecht, Netherlands: Kluwer.

Dionne, M., F. T. Short, and D. M. Burdick. 1999. "Fish Utilization of Restored, Created and Reference Salt-Marsh Habitat in the Gulf of Maine." *American Fisheries Society Symposium* 22:384–404.

DiQuinzio, D. A., P. W. C. Paton, and W. R. Eddleman. 2002. "Nesting Ecology of Saltmarsh Sharp-Tailed Sparrows in a Tidally Restricted Salt Marsh." *Wetlands* 22:179–85.

Eberhardt, A. L., D. M. Burdick, and M. Dionne. 2011. "The Effects of Road Culverts on Nekton in New England Salt Marshes: Implications for Tidal Restoration." *Restoration Ecology* 19:776–85.

Flewelling, L. R., and R. H. Lisante. 1982. *Soil Survey of York County, Maine*. USDA, Soil Conservation Service; Maine Agricultural Experiment Station, Maine Soil and Water Conservation Commission.

Fragosa, G., and T. Spencer. 2008. "Physiographic Control on the Development of *Spartina* Marshes." *Science* 322:1064.

Gjerdrum, C. C., S. Elphick, and M. Rebega. 2005. "Nest Site Selection and Nesting Success in Saltmarsh Breeding Sparrows: The Importance of Nest Habitat, Timing, and Study Site Differences." *Condor* 107:849–62.

Greenlaw, J. S., and J. D. Rising. 1994. "Sharp-Tailed Sparrow (*Ammodramus caudacutus*). No. 112 in *The Birds of North America*, edited by A. Poole and F. Gill. Philadelphia, PA: Academy of Natural Sciences; Washington, DC: American Ornithologists' Union.

Hodgman, T. P., W. G. Shriver, and P. D. Vickery. 2002. "Redefining Range Overlap between the Sharp-Tailed Sparrows of Coastal New England." *Wilson Bulletin* 114:38–43.

James-Pirri, M. J., C. T. Roman, and J. F. Heltshe. 2007. "Power Analysis to Determine Sample Size for Monitoring Vegetation Change in Salt Marsh Habitats." *Wetlands Ecology and Management* 15:335–45.

Kelley, J. T., D. F. Belknap, G. L. Jacobson Jr., and H. A. Jacobson. 1988. "The Morphology and Origin of Salt Marshes Along the Glaciated Coastline of Maine, USA." *Journal of Coastal Research* 4:649–65.

Kelley, J. T., W. R. Gehrels, and D. F. Belknap. 1995. "The Geological Development of Tidal Marshes at Wells, Maine." *Journal of Coastal Research* 11:136–53.

Kelly, M. W. 1978. *A History of Drakes Island 1630–1950*. Wells, ME: Drakes Island Improvement Association.

Konisky, R. A., and D. M. Burdick. 2003. "Analysis of Tidal and Storm Hydrology at Drakes Island Marsh (Wells, Maine) and Scenarios for Improvement." Unpublished report submitted as part of TRC 2005 permit application.

Maine Office of Geographic Information Systems (ME GIS). 2009 Maine GIS Data Catalog. http://www.maine.gov/megis/catalog/.

Nobeltec Corporation. 2004. *Nobeltec Tides and Currents* 3.3. Portland, OR.

Neckles, H. A., M. Dionne, D. M. Burdick, C. T. Roman, R. Buchsbaum, and E. Hutchins. 2002. "A Monitoring Protocol to Assess Tidal Restoration of Salt Marshes on Local and Regional Scales." *Restoration Ecology* 10:556–63.

Portnoy, J. W. 1999. "Salt Marsh Diking and Restoration: Biogeochemical Implications of Altered Wetland Hydrology." *Environmental Management* 24:111–20.

Redfield, A. C. 1972. "Development of a New England Salt Marsh." *Ecological Monographs* 42:201–37.

Rich, T. D., C. J. Beardmore, H. Berlanga, P. J. Blancher, M. S. W. Bradstreet, G. S.

Butler, D. W. Demarest, et al. 2004. *Partners in Flight North American Landbird Conservation Plan*. Ithaca, NY: Cornell Lab of Ornithology.

Roman, C. T., M. J. James-Pirri, and J. F. Heltshe. 2001. *Monitoring Salt Marsh Vegetation: A Protocol for the Long-Term Coastal Ecosystems Monitoring Program at Cape Cod National Seashore*. Technical Report. Narragansett, RI: USGS Patuxent Wildlife Research Center.

Roman, C. T., K. B. Raposa, S. C. Adamowicz, M. J. James-Pirri, and J. G. Catena. 2002. "Quantifying Vegetation and Nekton Response to Tidal Restoration of a New England Salt Marsh." *Restoration Ecology* 10:450–60.

Saltonstall, K. 2002. "Cryptic Invasion by a Non-native Genotype of the Common Reed, *Phragmites australis*, into North America." *Proceedings National Academy of Sciences* 99:2445–49.

Shelley, H. M. 1997. *Images of America: Beaches of Wells*. Portsmouth, NH: Arcadia Publishing.

Shriver, W. G., P. D. Vickery, T. P. Hodgman, and J. P. Gibbs. 2007. "Flood Tides Affect Breeding Ecology of Two Sympatric Sharp-Tailed Sparrows." *Auk* 124:552–60.

Spencer, W. D. 1973. *Pioneers on Maine Rivers with Lists to 1651: Compiled from Original Sources*. Baltimore, MD: Genealogical Publishing.

Stewart-Oaten, A., W. W. Murdoch, and K. R. Parker. 1986. "Environmental Impact Assessment: "Pseudoreplication in Time?" *Ecology* 67:929–40.

TRC. 2005. *Application for a [sic] NRPA Permit by Rule and ACOE Programmatic General Permit*, Drakes Island Marsh Restoration/Enhancement Project, Wells, ME February. Bremen, ME.

Turner, R. E. 2004. "Coastal Wetland Subsidence Arising from Local Hydrologic Manipulations." *Estuaries* 27:265–72.

US Fish and Wildlife Service. 2008. *Birds of Conservation Concern 2008*. Arlington, VA: US Department of Interior, Fish and Wildlife Service, Division of Migratory Bird Management. http://www.fws.gov/migratorybirds/.

US Geological Survey. 1891. *Kennebunk Quadrangle, Maine* [map]. 1:62500. 15 Minute Series. UNH Dimond Library, Documents Department & Data Center. http://docs.unh.edu/nhtopos/Kennebunk.htm.

US Geological Survey. 1979. *Kennebunk Quadrangle, Maine* [map]. 1:2400 7.5 Minute Series. Washington, DC.

Warren, R. S., P. E. Fell, A. H. Brawley, A. C. Orsted, E. T. Olson, V. Swamy, and W. A. Niering. 2002. "Salt Marsh Restoration in Connecticut: 20 Years of Science and Management." *Restoration Ecology* 10:497–513.

Chapter 20

Role of Science and Partnerships in Salt Marsh Restoration at the Galilee Bird Sanctuary, Narragansett, Rhode Island

Francis C. Golet, Dennis H. A. Myshrall,
Lawrence R. Oliver, Peter W. C. Paton,
and Brian C. Tefft

One of the earliest, and most extensive, collaborative salt marsh restorations in New England was launched east of the Port of Galilee, in Narragansett, Rhode Island, in the early 1990s (fig. 20.1). Once part of the largest coastal wetland complex in the state, by the 1950s these wetlands had been extensively altered by breachway construction, deposition of dredged material, and construction of docks, commercial facilities, houses, and roads (Lee 1980). In 1956, construction of the four-lane Galilee Escape Route, with only one 75 centimeter diameter culvert, severely restricted tidal flow from Bluff Hill Cove to the state-owned Galilee Bird Sanctuary and impounded freshwater runoff there. Over the next forty years, more than 80 percent of the 40 hectare Sanctuary salt marsh shifted from predominantly *Spartina* (cordgrass) to a mix of *Phragmites australis* (common reed) and freshwater or upland shrubs and trees (fig. 20.1).

Restoration planning began in 1989 with a series of meetings among the Rhode Island Division of Fish and Wildlife (RIDFW), the New England Division of the US Army Corps of Engineers (the Corps), and the US Fish and Wildlife Service (FWS). In 1991, the Rhode Island Department of Transportation (RIDOT) received approval from state and federal regulatory agencies to enhance tidal flow at the Galilee Bird Sanctuary (Applied Bio-Systems 1991) as mitigation for the filling of a 0.3 hectare *Phragmites* marsh in nearby North Kingstown. Through interagency collaboration, this narrowly focused mitigation project rapidly evolved into a comprehensive, proactive restoration effort that was designated as a Coastal America project early in 1992. Ultimately, partners included RIDFW, RIDOT, the Corps, FWS, US Environmental Protection Agency, National Fish and Wildlife Foundation, Ducks Unlimited, University of Rhode Island (URI), and Town of Narragansett.

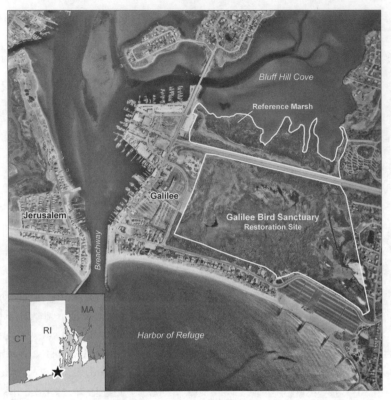

FIGURE 20.1. Setting of the Galilee Bird Sanctuary restoration site and Bluff Hill Cove reference site, Narragansett, Rhode Island. Photo was taken in 1995; restoration work began in 1996.

Supported by federal and state funding, the Corps and RIDFW assumed major responsibility for project oversight and compliance with federal and state regulations. The Corps also performed hydraulic modeling, culvert and gate design, and construction management, while RIDFW coordinated adaptive management efforts. RIDOT provided elevation surveys and shared in the cost of culvert construction. FWS funded URI's baseline ecological studies and monitoring.

Construction began in October 1996 and involved (1) removal of dredged material from a 3 hectare area in the northwestern part of the Sanctuary, (2) re-creation of a major tidal creek that had been filled with dredged material, (3) installation of two pairs of box culverts fitted with self-regulating tide gates, and (4) excavation of a network of channels to enhance tidal flushing throughout the central and western parts of the Sanctuary (fig. 20.2). The partners led a volunteer planting effort in the former dredged material disposal site just prior to enhancing tidal flow. Flow was restored in October 1997. This chapter explains how science and close collaboration among partners contributed to every aspect of the Galilee

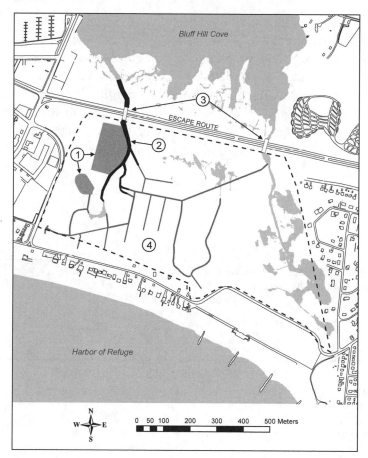

FIGURE 20.2. Phases of construction for the Galilee Bird Sanctuary salt marsh restoration.

Bird Sanctuary salt marsh restoration—planning, design, construction, monitoring, residential flood protection, evaluation of restoration success, and adaptive management.

Baseline Data Collection and Restoration Design

Establishment of baseline conditions within the Sanctuary and design of the restoration project were accomplished through an integrated series of steps.

Project Goal and Approach

The goal of this project was to convert more than 30 hectares of degraded coastal habitat (i.e., *Phragmites* and shrubs) back to productive salt marsh, channels, ponds, and mudflats. Prerequisite to designing and implementing a successful salt marsh restoration strategy was an understanding of the environmental conditions that supported, and limited, the distribution and abundance of both the degraded habitats and those targeted for reestablishment. For that reason, a comprehensive baseline ecological study was undertaken. Data gathered during that study (table 20.1), along with information gained from the scientific literature, were used to develop ecological targets for the Sanctuary restoration, to support hydrologic modeling efforts and related culvert and channel design, and to provide a basis for assessment of ecosystem response and restoration success.

TABLE 20.1

Ecological data gathered at the Galilee Bird Sanctuary (GBS) salt marsh restoration site and at the Bluff Hill Cove (BHC) reference site, Narragansett, Rhode Island, prior to, and after, tidal flow restoration in October 1997

Data type	Number of stations or plots	Frequency	Baseline survey	Postrestoration
Tide levels from automated gauges	1 at GBS, 1 at BHC	10 min. intervals	1992–97	1997–2000
Marsh groundwater levels	110	Biweekly-monthly	1992–97	—
Residential groundwater levels	12	Weekly-biweekly	1993–97	1999
Ground elevations	> 1000	Once	1992	—
Soil salinity	Up to 130	Weekly-biweekly in summer	1992–95	1999
Habitat mapping	Systemwide	Twice	1992	1999
Herb and shrub structure and cover by species	120–330	Annually	1992–94, 1997	1998–2001, 2005
Spartina alterniflora and *Phragmites* biomass	20–40	Annually	1997	1998–2001, 2005
Ruppia maritima cover	260	Annually	1997	1998–99
Habitat photographs	42	Annually	1995	1998–2002, 2005
Salt marsh snail density	180	Annually	1996–97	1998–99
Landbird numbers by species	31	3–4 times/summer	1996	1998–2008
Waterbird numbers by species	All of GBS and BHC	Twice weekly	1991–96	1998–2000
Saltmarsh sparrow breeding ecology	GBS only	Annually	1993–97	1998

Reference Site Selection

Selecting a reference site with environmental conditions and biotic communities like those targeted for the restoration area was an important first step in the baseline study. Reference site characteristics provide realistic targets for restoration and a basis for evaluating ecosystem response (Brinson and Rheinhardt 1996). The reference site was the 11.3 hectare Bluff Hill Cove marsh, where tidal flow is unrestricted (fig. 20.1). The reference marsh was dominated by short (less than 60 centimeters) *Spartina alterniflora* (smooth cordgrass), with small patches of tall (greater than or equal to 60 centimeters) *S. alterniflora* along creeks and in tidal guts, and a narrow band of salt meadow, dominated by *Spartina patens* (salt meadow cordgrass), *Distichlis spicata* (spikegrass), or *Juncus gerardii* (black rush) along the landward edge. *Phragmites*, *Iva frutescens* (hightide bush), *Baccharis halimifolia* (groundsel bush), and upland shrubs grew between the salt meadow zone and the Escape Route embankment.

Mapping Baseline Habitats

A RIDOT survey team established a grid of more than 250 wooden stakes placed at 60 meter intervals parallel and perpendicular to the Escape Route in the Sanctuary and at the reference marsh. Baseline habitat types were then mapped at both locations (table 20.2) through stereoscopic interpretation of 1:2,400-scale

TABLE 20.2

Prerestoration habitats at the Galilee Bird Sanctuary, Narragansett, Rhode Island

Habitat type	Description
Open water and mudflat	Subtidal or regularly flooded[1]; *Ruppia maritima* in subtidal areas
Tall *Spartina alterniflora*	≥ 60 cm tall; regularly flooded; soil salinity 31 ppt[2]
Short *Spartina alterniflora*	< 60 cm tall; irregularly flooded; soil salinity 29–31 ppt
Salt meadow	*Spartina patens, Distichlis spicata,* or *Juncus gerardii* dominant; irregularly flooded; soil salinity 17–24 ppt
Short *Phragmites*	*Phragmites* < 1.5 m tall with ≥ 30 percent cover; irregularly flooded; soil salinity 9–23 ppt; understory of salt meadow species
Medium *Phragmites*	*Phragmites* 1.5–2.5 m tall with ≥ 60 percent cover; irregularly flooded or nontidal; soil salinity 2–9 ppt; highly variable understory
Tall *Phragmites*	*Phragmites* > 2.5 m tall with ≥ 60 percent cover; mostly nontidal; soil salinity ≤ 1 ppt; understory plants fresh or lacking
Tall shrub	Shrubs 1.5–5.9 m tall with ≥ 60 percent cover; upland or freshwater wetland; soil salinity ≤ 1 ppt
Forest	Woody plants ≥ 6.0 m tall with ≥ 30 percent cover; upland; soil salinity 0 ppt

[1]Water regime definitions after Cowardin et al. (1979): subtidal—permanently flooded with tidal water; regularly flooded—flooded and exposed by tides daily; irregularly flooded—flooded by tides less than daily.
[2]Soil salinity values are means of samples collected June–September, 1992–93 or 1992–95.

false-color infrared aerial transparencies taken in November 1991, and entered into a geographic information system after extensive field checking in 1992. The map and grid system were used for orientation in the field, to randomly select and locate stations for sampling, and for assessment of habitat change after tidal flow restoration.

Elevation Surveys

In coastal wetlands, ground elevation controls the frequency and duration of tidal flooding and, in turn, the distribution of major plant zones (Bertness 2007). To better understand the water regimes of the habitats mapped in 1992, to identify target elevations for restoration, and to predict habitat area and distribution after tidal flow restoration, ground elevations (referenced to the 1929 National Geodetic Vertical Datum—NGVD 29) were collected at more than 750 points throughout the Sanctuary and reference marsh, including more than three hundred permanent vegetation sampling plots. Through this process the elevation range was established for each habitat type at the reference marsh and at the restoration site. The Corps used these data, along with more than six hundred spot elevations from channel and pond cross sections within the Sanctuary, to model site hydrology for culvert-sizing purposes.

Monitoring Tide Levels

A computerized tide gauge was deployed in a tidal pond within the Sanctuary and a second gauge was placed in permanent tidal water adjacent to the reference marsh to compare restricted and unrestricted tidal regimes prior to restoration, to compare pre- and postrestoration tide levels in the Sanctuary, and to facilitate hydrologic modeling of the Sanctuary and culvert design. Tidal flooding frequencies were calculated for individual habitat types using ground elevations obtained at vegetation sampling plots and tide data collected from May through September 1994 (Myshrall 1996).

Mapping Predicted Habitats

Assuming that the tidal regimes of restored habitats in the Galilee Bird Sanctuary would be approximately the same as at the reference site, and that ground elevation dictates both the tidal regime and habitat type, a map was created showing habitats predicted for the Sanctuary once restoration-induced changes had stabilized. It was estimated that more than 25 hectares of *Phragmites*, freshwater shrub swamp, and upland habitats would be converted to open water, mudflat, *Spartina alterniflora,* or salt meadow as a result of dredged material removal and reintro-duction of high-salinity (30 to 32 parts per thousand) tidal water (Myshrall and Golet 1996).

Locating Historic Channel for Reconstruction

The restoration process included re-creating a large tidal creek that had been filled north and south of the Escape Route (fig. 20.2). This historic channel was identified and delineated on a 1:14,400-scale, 1939 panchromatic aerial photo-graph, digitized and superimposed on the baseline habitats map. Channel loca-tion was flagged in the field for construction contractors to follow.

Hydrologic Modeling, Culvert Design, and Residential Flood Protection

Two steps critical to restoration success consisted of (1) determining the volume and elevation of tidal water needed to restore the former salt marsh and (2) de-signing a culvert system that would convey the water without flooding residential properties along the Sanctuary's southern border (fig. 20.2). The landward extent of southern New England salt marshes coincides roughly with the average eleva-tion of the highest monthly tides (Bertness and Ellison 1987; Lefor et al. 1987)—approximately mean high water of spring (new and full moon) tides (Niering and Warren 1980). As early as 1992, project partners had agreed that, to maximize the extent of salt marsh restored at the Sanctuary, new culverts should be large enough to allow the most distant reaches of the Sanctuary to flood to nearly the same elevation as the Bluff Hill Cove reference marsh during monthly astronom-ical high tides and to drain completely at low tide. At the same time, the Corps and RIDFW pledged to protect abutting residential properties from restoration-induced flooding.

   The Corps calculated the mean high water spring datum at Bluff Hill Cove to be 0.7 meters NGVD 29 and the maximum astronomical tide to be 1.0 meters (US Army Corps of Engineers 1994). Elevations obtained at residences along the southern edge of the Sanctuary indicated that surface water levels could reach about 1.2 meters before adversely affecting homes. Allowing 0.24 meters for stor-age of freshwater runoff, the Corps concluded that restoration tide levels should not be allowed to exceed 0.9 meters in the Sanctuary (US Army Corps of Engi-neers 1994). Partners then agreed that the new culverts should be fitted with self-regulating tide gates (SRTs), as well as slots for stoplogs, to ensure that water levels would not exceed that datum.

   Using a one-dimensional hydrodynamic model, the Corps analyzed the hy-drology of the Sanctuary under different design scenarios. On-site data required

for these analyses—or to predict the effects of different designs on restoration outcomes—included the type and extent of existing habitats in the Sanctuary, elevation ranges for reference marsh habitats, grid-point elevations and cross-sectional
profiles of major channels and ponds in the Sanctuary, tide levels recorded by the
Bluff Hill Cove gauge during periods of average and near-maximum tidal range,
tide levels required to maintain salt marsh habitat in the Sanctuary, and the tide
level above which residential areas would be flooded. During the modeling process, the Corps evaluated various culvert sizes and inverts, as well as different
channel widths and inverts north and south of the Escape Route. Ultimately, they
concluded that two pairs of 1.8 meter by 3.0 meter culverts—an eastern set located at the site of the existing, undersized culverts and a western set at the site of
the historic channel (fig. 20.2)—would provide sufficient saltwater exchange to
generate tide levels of 0.9 meters NGVD at the northern edge of the Sanctuary
and 0.8 meters at the southern edge when the maximum astronomical high tide of
1.0 meters occurred at Bluff Hill Cove. Each of the four box culverts was to be
equipped with a 1.8 meter by 1.5 meter SRT calibrated to close when tide levels
at Bluff Hill Cove reached 1.0 meters NGVD, thereby protecting residential
properties.

In response to concerns regarding possible flooding of residential properties,
groundwater levels were monitored weekly or biweekly in water-table wells at
each of five homes along the southern edge of the Sanctuary starting in the fall of
1993. Based on the first seven months of data, the US Army Corps of Engineers
(1994) estimated that the average groundwater level after tidal flow restoration
would be about 0.2 meters below the prerestoration average, primarily due to better soil drainage at low tide. Through water-table measurements obtained after
the tide gates were opened, we documented that this estimate was accurate
(Myshrall et al. 2000).

Mosquito Management

To ensure that the introduction of additional tidal water into the Sanctuary would
not increase mosquito breeding, a mosquito control strategy was developed based
partially on open marsh water management (OMWM) principles (Boyes 1997). It
was predicted that, after flow restoration, the increased tidal range in the Sanctuary would cause salt meadow, the primary breeding habitat for salt marsh mosquitoes, to redevelop closer to the upland edge, in areas dominated by shrubs or
Phragmites before restoration. The salt meadow zone was then targeted on the
predicted habitats map as the area most likely to support mosquito breeding after
restoration.

The first step in the Sanctuary mosquito management strategy involved enhancing the flow of tidal water to and from potential breeding sites by cleaning out selected mosquito control ditches dug in the 1930s and excavating new channels in strategic locations (Boyes 1997). After tidal flow had been restored, OMWM techniques were to be applied to specific locations where mosquitoes were breeding. To facilitate the first step, all of the ditches appearing on a 1939 aerial photograph were delineated and superimposed onto both the baseline and predicted habitats maps. Using these maps, a determination was made as to which of the old ditches to reopen and where new channels might be added.

Using two, low-ground-pressure, track excavators and a low-pressure bulldozer from FWS and the Connecticut Wetlands Restoration Project, RIDFW opened a network of channels (fig. 20.2) between August and October 1997; the Sanctuary was closed to tidal flow throughout that period (Boyes 1997). Creation of this channel network enhanced tidal flushing for mosquito control. It also maximized the extent of salt marsh restoration by efficiently conveying high-salinity tidal water to remote areas of the Sanctuary. OMWM activities undertaken after tidal flow restoration are described under "Adaptive Management."

Baseline Habitat Profiles

Between 1992 and 1997 detailed ecological profiles were developed for the major baseline habitats (all but open water and mudflats from table 20.2) using field data on vegetation structure and species composition, as well as environmental conditions (Myshrall and Golet 1993, 1996). These profiles were created to better understand the conditions controlling the presence and distribution of each habitat, to predict how individual habitats might change after tidal flow restoration, and to document those changes through comparisons with postproject data.

More than three hundred permanent vegetation sampling plots were established in herb, shrub, and forest communities. Percentage cover of herbs, by species, was visually estimated in 2.0 square meter quadrats and total shrub cover in 25.0 square meter quadrats. Ground elevations were obtained at each plot, and soil salinity and groundwater levels monitored in one third of the plots for four to five years prior to tidal flow restoration and for at least two years afterward (table 20.1). Changes in height, density, and aboveground biomass were closely examined in *Phragmites*, the primary species targeted for elimination, and *Spartina alterniflora*, the principal species targeted for reestablishment (Myshrall et al. 2000). From 1997 through 1999, cover of the submergent plant *Ruppia maritima* (widgeongrass) was also monitored in eleven ponds within the Sanctuary. Finally, in 1995 more than forty permanent photostations were established to provide visual documentation of restoration-induced changes in habitat structure and classification.

Baseline Animal Studies

A small number of animal species or groups were monitored to determine whether, and when, restoration-induced habitat changes had advanced sufficiently to affect habitat suitability. During baseline and postrestoration studies, *Melampus bidentatus* (salt marsh snail), landbirds, and waterbirds were monitored both in the Sanctuary and in the reference marsh (table 20.1; Myshrall et al. 2000). Snails were sampled in short *S. alterniflora*, salt meadow, and short and medium *Phragmites* habitats during one year prior to tidal flow restoration (1997) and for two years afterward (1998 and 1999). Landbirds were surveyed three to four times each summer at thirty-one 50 meter radius point-count stations during 1992–96 and 1998–99. Waterbirds were surveyed at high tide and low tide each week from the fall of 1991 through the spring of 2000 by walking fixed routes through the Sanctuary and the reference site. Waterbirds included mainly waders, waterfowl, shorebirds, and gulls. Population size, nest ecology, and nest success of the *Ammodramus caudacutus* (saltmarsh sparrow) were also monitored in remnant salt marsh patches within the Sanctuary for five years prior to tidal flow restoration (1993–97) and one year afterward (1998). Because salt marsh snails and saltmarsh sparrows are limited to salt marsh habitats, it was hoped that one or both might be a sensitive indicator of the quality of marsh habitats after tidal flow was restored. Although not reported on here, nekton (fish and decapod crustaceans) were sampled in the Sanctuary and reference marsh during prerestoration (1997) and postrestoration (1998 and 1999) periods (Raposa 2002).

Adaptive Management

During the first two years after tidal flow restoration at the Galilee Bird Sanctuary, several problems were encountered that prompted additional monitoring and remedial measures.

Inefficient Tidal Flushing

When the tide gates were opened in October 1997, peak tide levels varied locally within the Sanctuary. By comparing maximum daily levels observed at thirteen locations with those recorded by the Sanctuary tide gauge, one location was identified in the north-central part of the Sanctuary (fig. 20.2) where tide levels were markedly depressed due to the long, circuitous route that tidal water followed to reach that area during a typical tidal cycle. RIDFW solved this problem in 1998 by creating an auxiliary channel to link this area more directly to a major feeder channel.

Reduced Soil Salinity Due to Freshwater Inflow

Research by others (Sinicrope et al. 1990), as well as our own baseline habitat profiles (Myshrall and Golet 1996), indicated that vigorous growth of *Phragmites* occurs only at salinities below about 20 parts per thousand. Reintroduction of tidal water with a salinity of 30 to 32 parts per thousand caused severe stunting of *Phragmites* in most of the Sanctuary during the first two years and death shortly thereafter (see "Ecosystem Response"). However, in certain areas along the perimeter of the Sanctuary, where significant inflow of surface runoff or fresh groundwater occurred, soil salinity remained low enough after tidal flow restoration to allow *Phragmites* to persist in a vigorous condition. To address this problem, RIDFW excavated "perimeter ditches" at selected locations along the upland edge in 1998. These ditches effectively intercepted inflowing fresh water and allowed it to flow out with the tide. *Phragmites* dominated the upland side of these ditches, but salt marsh plants typically dominated the marsh side within two to three years. Perimeter ditching thus enhanced salt marsh restoration in areas where the introduction of high-salinity surface water alone was not sufficient.

Excessive Ground Elevations

Initially dredged material removal was confined to areas west of the reconstructed historic channel (fig. 20.2). After tidal flow was restored, one area of dredged material east of the channel was too high in elevation to support salt marsh, so in 1998 RIDFW bulldozed several centimeters from the soil surface. That area now supports short *S. alterniflora*, salt meadow, *Iva frutescens*, and short *Phragmites*.

Mosquito Breeding

Larval surveys conducted by the Rhode Island Department of Environmental Management's Mosquito Abatement Office early in 1998 identified one site along the southern edge of the Sanctuary where mosquito breeding was occurring due to a combination of ineffective tidal flushing and accumulation of freshwater runoff. OMWM techniques that were applied included reopening an old mosquito control ditch, constructing a permanent pond at the end of that ditch to serve as a reservoir for mummichogs (*Fundulus heteroclitus*), and creating a perimeter ditch to connect the pond to a feeder channel.

Self-Regulating Tide Gate Calibration and Maintenance

Accurate calibration of the self-regulating tide gates was necessary to maximize the effectiveness of the salt marsh restoration while also protecting abutting residential land from flooding. Calibration began shortly before the gates were opened in the fall of 1997 and continued until the spring of 1999—a protracted period necessary to encompass astronomical high tides and storm events. The restoration plan called for the SRTs to close whenever tide levels at Bluff Hill Cove reached or exceeded the maximum astronomic high tide of 1.0 meters NGVD 29. Project partners agreed to take a conservative approach to ensure protection of the residential properties. Initially, the SRTs were calibrated to close when the tide level reached 0.85 meters NGVD 29 at Bluff Hill Cove. When the gates closed, the maximum height of the tide achieved at our monitoring stations around the perimeter of the Sanctuary was checked to confirm that target elevations had not been exceeded. After a series of successively higher gate settings and checks within the Sanctuary, the four SRTs were ultimately set to close when Bluff Hill Cove tides reached between 0.95 meters and 1.0 meters NGVD 29.

Although they are "self-regulating," the gates must be checked periodically and cleaned of marine fouling organisms that might interfere with the action of the floats. The SRTs have been operating for more than ten years without serious mishap.

Ecosystem Response

The majority of ecosystem characteristics sampled prior to tidal flow restoration were monitored for at least a year postrestoration (table 20.1). Below are some examples of our findings on ecosystem response. A more comprehensive treatment of these findings may be found in Myshrall et al. (2000), DiQuinzio et al. (2002), and Shaffer (2007).

Changes in Tidal Regime and Soil Salinity

The increase in tidal range, frequency of flooding, and soil salinity at the Galilee Bird Sanctuary was immediate and pronounced after tidal flow restoration. During a two-week period (16–28 May 1994) before restoration, the daily tidal range at Bluff Hill Cove was 0.75–1.50 meters; in the Sanctuary it was only 0.15–0.25 meters. During a two-week period (1–13 August 1998) after restoration, the daily tidal range was 0.68–1.27 meters at Bluff Hill Cove and 0.56–1.03 meters in the Sanctuary. After restoration, the Sanctuary also drained more effectively at low tide; marsh ponds, which had been subtidal before flow restoration, became intertidal, and marsh soils were no longer waterlogged at low tide.

Before restoration of tidal flow, the frequency of tidal flooding ranged from 100 percent of the daily high tides for tall S. *alterniflora* to zero for tall *Phragmites*. After flow was restored, tall and short S. *alterniflora*, salt meadow, and all *Phragmites* habitats were flooded by more than 90 percent of the daily high tides. Before restoration, summer soil salinity during low tide ranged from 29 to 30 parts per thousand in S. *alterniflora* habitats to zero in most stands of tall *Phragmites* and some stands of medium *Phragmites*. Two years after the tide gates were opened, mean summer soil salinity levels exceeded 30 parts per thousand in all baseline S. *alterniflora*, salt meadow, and *Phragmites* habitats.

Early Postrestoration Habitat Dynamics

Updates made to the baseline habitats map through fieldwork in 1999, two years postrestoration, showed that the changes in tidal hydrology and soil salinity induced by tidal flow restoration had caused rapid changes in the classification and extent of Sanctuary habitats (table 20.3).

Woody vegetation died over almost 6 hectares of the Sanctuary. All of the short *Phragmites*, found along the landward edge of the salt meadow zone during baseline conditions, also died within the first two years. Most of the baseline medium and tall *Phragmites* was severely stunted by 1999 and reclassified as short *Phragmites*. All of the baseline short S. *alterniflora* had become tall (greater than 60 centimeters) in response to the increased frequency of tidal flooding and improved soil drainage at low tide. Mudflats increased by more than 3 hectares due to channel construction and dredged material removal. Overall, tidal flow restoration had significantly altered 39 hectares of Sanctuary habitats in only two years.

TABLE 20.3

Changes in the area (hectares) of habitat types within two years of tidal flow restoration at the Galilee Bird Sanctuary, Narragansett, Rhode Island

Habitat type	Baseline (1992)	Postrestoration (1999)	Change
Open water and mudflat	3.56	6.66	3.10
Tall *Spartina alterniflora*	0.25	2.41	2.16
Short *Spartina alterniflora*	2.16	0.00	−2.16
Salt meadow	1.15	3.91	2.76
Short *Phragmites*	2.76	14.74	11.98
Medium *Phragmites*	15.21	5.20	−10.01
Tall *Phragmites*	6.99	0.12	−6.87
Dead woody vegetation	0.00	5.99	5.99
Live woody vegetation	18.23	11.28	−6.95

Note: Tidal flow was restored in October 1997.

Decline of Phragmites

By the end of the first growing season postrestoration, *Phragmites* stem height and aboveground biomass had decreased significantly in all three height classes. Structural changes were most dramatic in the tall *Phragmites* habitat type, where soil salinity had increased from 0 to 31 parts per thousand. Median annual biomass production dropped from 740 grams per square meter to 164 grams per square meter in one year; during the next year it dropped below 100 grams per square meter (Shaffer 2007). Median stem height in that habitat type declined from 3.0 meters in 1997 to 1.7 meters in 1998 and 1.2 meters in 1999. As already noted, baseline short *Phragmites* was gone from the Sanctuary by 1999.

Although most restoration monitoring at Galilee ended early in 2000, tracking the response of *Phragmites* continued through 2005. Based on ground elevations obtained during the baseline study, it was predicted that the great majority of plots originally classified as *Phragmites* would become tall *S. alterniflora* if flooded daily after tidal flow restoration, or short *S. alterniflora* if flooded less often than daily (Myshrall et al. 2000). By 2001 *S. alterniflora* dominated 12 percent of the baseline short *Phragmites* plots and 3 percent of the medium *Phragmites* plots; none of the tall *Phragmites* plots had changed to *S. alterniflora*. By 2005, eight years postrestoration, the rate of conversion to *S. alterniflora* ranged between 22 percent and 27 percent for the three *Phragmites* height classes (fig. 20.3; Shaffer 2007).

Most medium and tall *Phragmites* plants either died after a total of four to five years or persisted as short *Phragmites* in elevated locations where the frequency and duration of tidal flooding were insufficient to support salt meadow or *S. alterniflora* (Shaffer 2007). If the plants died, the dead stems typically remained erect for another three to four years, then rotted off at ground level and were removed, along with *Phragmites* leaf litter, by the tides. Only then did salt marsh plants—first *Salicornia maritima* (slender glasswort) or *Suaeda maritima* (seablite) and later *S. alterniflora*—colonize the area and begin the conversion to salt marsh habitat.

During the reopening of mosquito control ditches and creation of new channels in the Sanctuary in 1997 and 1998, excavation equipment flattened belts of medium and tall *Phragmites* along the channels and covered them with a thin layer of soil removed from the channels. In 1999, twenty-seven additional sampling plots were established in those areas. The flattening process greatly accelerated the conversion from *Phragmites* to salt marsh (Shaffer 2007). By 1999, the pioneer forbs, *Salicornia* and *Suaeda*, dominated two thirds of the plots, and *S. alterniflora* was present in many of them. *S. alterniflora* dominated 30 percent of the plots by 2001 and 52 percent by 2005—more than twice the rate of conversion recorded in plots where the plants had not been flattened (fig. 20.3).

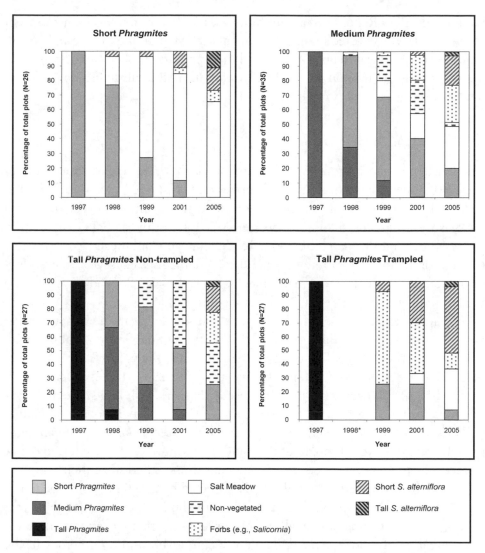

FIGURE 20.3. Changes in classification of *Phragmites* habitats due to tidal flow restoration at the Galilee Bird Sanctuary, Narragansett, Rhode Island, 1997–2005. Tidal flow was restored after the 1997 growing season. *Note:* Trampled tall *Phragmites* plots were not sampled in 1998.

Loss of Ruppia maritima

During the summer of 1997, immediately prior to tidal flow restoration, *Ruppia maritima* was found in all eleven ponds sampled in the Sanctuary; cover values ranged from less than 5 percent in three ponds to greater than 50 percent in three other ponds. Installation of enlarged culverts under the Escape Route permitted those ponds to drain at low tide, converting them from subtidal to intertidal habitats. By 1999 *Ruppia* was absent from two ponds, covered less than 1 percent of five ponds, and covered less than 5 percent of the remaining four ponds. No additional surveys have been conducted, but it appears unlikely that *Ruppia* remained in any of the ponds after 1999. Dewatering of the substrate at low tide was the most likely cause of *Ruppia* loss.

Response of the Animal Community

The frequency of occurrence of salt marsh snails in the Sanctuary dropped from 80 to 90 percent of quadrats in 1997 (prerestoration) to 50 percent in 1998 and zero in 1999 (postrestoration); density declined significantly between 1997 and 1998 as well. Baseline snail numbers were much lower in the Sanctuary than in the reference marsh, even in the preferred habitat type, salt meadow, presumably because of the nearly continuous waterlogging of the soils caused by tidal restriction. After the new culverts were installed, salt meadow was flooded every day, and the habitat became even less suitable for the pulmonate snails.

A major increase in the number of gulls and shorebirds in the Sanctuary was observed during the first three years (1998–2000) after tidal flow restoration; mudflats created when dredged material was removed from the northwestern corner of the Sanctuary were the principal habitat used. By 2001, however, salt marsh grasses had spread over more than half of the flats, and gull and shorebird numbers dropped nearly to prerestoration levels (F. Golet, pers. obs.). Migrating *Calidris minutilla* (least sandpiper) and breeding *Catoptrophorus semipalmatus* (willet) were observed throughout the Sanctuary within two years after tidal flow restoration, and their numbers have held steady or increased since then.

No change in waterfowl numbers was detected by 2000 that could be attributed to tidal flow restoration. Numbers of wading birds observed before and after restoration were comparable in all months except August; after restoration, *Ardea alba* (great egret) and *Egretta thula* (snowy egret) were more numerous in the Sanctuary and less numerous at the reference marsh in that month. The egrets may have simply moved across the Escape Route to take advantage of the high-quality habitat developing in the Sanctuary and to avoid disturbance caused by numerous clammers and other recreationists at Bluff Hill Cove.

Landbird species richness increased in the Sanctuary from twenty-six species per survey before restoration to thirty-two species afterward. Landbird numbers averaged 259 per survey before restoration and 523 afterward. This increase was due primarily to large numbers of *Sturnus vulgaris* (European starling), *Agelaius phoeniceus* (red-winged blackbird), and *Quiscalus quiscula* (common grackle) that foraged throughout the Sanctuary and roosted in the dead shrubs and *Phragmites* after restoration. Two species that bred in freshwater *Phragmites* stands prior to restoration, *Melospiza georgiana* (swamp sparrow) and *Cistothorus palustris* (marsh wren), were not observed afterward.

Population size and nest success of saltmarsh sparrows declined dramatically in the remnant salt marsh areas of the Sanctuary postrestoration. In southern New England, saltmarsh sparrows nest on the ground, primarily in salt meadow (DeRagon 1988; Benoit and Askins 1999). Once the tide gates were opened, the new regularly flooded water regime rendered this formerly key baseline habitat unsuitable for nesting. The annual nest success rate dropped from an average of 48 percent before restoration to 5 percent in 1998 (DiQuinzio et al. 2002). Between 1993 and 1997, flooding caused less than 33 percent of all nest failures; in 1998, the rate was 91 percent. These results were anticipated based strictly on the low elevation of the baseline salt meadow zone. Saltmarsh sparrow populations were also expected to recover—and even expand beyond baseline levels—once salt meadow had redeveloped at a higher elevation, in the zone that had been dominated by shrubs and high-elevation *Phragmites* stands prior to flow restoration. During bird surveys in 2007 and 2008, sizeable areas of salt meadow had begun to develop in the western and southern parts of the Sanctuary, where it had not occurred prior to restoration, and breeding saltmarsh sparrows were numerous in those areas.

Marsh Restoration in Stages

Research at Galilee has demonstrated that ecosystem response to tidal flow restoration occurs in discrete stages. These changes may be labeled short term (one to three years postrestoration), midterm (four to ten years postrestoration), and long term (more than ten years postrestoration). Variation in response between the Galilee Bird Sanctuary and other restoration sites will depend on differences in baseline habitats and the degree of change in tidal regime and soil salinity after flow restoration.

Short-term changes are prompted by the immediate increase in the frequency and duration of tidal flooding and in soil salinity. At Galilee, such changes involved (1) the loss of freshwater wetland plants and animals, as well as salt meadow animals such as the salt marsh snail and saltmarsh sparrow; (2) the loss of subtidal submergents such as *Ruppia maritima*; (3) severe stunting or death of *Phragmites*; (4) increased height of remnant *S. alterniflora*; (5) colonization of bare soil and trampled vegetation by salt marsh pioneers such as *Salicornia*; and (6) a sudden increase in saltwater fish, crabs, shorebirds, and wading birds.

Midterm changes included (1) death and collapse of *Phragmites* stems in areas with high soil salinity, followed by colonization by salt marsh pioneers; (2) full development of salt marsh vegetation in bare or trampled areas; (3) reemergence of salt meadow at higher elevations than before restoration; (4) emergence of salt marsh shrubs (*Iva frutescens* and *Baccharis halimifolia*) along the landward edge of the marsh; and (5) return of breeding saltmarsh sparrows.

Projected long-term changes include (1) full development of salt meadow, short and tall *S. alterniflora*, and salt marsh shrubs; and (2) continued increase, and eventual stabilization, of marsh-dependent animal populations.

Lessons Learned

Fourteen years of involvement in salt marsh restoration at the Galilee Bird Sanctuary have taught us some valuable lessons about restoration planning, partnerships, monitoring and adaptive management, and the need to keep the public informed. It is hoped that others will benefit from our experience.

Planning

- Preparation for on-the-ground restoration work (i.e., funding, planning, permitting, and design) can be a lengthy process. At Galilee, partners expected this phase to last about two years, but it took five. This prolonged period did allow us to develop an exceptionally strong, science-based design and a comprehensive baseline for assessing ecosystem response to tidal flow restoration; however, such an intensive effort may not be feasible, or necessary, for many restoration projects.
- Some degree of tidal restriction must persist after flow enhancement if bordering developed lands are to be protected from flooding. Self-regulating tide gates are an effective tool for achieving the necessary balance between restoration and protection.

Partnerships

- Close collaboration among major partners, whose roles have been clearly defined, is essential. At Galilee, frequent interaction at meetings and in the field among project scientists, engineers, and managers permitted development of a highly successful restoration plan that met ecological and management goals.
- Partnerships including a wide range of agencies assure that diverse expertise will be on hand for tackling the variety of technical and practical problems that restoration projects pose.

Monitoring and Adaptive Management

- Selection of a suitable, nearby reference marsh is highly desirable for development of appropriate ecological design criteria and evaluation of restoration success.
- Reintroduction of high-salinity water to tidally restricted wetlands greatly enhances prospects for restoration success. The high salinity (30 to 32 parts per thousand) of the tidal water reintroduced to the Galilee Bird Sanctuary assured elimination or severe stunting of *Phragmites* and reestablishment of target salt marsh habitats.
- Flattening *Phragmites* stems may accelerate the conversion to *S. alterniflora* or salt meadow, as long as the site experiences frequent flooding by high-salinity tidal water.
- If postrestoration monitoring is thorough, a need for adaptive management will almost certainly be identified, but additional construction costs can be minimized through application of sound science during restoration planning, design, and monitoring stages.

Communication with the Public

To guarantee public support for the restoration project, lead agencies need to inform local residents and public officials about project goals, methods, anticipated results, timelines, and precautions taken to protect bordering properties. The partners accomplished that at Galilee through a press conference, a public workshop, and a letter to neighbors of the Sanctuary before restoration; ad hoc field meetings with landowners over the course of the project; and a public celebratory event after tidal flow had been restored. Groundwater monitoring at residential properties ensured regular contact between project personnel and landowners and assured the latter group that protecting them from flooding was a high priority. Agency investment of more than $200,000 in self-regulating tide gates was the ultimate gesture to protect landowner interests and inspire their confidence in project success.

REFERENCES

Applied Bio-Systems. 1991. *Conceptual Design Report for the Galilee Bird Sanctuary Restoration/Enhancement Project*. Prepared for Rhode Island Department of Transportation, Division of Public Works, Providence, RI.

Benoit, L. K., and R. A. Askins. 1999. "Impact of the Spread of *Phragmites* on the Distribution of Birds in Connecticut Tidal Marshes." *Wetlands* 19:194–208.

Bertness, M. D. 2007. *Atlantic Shorelines: Natural History and Ecology*. Princeton, NJ: Princeton University Press.

352   COMMUNICATING RESTORATION SCIENCE

Bertness, M. D., and A. M. Ellison. 1987. "Determinants of Pattern in a New England Salt Marsh Plant Community." *Ecological Monographs* 57:129–47.

Boyes, D. 1997. "Phased Implementation of OMWM (Open Marsh Water Management) Principles in the Marsh Restoration Project at the Galilee Bird Sanctuary, Narragansett, RI." *Proceedings of the Northeastern Mosquito Control Association* 43:20–28.

Brinson, M. M., and R. Rheinhardt. 1996. "The Role of Reference Wetlands in Functional Assessment and Mitigation." *Ecological Applications* 6:69–76.

Cowardin, L. M., V. Carter, F. C. Golet, and E. T. LaRoe. 1979. *Classification of Wetlands and Deepwater Habitats of the United States.* FWS/OBS-79/31. Washington, DC: US Fish and Wildlife Service, Office of Biological Services.

DeRagon, W. R. 1988. "Breeding Ecology of Seaside and Sharp-tailed Sparrows in Rhode Island Salt Marshes." M.S. thesis, University of Rhode Island, Kingston, RI.

DiQuinzio, D. A., P. W. C. Paton, and W. R. Eddleman. 2002. "Nesting Ecology of Salt-marsh Sharp-tailed Sparrows in a Tidally Restricted Salt Marsh." *Wetlands* 22:179–85.

Lee, V. 1980. *An Elusive Compromise: Rhode Island Coastal Ponds and Their People.* Marine Technical Report 73. Narragansett: University of Rhode Island Coastal Resources Center.

Lefor, M. W., W. C. Kennard, and D. L. Civco. 1987. "Relationships of Salt-Marsh Plant Distributions to Tidal Levels in Connecticut, USA." *Environmental Management* 11:61–68.

Myshrall, D. H. A. 1996. "Influence of Tidal Hydrology on the Distribution and Structure of Salt Marsh Plant Communities at the Galilee Bird Sanctuary, Narragansett, RI." M.S. thesis, University of Rhode Island, Kingston, RI.

Myshrall, D. H. A., and F. C. Golet. 1993. *Baseline Inventory and Ecological Profiles of Coastal Habitats at the Galilee Bird Sanctuary, Narragansett, RI.* Final report prepared for Rhode Island Department of Environmental Management, Division of Fish, Wildlife, and Estuarine Resources, under Federal Aid in Wildlife Restoration Project No. W-23-R-32, Study No. III, Job 2. Wakefield, RI.

Myshrall, D. H. A., and F. C. Golet. 1996. *Baseline Ecological Survey of the Galilee Bird Sanctuary, Narragansett, RI.* Final report prepared for Rhode Island Department of Environmental Management, Division of Fish and Wildlife, under Federal Aid in Wildlife Restoration Project No. W-23-R-38/39, Study No. III, Job 4. Wakefield, RI.

Myshrall, D. H. A., F. C. Golet, P. W. C. Paton, and B. C. Tefft. 2000. *Salt Marsh Restoration Monitoring at the Galilee Bird Sanctuary, Narragansett, RI.* Final report prepared for Rhode Island Department of Environmental Management, Division of Fish and Wildlife, under Federal Aid in Wildlife Restoration, Rhode Island Avian Studies, Jobs 17–19. Wakefield, RI.

Niering, W. A., and R. S. Warren. 1980. "Vegetation Patterns and Processes in New England Salt Marshes." *BioScience* 30:301–7.

Raposa, K. 2002. "Early Response of Fishes and Crustaceans to Restoration of a Tidally Restricted New England Salt Marsh." *Restoration Ecology* 10:665–76.

Shaffer, M. 2007. "Mid-term Habitat Response to Tidal Flow Restoration at the Galilee Bird Sanctuary, Narragansett, RI." Master of Environmental Science and Management major paper, University of Rhode Island, Department of Natural Resources Science, Kingston.

Sinicrope, T. L., P. G. Hine, R. S. Warren, and W. A. Niering. 1990. "Restoration of an Impounded Salt Marsh in New England." *Estuaries* 13:25–30.

U.S. Army Corps of Engineers. 1994. *Galilee Salt Marsh Restoration: Feasibility (Section 1135) Report and Environmental Assessment.* Waltham, MA: US Army Corps of Engineers, New England District.

Chapter 21

Restoration of Tidally Restricted Salt Marshes at Rumney Marsh, Massachusetts

Balancing Flood Protection with Restoration by Use of Self-Regulating Tide Gates

EDWARD L. REINER

Low-lying coastal floodplain environments often experience extreme tidal events. When such high water conditions threaten improved property and highways, the property owners and government agencies often construct a system of earthen dikes, drainage pipes, and tide gates in an effort to exclude damaging tidal flood waters from the protected area while providing for the discharge of storm water from the upland watershed. Tide gates are simple gravity-operated mechanisms typically mounted on pipes through dikes and used to control the flow of water in coastal environments. Conventional tide gates have a weighted valve or one-way gate that is hinged at the top, allowing the discharge of runoff from a protected inner area when interior water levels are greater than exterior water levels, while preventing the return flow of incoming tides. When used in estuarine environments, these gates drain upstream wetlands, obstruct fish migration, increase sedimentation by blocking the scouring action of tides, and decrease salinity. This exclusion of the tide often leads to favorable conditions for the colonization of salt marshes by *Phragmites australis* (common reed). Dense stands of *Phragmites* pose fire hazards near developed property, can impair drainage, and result in increased freshwater flooding and mosquito breeding upstream of conventional tide gates.

Installing bidirectional-flow tide gates, such as the self-regulating tide gate (SRT), can improve drainage conditions, increase saline tidal flow, help control *Phragmites*, decrease fire hazards and mosquito breeding, restore normal salt marsh biotic assemblages, and protect adjacent development from flooding. The SRT has a top-hinged buoyant valve that opens with an incoming tide to allow saltwater to flow into the protected area and restore tidal wetlands. The buoyant valve is counterpoised with adjustable back floats whose greater buoyancy counteracts the buoyant valve so as to close the gate at a predetermined adjustable

355

FIGURE 21.1. Self-regulating tide gate (SRT), Route 1A in Revere, Massachusetts, adjacent to the Pines River. Unlike conventional flapper tide gates, which provide drainage flow only through a top-hinged structure, the SRT is a float-actuated water control valve that uses a bottom float to open the gate on incoming tides. A second set of adjustable floats (the round floats in this photo) on the valve allows the gate to close for flood protection. (Photo courtesy of Edward Reiner).

water level for flood protection of interior areas. Similar to conventional tide gates, on ebb tide the hydraulic force of the water pushes the gate open allowing for upstream drainage (fig. 21.1).

Prior to restoration actions, Rumney Marsh, located in Revere, Saugus, and Lynn, Massachusetts, had twenty-one missing, poorly functional, or nonfunctional tide gates with up-gradient wetlands. The tide gates adversely affected more than 45 hectares of wetlands. To restore or enhance salt marsh ecology, as well as provide flood protection, eleven SRTs were installed between 1997 and 2001 to provide controlled tidal flow to approximately 32 hectares of wetlands. While flood protection has improved with the installation of these new tide gates, numerous problems such as lack of maintenance of the tide gates and culverts, engineering errors, vandalism, and noncompliance with permit conditions have limited the success of these efforts. Continued effort is needed to correct deficiencies and achieve the desired flood protection and marsh restoration benefits. The lessons learned from the successes and failures of the Rumney Marsh project will help direct future restoration efforts, especially those in urban settings.

Tide gate installation efforts are part of a comprehensive salt marsh restoration plan that also includes fill removal and excavation, culvert and bridge replacement, open marsh water management for mosquito control, and other potential tide gate modification projects (EOEA 2002). The Rumney Marsh case study describes completed restoration projects involving the replacement of conventional tide gates with SRTs and highlights other tidal restrictions in the marsh.

Rumney Marsh Background

Situated at the lower end of the 122 square kilometer watershed of the Saugus and Pines Rivers, Rumney Marsh is the largest remaining salt marsh in the metropolitan Boston area. The mean tidal range is 2.9 meters, and the mean spring tide range is 3.4 meters. On its east side, the marsh is bounded by Revere Beach, a developed barrier beach. The Saugus Iron Works National Historic Site is located at the upper limit of the estuary, approximately 7.2 kilometers from the mouth of the Saugus River. America's first successful ironworks, the Saugus Iron Works, operated from 1646 to 1668. With more than 350 years of development history since the arrival of English colonists in the seventeenth century, Rumney Marsh provides a good case study for the examination of how the wetlands in this urban watershed have been impacted and the challenges and difficulties affecting restoration projects.

Despite its highly urbanized setting, Rumney Marsh supports an abundance of natural resources, including resident and migratory fish, shellfish, shorebirds, wading birds, and waterfowl. In recognition of the importance of protecting these resources, the Commonwealth of Massachusetts in 1988 designated the majority of the marsh an Area of Critical Environmental Concern. A comprehensive evaluation characterized the estuary as containing approximately 668 hectares of wetlands, including intertidal and subtidal areas. Vegetated wetlands totaled approximately 433 hectares, including 383 hectares of high salt marsh and 46.5 hectares of low salt marsh. Approximately 57 hectares or 15 percent of the high salt marsh was vegetated with *Phragmites australis*, largely due to tidal restrictions caused by rail and roadway construction (USACE 1988). Approximately 38 hectares or 67 percent of the *Phragmites* areas were in wetlands impacted by tide gates.

Historical Impacts

Construction of roads and railroads in the nineteenth century as well as more recent road projects segmented the marsh and affected tidal hydrology. More than 120 hectares at the mouth of the estuary were filled for the Lynn Harbor Improvement and Nuisance Abatement project permitted in 1928. Landfills in Saugus

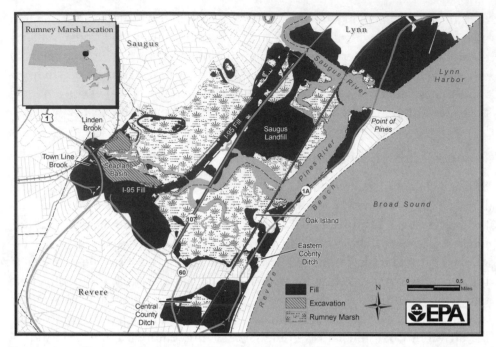

FIGURE 21.2. Rumney Marsh historical wetland impacts 1803–2007. Approximately 509 hectares, 43 percent of the marsh, has been filled, as shown in black. Cross-hatched area depicts 35 hectares of wetland excavation where salt marsh was converted to intertidal habitat. The upper estuary area is not depicted. An estimated 10 hectares of I-95 fill removal for restoration projects undertaken since 1990 are not included. Note: A detailed and informative color aerial map of the site is located on the US Environmental Protection Agency Rumney Marsh Information page (http://www.epa.gov/region1/eco/rumneymarsh/).

filled approximately 80 hectares of marsh. Figure 21.2 shows Rumney Marsh and the historical impacts described here. More specific information about some of the historical impacts and effects on tidal hydrology of the marsh is explained in this chapter, depicted in figure 21.3, and documented in tables 21.1 and 21.2.

Road and Rail Construction

In 1803, the Salem Turnpike (Rt. 107) was constructed across the marsh providing stagecoach service, later a horse railway, and in 1860 an electric trolley car between Salem and Boston. This road roughly bisected the marsh in a north–south direction. A 3.7 kilometer embankment restricts tidal flows across the marsh by three relatively small bridges at the major waterways and creeks of the Pines River.

The marsh was bisected again by embankment fill for the construction of the Eastern Railroad in the 1830s. This prevented tidal flow across the marsh except at a small granite culvert at Diamond Creek in Revere. An embankment fill con-

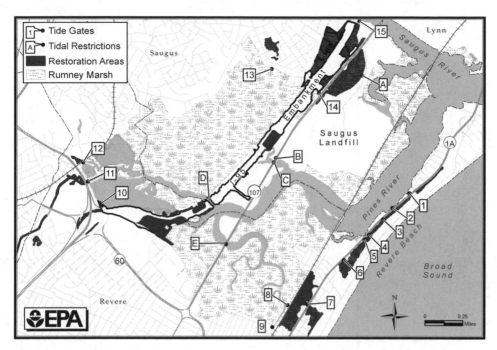

FIGURE 21.3. Tidal restrictions and associated restoration areas in Rumney Marsh. Tide gates and their associated tidally restricted wetlands shown at fifteen numbered locations as listed in table 21.1. Details of the additional tidal restrictions A–E refer to table 21.2. Approximately 10 hectares of I-95 fill removal restoration areas are also depicted. The Central County ditch wetland associated with tide gate 1 is not depicted (see fig. 21.2).

nection between the Eastern Railroad and the Boston, Revere Beach, and Lynn Railroad, which operated along Revere Beach from the 1870s to 1930s, created an additional tidal restriction near Oak Island along the Pines River. The construction of the Saugus Branch Railroad in the 1850s also created a tidal restriction at the western edge of the marsh by Linden Brook (fig. 21.2).

Tide Gates and Drainage Projects

Adverse effects of tidal restrictions from railroad construction were compounded by the installation of conventional tide gates in the early 1900s on railroad culverts. Excavation of creeks or ditches behind these tide gates drained wetlands, which were filled and sold for residential and commercial development, attracted to the area by the convenient railroad access. Approximately 74 percent of the wetlands associated with the Eastern and Central County Ditch drainage projects upstream of the Eastern Railroad tide gate were filled. The remaining wetlands account for approximately 20 hectares of the *Phragmites*-dominated, tidally

TABLE 21.1

Wetlands affected by tide gates in Rumney Marsh

Site		Wetland hectares	Prerestoration gate condition	Culvert size	Restoration action year
Route 1A, 6 locations (map 1–6)[1]	1–6	6	Three missing, three nonfunctional	61 cm dia.	6-SRTs installed 2000
Eastern County Ditch (Rte 1A)	7	2.8	Leaking	183 × 122 cm elliptical	213 cm × 107 cm twin box culvert and tide gate removed 2001
Eastern County Ditch	8	7.7	Leaking	142 × 152 cm	SRT on new 183 cm culvert 2003
Central County Ditch	9	9.3	One missing, one broken (in tandem)	152 cm dia.	SRT installed 1997
Copeland Circle	10	.97	Leaking	107 cm dia.	None
Route 1 Town Line Brook	11	6.9	Two broken, four leaking, ten leaking stop-log culvert bays	305 × 152 cm 8 culverts with 16 openings	Three-SRTs installed 2001; eight new tide gates; five stop-log bays
Linden Brook	12	.3	Leaking	137 × 122 cm	None
Seagirt Avenue	13	1.5	Missing	61 cm dia.	None
Bristow Street	14	4.2	Missing	221 × 163 cm	Temporary blocked
Ballard Street	15	12	Leaking	122 cm dia.	SRT proposed

Note: Fifteen wetland areas affected by tide gates and their condition in 1995 are listed with estimated wetland area, culvert size, and restoration action if any.

[1] Map number corresponds to numbered locations in figure 21.3.

TABLE 21.2

TABLE 21.2

Other tidal restrictions in Rumney Marsh

Location	Wetland hectares	Condition size	Planned improvements	Restoration action
Crescent Marsh (map A)[1]	9.3	Nonfunctional 122 cm CMP[2,3]	New culvert (RCP)	Future project
Route 107 drainage (map B)	.2	Nonfunctional 122 cm CMP	Culvert replacement	Complete
Rt. 107 E. Branch Pines River (map C)	> 16	6.7 m bridge	New bridge–13.7 m at MHW	Complete
I-95 embankment Pines River (map D)	180	48 m channel at MHW[4]	None	None
Rt. 107 Diamond Creek (map E)	> 40	7.6 m bridge	New Bridge–13.7 m at MHW	Complete

Note: Tidal restrictions, other than tide gates, in Rumney Marsh include nonfunctional culverts, bridges on Route 107, and the river channel embankment. Estimated wetland area affected and planned improvements, if any, are listed.
[1]Map letter corresponds to lettered locations in figure 21.3.
[2]Nonfunctional = significant obstructions due to collapse observed or suspected.
[3]CMP, corrugated metal pipe; RCP, reinforced concrete pipe.
[4]Dimensions from Corps permit file, revised plan dated September 1966. 48 meters wide at mean high water and 23 meters wide at mean low water.

restricted wetlands in Rumney Marsh. Upstream of the Saugus Branch Railroad tide gate at Linden and Town Line Brook approximately 85 percent of the former wetlands were also filled, leaving only about 4 hectares of *Phragmites*-dominated wetlands in this area (fig. 21.3).

Interstate 95 Project

Between 1967 and 1969, approximately 49 hectares of marsh were filled, creating an embankment nearly 3.9 kilometers long for the Interstate 95 (I-95) project, which was never completed (Commonwealth of Massachusetts 1995; fig. 21.2). The original meandering Pines River channels and the east branch of the Pines River were completely filled and replaced with a rock-armored Pines River channel (fig. 21.3; location D). With the exception of this new Pines River channel, the I-95 embankment formed a continuous barrier across the marsh, affecting tidal flow and drainage patterns and causing a tidal restriction to approximately 180 hectares of wetlands located west of the embankment (MDC 1987; USACE 1989a). At its northern end, approximately 12 hectares of salt marsh near Ballard Street are tidally restricted by tide gates and are now dominated by *Phragmites*.

Since the 1980s, several construction projects have been allowed to remove portions of the abandoned I-95 fill for beneficial reuse of the gravel material. Four projects, completed in the 1990s as mitigation for aquatic resource impacts outside of the marsh, restored approximately 10 hectares of salt marsh or intertidal habitat by fill removal and grading. While the height and width of the embankment have been reduced with these efforts, full removal or breaching of the embankment may not occur due to the belief that it provides some flood protection (USACE 1989b).

Quantifying Wetland Loss

According to studies by the US Army Corps of Engineers (the Corps), the ongoing development of Lynn, Saugus, and Revere between 1951 and 1971 resulted in a total of 248 hectares of salt marsh loss, about 33 percent of salt marsh resources over that twenty-year period (USACE 1989a). The estimated cumulative wetland loss in the estuary for the years between 1803 and 2007 is depicted in figure 21.2. Fill areas encompass approximately 509 hectares of salt marsh and intertidal and subtidal area or 43 percent of the total estimated acreage of the 1177 hectare estuary area. An additional 35 hectares of salt marsh were altered by excavation for gravel and construction of the Seaplane Basin, adjacent to the Revere Airport (now part of Northgate Shopping Center), and to relocate the Pines River and construct drainage channels associated with the abandoned I-95 project.

Massachusetts enacted the first wetland protection legislation in the nation in 1963. Coastal wetland filling, however, was commonly permitted until the regulations for the protection of coastal wetlands became effective in 1978. These regulations have nearly halted legally permitted losses of salt marsh.

Tide Gates, Flood Protection, and Marsh Restoration

The February Blizzard of 1978 caused extensive flood damage in the Saugus and Pines Rivers estuary, in part due to storm surges conveyed up the estuary and waves overtopping a seawall along Revere Beach. Measures to reduce flood damage in the estuary included a proposed Regional Saugus River Floodgate plan, which included tidal floodgates at the mouth of the Saugus River (USACE 1989b). A component of this plan included maintenance of tide gates throughout the estuary (fig. 21.3). The Environmental Protection Agency (EPA) specifically recommended consideration of alternative tide gates such as the SRT in several areas of the marsh to protect and restore up-gradient wetlands. Table 21.1 lists fifteen sites as shown in figure 21.3, where conventional tide gate repair or maintenance would have affected more than 45 hectares of salt marsh, and the restoration actions completed to date. Restoration actions involving additional tidal restrictions in the marsh are documented in table 21.2.

Central County Ditch Tide Gate

The Central County Ditch tide gate (fig. 21.3, location 9) illustrates several difficulties balancing flood control and marsh restoration. In 1979, its poorly functioning condition allowed sufficient tidal flushing to stunt *Phragmites* in the brackish Central County Ditch wetland (fig. 21.2). The conditions of a 1981 Corps permit for a large warehouse development project included grading 9.9 hectares to elevations suitable for salt marsh growth and a requirement to maintain the tide gate in an open position for tidal flow. When the warehouse construction was complete, the city of Revere was concerned about potential flooding and requested that the Corps allow the tide gate to operate conventionally, for drainage only. The Corps agreed with the request. With the conventional tide gate operating, salt marsh restoration was not possible, and the excavated wetland area colonized with *Phragmites* (Reiner 1989).

During October 20–21, 1996, more than 25 centimeters of rain fell in a thirty-six-hour period causing extensive interior flooding behind several tide gates in Revere on the County Ditch drainage system. EPA prepared a report describing the problems, which included undersized culverts, debris such as *Phragmites* stems, tires, and other refuse at trash racks (grates to protect culvert inlets from debris), and drainage channels obstructed by *Phragmites*. EPA recommended clearing the channel of debris and dense *Phragmites* and replacement of tide gates with SRTs to increase salinity and control *Phragmites*, improve drainage and flood control, and restore the salt marshes. An inspection revealed that a tide gate at the Central County Ditch outlet had a broken hinge that required immediate attention. Rather than repairing the tide gate, EPA, working along with the Corps, the City of Revere, and the landowner of the Central County Ditch wetland, developed a plan to replace the broken tide gate with a Waterman/Nekton SRT. Invented by Thomas J. Steinke and first used in Connecticut in 1980, this invention had restored tidal flow and provided flood protection at numerous installations in Connecticut and has since been used in several other states (Steinke 1995). The SRT, which was the first in Massachusetts, was installed by the City of Revere in October 1997 on the outlet of the Central County Ditch tide gate vault (fig. 21.3, location 9).

The damaged tide gate allowed bidirectional tidal flow to the wetland, resulting in significant *Phragmites* die-off due to the increased salinity. By the summer of 1997, half of the wetland consisted of unvegetated areas. During the first and second growing seasons following installation of the SRT, *Salicornia* sp. (glasswort) colonized most of the unvegetated areas where the *Phragmites* had died. By the third growing season, those areas started colonizing with *Spartina* sp. (cordgrass).

Although this initial success was promising, half of the wetland area was still dominated by stunted *Phragmites*. EPA recommended adjustments to the SRT closure setting to achieve a higher tidal water level; unfortunately, City staff was rarely available for such efforts due to other priority work. In June 2004, the bottom float of the SRT was damaged by an accumulation of loose riprap stones beneath the gate, which had apparently been thrown at the structure. A new float was provided; however, the City did not want to install it until they removed the riprap at the SRT. Four years later, the SRT continued to be operated as a conventional tide gate without the bottom float. In 2008, the aluminum parts of the SRT were stolen and sold for scrap metal—a challenge of restoration in an urban setting.

The potential for successful restoration of salt marsh at the Central County Ditch wetland was first diminished by the municipality's resistance to adjust the SRT beyond the initial conservative closure setting and later impacted by vandalism and theft, as well as failure to promptly address repair and maintenance. In the seven years that the SRT has not been used as intended for bidirectional tidal flow, *Phragmites* has once again taken over the wetland area. The City, encouraged by EPA to repair the SRT, contracted for the purchase and installation of replacement parts in 2009. Repairs were completed in 2010; however, the City was still operating the tide gate without a bottom float in 2012. The weight of the back floats create a slightly buoyant valve cover on an incoming tide, allowing some upstream tidal water flow. The hydraulic forces of an incoming tide close the gate before the flotation effect of the back floats would ordinary close the gate as intended by the manufacturer.

Route 1A Tide Gates

Following the October 1996 flood in Revere, local and state interest in repairing or replacing tide gates increased. Repairs to tide gates on six 61 centimeter culverts, which would affect 6 hectares of wetlands between Revere Beach and Route 1A, were proposed by the Massachusetts Highway Department (MassHighway) at the request of the City (fig. 21.3, locations 1–6). To fully consider the potential impacts to wetlands from tide gate repair or replacement, EPA documented that three of the six tide gates were missing and three were not functioning due to sediment or debris obstructions. Since existing salt marsh would have been adversely impacted if the missing or nonfunctioning tide gates were replaced with new conventional tide gates, at the request of the Corps, National Marine Fisheries Service, and EPA, MassHighway modified the project to incorporate SRTs.

Permit conditions required that all six new SRTs close at a specified elevation, which was approximately 60 centimeters above mean high water (MHW). This would ensure no adverse impacts to the existing salt marsh, restore a salt marsh overgrown by *Phragmites*, and protect adjacent developed property. The state and federal permits required monitoring SRT operation and performance relative to achieving the specified tidal heights in the marsh. To ensure adequate property flood protection the permit allowed for manual closure of the SRTs in advance of predicted storms. Installation of the new SRTs was completed in 2000.

The project has improved flood protection for the developed property surrounding the marsh and successfully protected the existing salt marsh for one section of marsh that is drained by four culverts with SRTs (fig. 21.3; locations 1–4). The expected restoration of salt marsh in another section of the marsh drained by two culverts with SRTs (fig. 21.3; locations 5–6) has not been as successful for a variety of reasons. MassHighway did not comply with the permit monitoring and maintenance requirements. A 2005 vandalism incident, which required repairs of a broken SRT float arm, had not been addressed until 2009 after several meetings between the City of Revere, MassHighway, EPA, and the Corps. The initial SRT closure elevation for the section of marsh with only two drainage culverts was lowered from approximately 60 centimeters above MHW to approximately 60 centimeters below MHW in order to avoid flooding several low-lying properties, which were not identified during permit review. The amount of tidal water inundating this marsh rarely exceeded the creek bank elevations and was not sufficient to restore the salt marsh, which was still dominated by *Phragmites* in 2012.

Inspections by MassHighway and EPA in November 2008 revealed several culvert obstruction problems that needed to be addressed. One culvert (fig. 21.3; location 2) was obstructed by riprap, which appeared to have fallen from its position on the slope at the inlet. A second culvert (fig. 21.3; location 3) had a crushed pipe outlet (located on private property), which was not repaired during construction. A third culvert (fig. 21.3; location 6) appeared to be completely nonfunctional with no indication of flow and a buried outlet. During the Corps permit review for the project, EPA, having observed similar culvert obstructions in 1997, had recommended consideration of installing larger culverts with SRTs for tidal flow and marsh drainage. Replacing any of these culverts would likely be difficult and costly to construct due to the presence of buried utilities. Use of the existing culverts, which MassHighway thought could be repaired or cleaned of debris, was less expensive and avoided causing traffic delays. These are additional challenges of restoration in an urban setting. In 2011, further vandalism was reported to MassHighway; the alumimum grates of the four concrete vaults were stolen, and the SRT back floats were broken or stolen from two units. The intended salt marsh restoration and flood control benefits will not be fully achieved at this marsh area until culvert obstructions are removed, faulty culverts are replaced, and necessary repairs and maintenance are implemented.

Town Line Brook

The October 1996 flood also prompted efforts to repair six leaking, nonfunctional tide gates and ten leaking "stop log" structures at Town Line Brook (fig. 21.3, location 11). Similar to the repair project at Route 1A, replacement of the failed structures with conventional tide gates (initially proposed by MassHighway) would have adversely impacted upstream salt marsh areas. MassHighway prepared a coastal wetland inventory and self-regulating tide gate analysis for the project in 1998 and agreed to plan for three new SRTs to sustain upstream salt marsh and eight new tide gates for increased drainage. The state and federal permits issued to MassHighway and Massachusetts Department of Conservation and Recreation required both short- and long-term monitoring in order to calibrate the SRT settings and determine their effectiveness at restoring salt marsh and controlling flood levels. The permit included specific summer and winter settings for the SRT closure elevations that could be modified based on the monitoring.

For numerous reasons, the anticipated ecological restoration benefit of this project has not been fully achieved. The initial monitoring and reporting for the first thirty days of operation to calibrate the SRT operation were not performed. Occasional inspections by EPA documented that the SRTs were closing before water levels on the interior flooded the wetland areas outside of the immediate channel bank. The long-term monitoring required annual reporting for three years to document any changes in the characteristics of the aquatic, wetland, and floodplain conditions in the project area as compared to the 1998 baseline report, but this monitoring was not performed.

In 2003 EPA discovered that one of the three SRTs was broken and a second was likely damaged due to a manufacturing defect. To protect upstream areas from flooding, "stop logs" were installed in two SRT culverts and the third SRT was modified to function for drainage flow only by removal of the bottom float. Prompt repairs to the defective SRTs were not possible because the manufacturer had filed for bankruptcy. The manufacturer emerged from bankruptcy with a new owner and delivered the three replacement SRT parts to the Revere Department of Public Works in 2008 at no charge. In 2009 MassHighway contracted for the installation of the replacement parts and other necessary repairs. The three SRTs were operational in August 2009, but in response to concerns of the City of Revere for adequate flood protection the SRTs' initial closure elevation was adjusted to a more conservative lower level. MassHighway agreed to perform appropriate monitoring, and the Department of Conservation and Recreation will develop a formal agreement for operation and maintenance with the City. At the adjusted SRT closure elevations, any salt marsh restoration beyond the channel of the brook will likely require excavation to lower the elevations within the drained wetland/floodplain. A bottom float on one SRT was lost in 2011. Recommended measures to maintain, adjust, and monitor tide gate settings have yet to be implemented.

Eastern County Ditch Tide Gate

The 11th SRT in Rumney Marsh was installed by the City of Revere to replace the Eastern County Ditch Diamond Creek tide gate (fig. 21.3; location 8) as part of the Oak Island Salt Marsh Restoration project. This project, which was conceived in 1996, permitted in 2000–01, and constructed in 2003 at a cost of more than $1 million, was intended to improve drainage and restore up to 10.5 hectares of salt marsh associated with the Eastern County Ditch. Funding for this project was provided partly by grants from several state and federal agencies. The most difficult and expensive part of the project involved installation of a new 183 centimeter diameter culvert, which had to be hydraulically jacked under an active railroad. Although the agencies involved had expected another successful project, several significant problems have prevented that success.

Without fully understanding the SRT and its design requirements, the engineering company designed a concrete vault outlet structure for the new SRT. During construction, it was discovered that the SRT would not fit in the vault. The vault and the SRT were modified to fit, but problems resulted. The SRT, an elegant design, could not be set for the closure elevation required to avoid property flooding due to a lack of any lateral clearance for side-mounted SRT floats.

The City of Revere replaced the SRT in 2011 with an electrically operated automatic combination sluice-flap gate. Due to the presence of low-lying developed property surrounding this marsh, the level of tidal inundation will need to be carefully controlled. Water level monitoring is being collected to determine appropriate settings for the automatic tide gate. While decades of drainage from the conventional tide gate at this site caused some wetland subsidence as the organic matter decomposed in the oxidized soils, higher areas within the wetland will need to be graded to suitable elevations to promote salt marsh plant establishment. Such efforts are intended to be accomplished in part through off-site wetland mitigation agreements with other parties. Water level monitoring data gathered since the tide gate became operational will be used to determine appropriate grading plans for future marsh restoration.

Lessons Learned and Recommendations

The installation of 11 SRTs to replace missing or nonfunctional conventional tide gates in Rumney Marsh provided innovative technology, the goal of which was to balance flood control needs with salt marsh restoration at approximately 32 hectares of wetlands. Flood control has been improved over the previous broken or missing tide gate situations at all the installations. The ability to provide bi-directional tidal flow as compared to the continued use of conventional tide gates is a major accomplishment and success story for Rumney Marsh. The key justification for restoration, which initially garnered municipal acceptance of the SRTs, was the benefit of saline tidal flow to control *Phragmites* growth, which had repeatedly caused both fire and flooding problems in the marshes. The first SRT installation at Central County Ditch was originally quite successful in controlling *Phragmites* and reducing the fire threat and flooding problems. Further ecological improvements in the marsh could have been made by adjusting the SRT closure heights to increase tidal water inundation in the marsh using an adaptive management approach. However, where developed property is at risk of flooding, flood control will always be a priority over marsh restoration at the municipal level.

The SRT design incorporates several features that help to alleviate the dichotomy between wetland restoration and flood control. In storm conditions the associated low barometric pressure causes the ocean water level to rise more quickly than normal. This causes the SRT to close sooner having passed less ocean water than normal, thereby providing increased volume for storing stormwater runoff during the coincident high tide and heavy rain associated with the coastal storm. Additional interior flood storage can also be provided in advance of storms by pulling a door-float release rod to convert the gate to a one-way flap valve. If set to a proper elevation for closure and if maintained at this setting, the SRT can achieve flood protection and provide adequate tidal flow to sustain/restore salt marsh habitat.

Successful use of SRTs requires an understanding of all critical field and design factors and operational requirements of this new technology. Regulatory and funding agencies, as well as applicants and consultants, should thoroughly examine permit application plans and specifications to ensure that these factors and requirements are evaluated. Installations should be appropriately designed to accommodate the full function and adjustability of the SRTs.

Like other mechanisms operating in a harsh environment, SRTs require periodic inspection and cleaning to ensure proper functioning. The need for SRT inspections, maintenance, and adjustments requires a commitment from the applicant or municipality expected to maintain them. Operation and maintenance plans for these projects should help define these responsibilities, and dedicating staff to such efforts is recommended. Use of instruments for recording water level would help in the evaluation of SRT performance and the adjustment of the closure elevations and would reduce the need for frequent inspections. Small-size riprap near culverts and tide gates should be avoided since the material is easily dislodged or vandalized and can block culverts as well as damage or interfere with tide gate operation. Special measures such as locked fences or grates on access vaults are recommended for urban environments to deter vandalism or theft of aluminum components. Culvert obstructions, vandalized or broken tide gates all require prompt repairs or replacement. Having extra SRT parts on hand and budgeting for repairs are recommended.

For systems where development of properties and infrastructure is adjacent to tide-restricted marshes, SRTs provide a valuable tool to aid in restoration while protecting improved property. A successful project is possible, but only with diligent measures by all.

Acknowledgments

The author gratefully acknowledges the assistance of Dianne Gould of EPA Region 1, Kenneth Finkelstein, of the National Oceanographic and Atmospheric Administration, Mary Anne Thiesing of EPA Region 10, and Mr. Thomas Steinke for their contributions to this chapter.

This chapter was prepared by the author as part of his official duties for the US EPA. However, it has not been reviewed from a policy standpoint, and it does not necessarily express any official position of the EPA.

REFERENCES

Commonwealth of Massachusetts. 1995. *Central Artery (I-93)/Tunnel (I-90) Project, Rumney Marsh Intertidal Mitigation Site, Supplemental 25 % Mitigation Design Report.*

Executive Office of Environmental Affairs (EOEA). 2002. *Rumney Marshes Area of Critical Environmental Concern Salt Marsh Restoration Plan.* Commonwealth of Massachusetts: Executive Office of Environmental Affairs.

Metropolitan District Commission (MDC). 1987. *Final Environmental Impact Report Revere Beach Erosion Control Project.*

Reiner, E. L. 1989. "The Biological and Regulatory Aspects of Salt Marsh Protection, Restoration and Creation, in Massachusetts." University Microfilms International no. 1338978. M.S. thesis Northeastern University, Boston, MA.

Steinke, T. 1995. "The Self-Regulating Tidegate." Unpublished literature provided to the Corps of Engineers. Fairfield, CT: Nekton.

US Army Corps of Engineers (USACE). 1988. *Final Report Wetland-Estuary Assessment for the Saugus River and Tributaries, Flood Damage Reduction Study.* Prepared by IEP, Inc., Northborough, MA.

US Army Corps of Engineers (USACE). 1989a. *Feasibility Report and Draft Environmental Impact Statement/Report Water Resources Investigation Saugus River and Tributaries, Lynn, Malden, Revere, and Saugus, Massachusetts Flood Damage Reduction.* Vol.

8, Appendix K–Environmental. Waltham, MA: US Army Corps of Engineers, New England District.

US Army Corps of Engineers (USACE). 1989b. *Feasibility Report and Final Environmental Impact Statement/Report Water Resources Investigation Saugus River and Tributaries, Lynn, Malden, Revere, and Saugus, Massachusetts Flood Damage Reduction.* Waltham, MA: US Army Corps of Engineers, New England District.

Summary

Chapter 22

Salt Marsh Responses to Tidal Restriction and Restoration

A Summary of Experiences

DAVID M. BURDICK AND CHARLES T. ROMAN

People have had many reasons to build structures across tidal marshes that restrict tides, including agriculture, pasture, salt works, flood prevention, transportation, and access to uplands. Impacts to tidal marshes caused by tidal restrictions vary by type of restriction and its severity, as well as the geomorphology of the system. In most cases tidal flow can be restored, at least partially, and through monitoring, the physical and biological responses of the recovering marshes have been found to be predictable. Restoration of tidal exchange is important to restore various ecosystem services, such as essential fish habitat, and to reestablish a self-maintaining system. Unfortunately, tidal restrictions have enabled landowners and developers to build dwellings and infrastructure within areas landward of barriers that were once flooded by the highest tides. As documented by Reiner (chap. 21, this volume), as well as Adamowicz and O'Brien (chap. 19, this volume), restoration of such systems has been only marginally successful due to development within or adjacent to the marsh.

This concluding chapter reflects on what we have learned about marsh functions from research, especially research associated with tidal restoration, and discusses research avenues to aid in the protection and restoration of tidal marshes into the future. Drawing from case studies of specific projects and programs, we discuss lessons learned from the practice of tidal restoration in New England and Atlantic Canada, from programmatic structure to partnering to communication. The experiences of planners, managers, and scientists are discussed to guide development of salt marsh tidal restoration programs in other regions of the world.

Tidal Marsh Development and Persistence

Tidal marshes are a product of physical and biological processes dominated by hydrology (flooding, with salts and sediments) and perennial grasses (primarily *Spartina* species along the Atlantic coast). Early descriptive studies of the structure and function of tidal salt marshes (Miller and Egler 1950; Chapman 1960; Teal 1962; Ranwell 1972; Redfield 1972) helped ecologists focus on the underlying coastal processes that result in these specific, recognizable ecosystems. Redfield's (1972) description of marsh development in postglacial New England especially led to research devoted to understanding marsh development and persistence. Scientists regard tidal marshes as poised systems that depend upon vascular plants to mediate dynamic physical forces and result in an ecosystem that can sustain itself. Typically located between relatively steep uplands and tidal flats, tidal marshes maintain elevation of the marsh plain as sea level rises through a negative feedback system (fig. 22.1). Increased flooding leads to more sediment trapping and peat accumulation, resulting in marsh elevation growth relative to sea level. The self-maintenance ability of tidal marshes is remarkable and has emerged as the dominant paradigm as our understanding of processes leading to marsh stability and persistence has increased. Ideas that marshes are temporary stages in hydrarch succession such as the sequences proposed by Chapman (1960) have been displaced, though such sequences do occur infrequently (e.g., along uplifting coasts; Ranwell 1972).

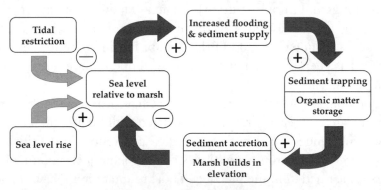

Salt Marsh Self-Maintenance Model
(a negative feedback system)

FIGURE 22.1. Feedback system regulating tidal marsh persistence in the face of two drivers: sea level rise (up to 5 millimeters per year) results in greater (positive) sea level relative to the marsh surface, whereas a tidal restriction typically reduces sea level relative to the marsh surface. Lower sea levels associated with tidal restrictions lead to less flooding and sediment delivery resulting in less sediment trapping and organic matter storage (typically a net loss of organic matter storage) and therefore less accretion (typically elevation loss, called subsidence).

Marshes are a bit more complex than presented in figure 22.1. Low marsh areas typically dominated by *Spartina alterniflora* (smooth cordgrass) that form at low elevations receive greater flooding. If they are protected from physical exposure and supplied with ample suspended sediment, low marshes build rapidly in elevation and become high marsh, typically dominated by *Spartina patens* (salt meadow cordgrass). On the other hand, if sea level rises quickly and sediment supply is low, high marsh can revert back to low marsh (Donnelly and Bertness 2001) or tidal flat (Fagherazzi et al. 2006). Fringing marshes are common in New England (Morgan et al. 2009), and many appear to migrate landward without leaving behind a record of peat deposits as erosion at the seaward face is balanced by transgression. Fringing marshes have rarely been the focus of tidal restoration projects, but they should be considered as special cases that typically endure greater physical exposure and may not follow the conceptual model in figure 22.1.

A primary driving force in development of tidal restoration programs is the recognition that the rich ecosystem services provided by tidal marshes (as well as their ability to perpetuate themselves), depend upon tidal flooding. On a decadal scale, sea level rise provides for greater tidal flooding and development of peat deposits. Early research in tide-restricted marshes has shown that elimination of flooding and soil waterlogging actually results in decreased elevation, termed subsidence (Roman et al. 1984; Frenkel and Morlan 1991) as explained by Anisfeld (chap. 3, this volume). Loss in elevation due to tidal restriction puts these systems at greater risk of drowning as sea level rises.

Tidal Marsh Impacts from Restrictions and Responses to Hydraulic Restoration

Research associated with restoration projects has documented several effects of restricting hydrology on tidal marsh ecology. First and foremost, restrictions reduce or eliminate tidal flooding, directly interfering with marsh maintenance processes (fig. 22.1). Restrictions typically reduce flood tides, especially the larger spring tides that periodically flood the marsh surface. Oftentimes the restriction also impedes drainage of freshwater from upstream, and rainfall and spring melt events can lead to flooding and damage of private property and infrastructure. Increased tidal prism following removal of the restriction will typically drive water circulation, promote drainage, expand creeks, and enhance marsh accretion.

Tidal marshes encompass complex hydraulic systems and pose significant challenges to engineers; however, a variety of models can describe and help managers select hydraulic solutions for restoration (McBroom and Schiff, chap. 2, this volume). Both simple and complex models can predict hydraulic responses to specific restoration solutions, but quality data are needed to support the models, and there is no substitute for careful on-site observations. Oftentimes the results from hydrodynamic models will reveal that tides can be restored to a tide-restricted marsh, but controls on the amount of flow are required to prevent flooding of infrastructure within the marsh floodplain, or to manage excessive flooding of the marsh to be restored. Glamore (chap. 17, this volume) and several case studies in this volume (Portnoy, chap. 18; Adamowicz and O'Brien, chap. 19; Golet et al., chap. 20; Reiner, chap. 21) describe the interaction of hydrologic modeling and the appropriate design of new or modified tide control structures that will restore tidal flushing while accommodating property flooding and ecological concerns.

Early work showed that restrictions cause an obvious lack of tidal flooding and result in decreased soil salinity (Roman et al. 1984). Moreover, as drained soils oxidize, organic matter declines and sulfur oxidizes to acid, leading to low pH and associated toxicity (Portnoy 1999; Anisfeld, chap. 3, this volume). The oxidation of organic matter can result in substantial subsidence, with comparable portions of marsh depressed 60 centimeters or more behind tidal barriers. Following restoration, tidal floodwaters bring salinity and sediments to the marsh.

Natural tidal marshes are typically dominated by emergent perennial graminoids that form several zones along an elevation gradient. The low elevation of New England marshes is dominated by S. alterniflora, which gives way to S. patens (salt meadow cordgrass), at the mean high tide line. Higher in elevation, Juncus gerardii (black rush) may prevail or be absent, but the upper edge of the marsh is typically vegetated by a variety of taller grasses and sedges or even shrubs (e.g., Iva fructescens, Baccharis halimifolia). Following the flush of freshwater associated with spring melt in temperate climates, severe tidal restrictions often result in draining over much of the growing season. Native halophytes are replaced by less salt tolerant grasses and weedy upland species. Less severe restrictions and those impounding freshwater often lead to Typha (cattail) and other brackish emergent plants. Some barriers result in hypersaline systems (Kelts 1979), but typically soil salinity falls and native halophytes are replaced by less salt tolerant weedy species, especially invasive exotic species like Phragmites australis (common reed), Lythrum salicaria (purple loosestrife), and most recently, Lepidium latifolium (broad-leaved pepperweed).

Dominance by exotic Phragmites (there are also noninvasive native forms; Saltonstall 2002) in tide-restricted marshes provides a clear indicator of impact. Yet this invader has proven difficult to turn back using tidal restoration alone because it not only is a great competitor, it also modifies the habitat (Chambers et al., chap. 5, this volume). Thus, while some restoration projects can increase hydroperiod and salinity enough to eliminate exotic Phragmites, many appear to only reduce and control its dominance.

Commensurate with the objectives of most tidal restoration projects, restoration is expected to result in increased halophyte cover and decreased cover of flood- and salt-intolerant plants. However, vegetation responses can be quite variable, with initial declines in halophytes common (Konisky et al. 2006) and convergence on reference communities sometimes elusive (Garbutt and Wolters 2008). In some cases, trees that colonized restricted marsh levees and plains have died following tidal restoration (Portnoy and Reynolds 1987; Diers and Richardson, chap. 11, this volume). Smith and Warren (chap. 4, this volume) have outlined a set of thirteen important site factors, ranging from the duration and magnitude of tidal restriction to herbivory, that may influence the development of plant communities on restored marshes and as such explain the variable responses to tidal restoration.

The impact of tidal restriction on nekton (fish, crabs, and shrimp) can be dramatic, but it varies with the severity of restriction (Raposa and Roman 2003) and also the type of restriction (restricted and drained versus restricted and impounded; Raposa and Talley, chap. 6, this volume). Some restrictions have impounded pools of water in the tidal creek immediately upstream of the restriction, a subtidal refuge where large populations of resident fish remain during low tide. However, even with high fish density, such marshes may fail to support higher trophic levels through the trophic relay as described by Kneib (2002). Passage rates through water control structures have been shown to be significantly depressed and may be eliminated at most tidal restrictions since small fish tend to avoid high current velocities associated with culverts (Eberhardt et al. 2011). Raposa and Talley (chap. 6, this volume) show that restoring hydrology leads to rapid improvements in the nekton community, but they stress the need for monitoring to document longer-term structural changes using standardized gear and protocols as well as assessment of functional benefits (e.g., reestablishing trophic export and food web support).

Besides marine export from fish and tides, a variety of birds use and export energy from tidal marshes throughout the year. Shriver and Greenberg (chap. 7, this volume) point out that little is known about avian responses to tidal restriction or restoration, perhaps due to the small size of most projects, short monitoring periods (typically less than five years), the high mobility of birds, and the inherent variability of sampling. One study concluded that generalist species (e.g., egrets, herons) returned relatively quickly following restoration, but marsh specialists such as *Ammodramus caudacutus* (salt marsh sparrow) only recovered after fifteen years (Warren et al. 2002). Shriver and Greenberg (chap. 7, this volume), Golet et al. (chap. 20, this volume), and others report that specialist species recolonize sites following recovery of vegetation, which can take many years. They also stress that the specific configuration of subhabitats (e.g., creeks, mudflats, pools, and pannes) as well as prey populations developed during recovery will help determine types of birds (shorebirds, waders, etc.) and their use of the system.

Responses to Accelerated Rates of Sea Level Rise

Hydrology is the key to a sustainable marsh, with the tidal floodwaters providing salt and suspended sediments to allow marshes to build seaward as well as landward during times of slow to moderate (0 to 3 millimeters per year) sea level rise (Redfield 1972). However, due to global climate change, eustatic sea level rise is surpassing 3 millimeters per year, and coastal managers are faced with the prospect of increased flooding of salt marshes that will lead to vegetation change and significant loss of marsh area. Indeed, local rates of sea level rise are reaching 6 millimeters per year in the Chesapeake area (http://tidesandcurrents.noaa.gov /sltrends/sltrends_station.shtml?stnid=8638863), and it is not known whether marsh-building processes can keep this pace. Clearly, some systems are not keeping pace (Blackwater National Wildlife Refuge, Stevenson et al. 1985; Jamaica Bay, Hartig et al. 2002), but human interference in the natural hydrology and sediment supply of these systems prevents clear inferences pointing to climate change alone. Research by Morris et al. (2002) has suggested 5 millimeters per year as an upward limit of the ability of marshes in the southeastern United States to build in response to rising sea levels. This threshold could vary in other regions, such as New England, with different sediment supplies (Kirwan et al. 2010).

With accelerated rates of sea level rise there should be an urgency to restore tide-restricted salt marshes. Even if marshes cannot keep pace with sea level rise, vegetated tidal marshes will build in elevation (Kirwan et al. 2010) and slow shoreline erosion (Morgan et al. 2009), so removing tidal barriers will increase resistance to marine transgression and also ameliorate catastrophic losses as tidal barriers are inevitably breached by storms. In many coastal areas of the northeastern United States, especially those with a high tidal range, marsh elevation is well above the lower limit of S. *alterniflora* growth and will provide many decades of function. Even with dramatic sea level rise (perhaps 10 millimeters per year), marsh areas only 50 centimeters above the lower growth limit should survive fifty years without any net growth from accretion. This concept is termed elevation capital (Cahoon and Guntenspergen 2009). Unfortunately the concept that uncontrolled tidal flooding of marshes will reduce flooding losses in the future is counterintuitive, so scientists and managers must work especially well together to communicate the importance of tidal restoration.

Without seawalls, berms, and other barriers to marsh transgression at the marsh-to-upland interface, sea level rise will allow marshes to spill over low-lying uplands, hopefully replacing the marsh area lost at the lower elevation edges. Further, without tidal restrictions that interfere with marsh accretion and organic matter accumulation processes, marshes will grow in elevation in response to sea-level rise as they migrate landward. However, homeowners, businesses, and those charged with maintaining infrastructure will want to erect barriers to marine transgression as sea level rise is punctuated by storm events. Such barriers will lead to coastal squeeze, where marshes erode and submerge at lower elevations, without their ability to migrate inland (Pethick 2002), leading to dramatic losses of these important ecosystems.

Where Should Marsh Restoration Research Be Focused in the Future?

If we are to manage tidal marshes and mangroves to provide local communities and economies with essential ecosystem services, detrimental human impacts, such as tidal restrictions, need to be ameliorated. With global change anticipated to threaten many coastal resources, most prominently through accelerated rates of sea level rise (Cahoon and Guntenspergen 2009; Kirwan et al. 2010), it is important to reestablish the natural processes, including hydrology and sediment flux, to promote resilience in tidal wetlands (Langley et al. 2009; Huxham et al. 2010; Kirwan et al. 2010). Enhanced understanding and modeling of the accretionary (Kirwan et al. 2010) and erosive (Fagherazzi et al. 2006) processes involved will allow coastal scientists, planners, and managers to better plan for sea level rise and promote tidal restoration as the first step in protection of coasts. Model outputs are also valuable education and communication tools in promoting tidal restoration (Konisky, chap. 16, this volume), but need to be continually updated as our understanding of feedback processes grows through research (e.g., fig. 22.1; Langley et al. 2009; Huxham et al. 2010). We also need better science to support the adaptive management process in salt marsh restoration, which requires a thorough understanding of how marshes will respond to various scenarios of tidal restoration (Buchsbaum and Wigand, chap. 14, this volume).

Justification of restoration costs and assessment could be improved if the values of ecosystem services from tidal marshes (and losses associated with human impacts) could be determined easily. Better indicators of function (rather than relying on measures of structure as a proxy for function), or better yet the specific services themselves, should be developed (Chmura et al., chap. 15, this volume).

Transferring Restoration Science and Practice to Enhance Salt Marsh Conservation

Coastal resource managers have found that a significant portion of tidal marshes have already been destroyed for development. Many of the remaining marshes have been affected by a variety of human actions, including hydrologic restrictions from transportation corridors, agricultural projects and dams, dredge spoil disposal, ditching for mosquito control, and spread of invasive species. Partnering with landowners, transportation agencies, and others, coastal managers in the New England states (chaps. 8–12, this volume) have shown great success in removing or reducing tidal restrictions to help restore ecosystem services. Furthermore, developing partnerships with staff from key state and federal agencies can reduce costs dramatically. Funding from state contributions and trust funds is very important in getting projects started and providing match for larger federal contributions.

Even within our limited perspective of the Canadian Maritime Provinces and the New England states, variation in tidal range and climate has led to different considerations regarding tidal restoration (Bowron et al., chap. 13, this volume). Developing successful tidal restoration programs along other coasts of the world will need to recognize and react to significant differences in hydrology and geomorphology affecting marshes (and mangroves) as well as human needs and cultural values.

Long-term monitoring must be incorporated into restoration projects, with the data being used to guide the adaptive management process (Buchsbaum and Wigand, chap. 14, this volume). All authors in this edited volume have identified the need for long-term monitoring to fill gaps in our understanding of how various ecosystem components and processes will respond to restored tidal flow. Few studies have evaluated responses over multidecade timescales (e.g., Warren et al. 2002), but these kinds of datasets are essential to improving our understanding of trends or trajectories of marsh response toward a restored condition. Given the uncertainties of how tidally restored marshes will respond to accelerated rates of sea level rise, it is especially imperative that targeted research and long-term multidisciplinary monitoring programs be maintained, providing relevant information to support predictive models and adaptive management decisions.

Regulation of human activities, including restoration, is increasing worldwide as competition for coastal resources increases. Restoration practice is becoming more complex and requires interdisciplinary teams to navigate social and regulatory issues as well as technical ones. Planning has an integral role in coastal restoration; whether a program is centrally managed or is based on opportunistic approaches to identify and restore tidal restrictions. Planning for tidal restoration in states and provinces has benefited from geographic inventories of degraded marshes, identification of human impacts, and discussion of potential restoration approaches (chaps. 8–13, this volume).

Restoration leaders should determine what processes are affected by tidal restriction and what impacts to tidal ecosystems are apparent. By examining the role of tidal hydrology in supporting reference ecosystems, impacts to tide-restricted systems can be inferred and the potential benefits from tidal restoration can be predicted and communicated to the public. Managers and planners should focus on the roles of natural hydrology and coastal processes as the fundamental mechanisms to initiate change along a trajectory toward a restored condition. By establishing the natural processes needed to maintain salt marsh, most practitioners have found that the marsh will self-organize or develop on its own, with minimal active management.

Acknowledgments

We thank the Marine Program and the Class of 1937, University of New Hampshire, for their support of David Burdick with a professorship in marine biology. Jackson Estuarine Laboratory contribution no. 507.

References

Cahoon, D. R., and G. R. Guntenspergen. 2009. "Climate Change, Sea-Level Rise, and Coastal Wetlands." *National Wetlands Newsletter* 32:8–12.

Chapman, V. J. 1960. *Salt Marshes and Salt Deserts of the World.* New York: Interscience.

Donnelly, J. P., and M. D. Bertness. 2001. "Rapid Shoreward Encroachment of Salt Marsh Cordgrass in Response to Accelerated Sea-Level Rise." *Proceedings of the National Academy of Sciences* 98:14218–23.

Eberhardt, A. L., D. M. Burdick, and M. Dionne. 2011. "The Effects of Road Culverts on Nekton in New England Salt Marshes: Implications for Tidal Restoration." *Restoration Ecology* 19:776–85.

Fagherazzi, S., L. Carniello, L. D'Alpaos, and A. Defina. 2006. "Critical Bifurcation of Shallow Microtidal Landforms in Tidal Flats and Salt Marshes." *Proceedings of the National Academy of Sciences* 103:8337–41.

Frenkel, R. E., and J. C. Morlan. 1991. "Can We Restore Our Salt Marshes? Lessons from the Salmon River, Oregon." *Northwest Environmental Journal* 7:119–35.

Garbutt, A., and M. Wolters. 2008. "The Natural Regeneration of Salt Marsh on Formerly Reclaimed Land." *Vegetation Science* 11:335–44.

Hartig, E. K., V. Gornitz, A. Kolker, F. Mushacke, and D. Fallon. 2002. "Anthropogenic and Climate-Change Impacts on Salt Marshes of Jamaica Bay, New York City." *Wetlands* 22:71–89.

Huxham, M., M. P. Kumara, L. P. Jayatissa, K. W. Krauss, J. Kairo, J. Langat, M. Mencuccini, M. W. Skov, and B. Kirui. 2010. "Intra- and Interspecific Facilitation in Mangroves May Increase Resilience to Climate Change Threats." *Philosophical Transactions of the Royal Society B* 365:2127–35.

Kelts, L. J. 1979. "Ecology of a Tidal Marsh Corixid, *Trichocorixa verticalis* (Insecta, Hemiptera)." *Hydrobiologia* 64:37–57.

Kirwan, M. L., G. R. Guntenspergen, A. D'Alpaos, J. T. Morris, S. M. Mudd, and S. Temmerman. 2010. "Limits on the Adaptability of Coastal Marshes to Rising Sea Level." *Geophysical Research Letters* 37. doi:10.1029/2010GL045489.

Kneib, R. T. 2002. "Salt Marsh Ecoscapes and Production Transfers by Estuarine Nekton in the Southeastern United States." Pp. 267–91 in *Concepts and Controversies in Tidal Marsh Ecology*, edited by M. Weinstein and D. Kreeger. Boston: Kluwer Academic.

Konisky, R. A., D. M. Burdick, M. Dionne, and H. A. Neckles. 2006. "A Regional Assessment of Salt Marsh Restoration and Monitoring in the Gulf of Maine." *Restoration Ecology* 14:516–25.

Langley, J. A., K. L. McKee, D. R. Cahoon, J. A. Cherrve, and J. P. Megonigal. 2009. "Elevated CO_2 Stimulates Marsh Elevation Gain, Counterbalancing Sea-Level Rise." *Proceedings of the National Academy of Sciences* 106:6182–86.

Miller, W. R., and F. E. Egler. 1950. "Vegetation of the Wequetequock-Pawcatuck Tidal Marshes, Connecticut." *Ecological Monographs* 20:145–70.

Morgan, P. A., D. M. Burdick, and F. T. Short. 2009. "The Functions and Values of Fringing Salt Marshes in Northern New England, USA." *Estuaries and Coasts* 32:483–95.

Morris, J. T., P. V. Sundareshwar, C. T. Nietch, B. Kjerfve, and D. R. Cahoon. 2002. "Responses of Coastal Wetlands to Rising Sea Level." *Ecology* 83:2869–77.

Pethick, J. 2002. "Estuarine and Tidal Wetland Restoration in the United Kingdom: Policy versus Practice." *Restoration Ecology* 10:431–37.

Portnoy, J., and M. Reynolds. 1997. "Wellfleet's Herring River: The Case for Habitat Restoration." *Environment Cape Cod* 1:35–43.

Portnoy, J. W. 1999. "Salt Marsh Diking and Restoration: Biogeochemical Implications of Altered Wetland Hydrology." *Environmental Management* 24:111–20.

Ranwell, D. S. 1972. *Ecology of Salt Marshes and Sand Dunes*. London: Chapman and Hall.

Raposa, K. B., and C. T. Roman. 2003. "Using Gradients in Tidal Restriction to Evaluate Nekton Community Responses to Salt Marsh Restoration." *Estuaries* 26:98–105.

Redfield, A. C. 1972. "Development of a New England Salt Marsh." *Ecological Monographs* 42:201–37.

Roman, C. T., W. A. Niering, and R. S. Warren. 1984. "Salt Marsh Vegetation Change in Response to Tidal Restriction." *Environmental Management* 8:141–49.

Saltonstall, K. 2002. "Cryptic Invasion by a Non-native Genotype of the Common Reed, *Phragmites australis*, into North America." *Proceedings of the National Academy of Sciences* 99:2445–49.

Stevenson, J. C., M. S. Kearney, and E. C. Pendleton. 1985. "Sedimentation and Erosion in a Chesapeake Bay Brackish Marsh System." *Marine Geology* 67:212–35.

Teal, J. M. 1962. "Energy Flow in the Salt Marsh Ecosystem of Georgia." *Ecology* 43:614–24.

Warren, R. S., P. E. Fell, R. Rozsa, A. H. Brawley, A. C. Orsted, E. T. Olson, V. Swamy, and W. A. Niering. 2002. "Salt Marsh Restoration in Connecticut: 20 Years of Science and Management." *Restoration Ecology* 10:497–515.

Charles T. Roman leads the National Park Service's North Atlantic Coast Cooperative Ecosystem Studies Unit. His research has focused on salt marsh ecology with a concentration on habitat restoration for over three decades. He is also professor-in-residence at the University of Rhode Island's Graduate School of Oceanography.

David M. Burdick is research associate professor of coastal ecology and restoration in the Department of Natural Resources and the Jackson Estuarine Laboratory at the University of New Hampshire, where he has taught wetlands courses over the past twelve years. His study of coastal science spans more than thirty years, concentrating on coastal habitats, primarily salt marsh and seagrass beds, assessing human impacts and habitat restoration.

Susan C. Adamowicz is the land management research and demonstration biologist with the US Fish and Wildlife Service, specializing in salt marsh ecology and restoration. She is currently engaged in salt marsh management and restoration projects from Maine to Virginia.

Shimon C. Anisfeld is a senior lecturer and research scientist at the Yale University School of Forestry and Environmental Studies. He has an active research program investigating biogeochemical processes in degraded and restoring tidal marshes of Long Island Sound.

Tony M. Bowron is the director and coastal wetland ecologist with CB Wetlands and Environmental Specialists Inc., a consulting firm serving Atlantic Canada.

He has over a decade of experience designing and monitoring estuarine wetland restoration projects.

Robert N. Buchsbaum is the coastal ecologist with the Massachusetts Audubon Society. He has been active in monitoring salt marsh restoration projects in Massachusetts.

Marc Carullo is the coastal GIS and habitat analyst with the Massachusetts Office of Coastal Zone Management.

Caitlin Chaffee is a coastal policy analyst with the Rhode Island Coastal Resources Management Council.

Randolph M. Chambers is a professor of biology and the director of the W. M. Keck Environmental Field Laboratory at the College of William and Mary. Much of his research has focused on biogeochemistry of *Phragmites* wetlands.

Gail L. Chmura is an associate professor in the Department of Geography and Centre for Climate and Global Change Research at McGill University. Her research interests related to salt marshes include diked marsh restoration, paleoecology to address issues of climate change and anthropogenic stress, carbon sequestration, and quantification of ecosystem services.

Ted Diers is the administrator of the Watershed Management Bureau, New Hampshire Department of Environmental Services. He has over thirteen years of experience planning and managing coastal wetland restoration projects in New Hampshire.

Hunt Durey is with the Department of Fish and Game, Division of Ecological Restoration, Commonwealth of Massachusetts, and is engaged with planning and implementation of coastal wetland restoration projects.

Marci Cole Ekberg is a coastal ecologist with Save The Bay, Rhode Island. She is engaged in salt marsh and eelgrass restoration programs, water quality restoration, and development of nutrient reduction strategies for Narragansett Bay.

Wenley Ferguson is the director of habitat restoration for Save The Bay, Rhode Island. She is involved with planning and implementation of wetland restoration projects throughout the Narragansett Bay watershed.

William C. Glamore is a senior research fellow with the Water Research Laboratory, School of Civil and Environmental Engineering, University of New South Wales (Australia), with active research programs in wetland hydraulics, modeling, and engineering. He is a Churchill Fellow in international wetland engineering practices.

Francis C. Golet is an emeritus professor of wetland ecology in the Department of Natural Resources Science, University of Rhode Island. Since 1973 he has taught courses and conducted research on the ecology and management of both inland and coastal wetlands.

Russell Greenberg is the head of the Migratory Bird Center of the Smithsonian Conservation Biology Institute. His studies are focused on the ecology and evolution of migration, habitat selection, and interspecific interaction, with a keen interest in coastal marsh birds.

Elizabeth Hertz is the Land Use Team manager with the Maine State Planning Office.

W. Gregory Hood is a senior research scientist with the Skagit River System Cooperative located within the Puget Sound watershed. His research on estuarine geomorphology and vegetation distribution is conducted in support of habitat restoration design.

Jon Kachmar, formerly with the Maine State Planning Office, is now the Southeast Massachusetts program director for The Nature Conservancy.

Raymond A. Konisky is the director of Marine Science and Conservation for The Nature Conservancy, New Hampshire Chapter. He has developed marsh restoration simulation models and been involved with a regional synthesis of marsh restoration monitoring in the Gulf of Maine.

Kimberly L. Dibble is a doctoral candidate in the Department of Natural Resources Science, University of Rhode Island. Her research is on the response of nekton communities to tidal restoration of *Phragmites*-dominated marshes.

Jeremy Lundholm is an associate professor in the Department of Biology, Saint Mary's University, Halifax. As a plant ecologist he has been involved with baseline vegetation classification and postrestoration monitoring at salt marsh restoration sites in the Bay of Fundy and throughout Nova Scotia.

James G. MacBroom is the senior vice president of Milone & MacBroom Inc., a civil and environmental engineering firm in Connecticut, and lecturer at Yale University. He has over forty years of experience in restoration of tidal systems and other watershed management practices.

Laura A. Meyerson is an associate professor of habitat restoration ecology in the Department of Natural Resources Science, University of Rhode Island. Her research focuses on invasive species, restoration ecology, and environmental policy.

Gregg E. Moore is a research assistant professor in the Department of Biological Sciences and Jackson Estuarine Laboratory at the University of New Hampshire. He conducts research on coastal wetland restoration, with special interests in the biogeochemistry of natural and disturbed salt marshes and the ecology of invasive plant species.

Dennis H. A. Myshrall received his master's degree from the Department of Natural Resources Science, University of Rhode Island. He coordinated the ecological monitoring program at the Galilee marsh restoration site.

Nancy Neatt is the director and coastal marine ecologist with CB Wetlands and Environmental Specialists Inc., a consulting firm serving Atlantic Canada. She has extensive experience in monitoring salt marsh restoration projects throughout Atlantic Canada.

Kathleen M. O'Brien is a biologist with the Rachel Carson National Wildlife Refuge, US Fish and Wildlife Service, Wells, Maine. She is involved in all aspects of natural resource management, including rare species monitoring, invasive species control, and habitat restoration, including salt marshes.

Lawrence R. Oliver is an environmental biologist and study manager with the US Army Corps of Engineers, New England Division. He has been engaged in the design, planning, implementation, and monitoring of salt marsh tidal restoration projects throughout New England.

Peter W. C. Paton is a professor of wildlife ecology at the University of Rhode Island. His research focuses on the conservation of avian and amphibian populations in coastal environments.

John W. Portnoy, retired, was an ecologist at Cape Cod National Seashore, National Park Service, for over twenty-nine years. His salt marsh restoration interests include biogeochemical cycling, hydrologic modeling, long-term monitoring, and policy development.

Kenneth B. Raposa serves as the research coordinator of the Narragansett Bay National Estuarine Research Reserve. He maintains an active research focus on estuarine nekton ecology, particularly on nekton habitat use, community ecology, and responses to habitat restoration.

Edward L. Reiner is a senior wetland scientist at the US Environmental Protection Agency–Region One, Office of Ecosystem Protection, Wetland Protection Section. He has been with EPA for over thirty years doing federal permit review and supporting salt marsh restoration activities throughout New England.

Frank D. Richardson is a coastal ecologist with the Wetlands Bureau of the New Hampshire Department of Environmental Services, with more than thirty years of experience.

Ron Rozsa is retired from the Office of Long Island Sound Programs of the Connecticut Department of Environmental Protection. He formulated tidal wetland restoration policy in Connecticut.

Roy Schiff is a water science specialist and engineer with Milone and MacBroom Inc., a civil and environmental engineering firm in Vermont. He specializes in stream restoration and aquatic ecosystem studies.

W. Gregory Shriver is an assistant professor of wildlife ecology, Department of Entomology and Wildlife Ecology, University of Delaware. His research interests are in conservation biology, landscape ecology, and avian ecology, with a focus on coastal and salt marsh breeding birds.

Charles A. Simenstad is a research professor at the School of Aquatic and Fishery Sciences, University of Washington. He studies food web structure and restoration ecology of estuarine and coastal marine ecosystems along the Pacific Northwest coast.

Stephen M. Smith is a plant ecologist at Cape Cod National Seashore, National Park Service. His studies focus on understanding the processes and factors that shape coastal plant communities.

Timothy Smith, formerly with the Massachusetts Wetlands Restoration Program, is now a restoration ecologist with Cape Cod National Seashore, National Park Service. He has extensive experience with the design, planning, implementation, and monitoring of salt marsh restoration projects.

Drew M. Talley is an assistant professor in the Department of Marine Science and Environmental Studies, University of San Diego. He has active research on marsh restoration on both the Pacific and Atlantic coasts of the United States, focusing on issues of habitat connectivity and wetland fishes.

Brian C. Tefft is a wildlife biologist with the Division of Fish and Wildlife, Rhode Island Department of Environmental Management.

Danika van Proosdij is an associate professor of geography at Saint Mary's University, Halifax. Her current studies focus on geomorphology and biogeography of intertidal marine systems, with a particular interest in climate change research and restoration of salt marshes.

R. Scott Warren is an emeritus professor of botany at Connecticut College and has been studying salt marsh ecology for over thirty-five years, with a focus on vegetation dynamics and habitat restoration.

Cathleen Wigand is a research ecologist with the US Environmental Protection Agency, Atlantic Ecology Division, Narragansett, Rhode Island. Her recent studies have been evaluating the response of salt marsh vegetation communities to nutrient enrichment, among other topics.

THE SCIENCE AND PRACTICE
OF ECOLOGICAL RESTORATION

Restoring Natural Capital: Science, Business, and Practice, edited by James Aronson, Suzanne J. Milton, and James N. Blignaut

Old Fields: Dynamics and Restoration of Abandoned Farmland, edited by Viki A. Cramer and Richard J. Hobbs

Ecological Restoration: Principles, Values, and Structure of an Emerging Profession, by Andre F. Clewell and James Aronson

River Futures: An Integrative Scientific Approach to River Repair, edited by Gary J. Brierley and Kirstie A. Fryirs

Large-Scale Ecosystem Restoration: Five Case Studies from the United States, edited by Mary Doyle and Cynthia A. Drew

New Models for Ecosystem Dynamics and Restoration, edited by Richard J. Hobbs, and Katharine N. Suding

Cork Oak Woodlands in Transition: Ecology, Adaptive Management, and Restoration of an Ancient Mediterranean Ecosystem, edited by James Aronson, João S. Pereira, and Juli G. Pausas

Restoring Wildlife: Ecological Concepts and Practical Applications, by Michael L. Morrison

Restoring Ecological Health to Your Land, by Steven I. Apfelbaum and Alan W. Haney

Restoring Disturbed Landscapes: Putting Principles into Practice, by David J. Tongway and John A. Ludwig

Intelligent Tinkering: Bridging the Gap between Science and Practice, by Robert J. Cabin

Making Nature Whole: A History of Ecological Restoration, by William R. Jordan and George M. Lubick

Human Dimensions of Ecological Restoration: Integrating Science, Nature, and Culture, edited by Dave Egan, Evan E. Hjerpe, and Jesse Abrams

Tidal Marsh Restoration: A Synthesis of Science and Management, edited by Charles T. Roman and David M. Burdick